“十四五”高等职业教育计算机教育系列教材

Network Security Technology and Application

刘　琤　杨　阳◎主　编
关　剑　隋玮玥　李　粟　张运亨◎副主编
张　良　于　洋　任雁汇
沈　婷　韩美琪　王占云◎参　编

中国铁道出版社有限公司
CHINA RAILWAY PUBLISHING HOUSE CO., LTD.

Introduction

This book provides a detailed introduction to the main technologies and application areas of modern network security. It covers multiple topics such as authentication, authorization, disaster recovery planning, data monitoring and analysis, operating system and host security, helping readers understand how to safeguard network security through various technical means. The content ranges from basic concepts to specific technical implementations, from theoretical knowledge to practical operations, comprehensively covering all aspects of network security.

Additionally, the book introduces the latest network security technologies and trends, such as Software-Defined Security (SDSec), ensuring the content is both timely and advanced.

This book is suitable for students majoring in network security, network administrators, and security engineers. It can serve as a textbook for relevant courses in higher education institutions or as a reference book for practitioners, helping them enhance their network security knowledge and skills.

图书在版编目（CIP）数据

网络安全技术与应用 = Network Security Technology and Application / 刘琤，杨阳主编. 北京：中国铁道出版社有限公司，2025. 6. --（"十四五"高等职业教育计算机教育系列教材）. -- ISBN 978-7-113-31538-2

Ⅰ. TP393.08

中国国家版本馆CIP数据核字第20257VM440号

书　　名： Network Security Technology and Application
作　　者： 刘　琤　杨　阳

策　　划： 祁　云　　　**编辑部电话：**（010）63549458
责任编辑： 祁　云　绳　超
封面设计： 刘　颖
责任校对： 刘　畅
责任印制： 赵星辰

出版发行： 中国铁道出版社有限公司（100054，北京市西城区右安门西街8号）
网　　址： https://www.tdpress.com/51eds
印　　刷： 天津嘉恒印务有限公司
版　　次： 2025年6月第1版　2025年6月第1次印刷
开　　本： 850 mm × 1 168 mm　1/16　**印张：** 21.75　**字数：** 472千
书　　号： ISBN 978-7-113-31538-2
定　　价： 79.80元

Preface

With the rapid development of information technology, network security has become a crucial infrastructure across various industries. This textbook provides comprehensive theoretical and practical guidance for network security courses, aiming to equip students and professionals with the ability to handle various network security threats and lay a solid foundation for their future career development.

The writing of this book is based on current network security teaching standards and the latest technological developments. The content includes both foundational theoretical knowledge and practical operational skills, striving to help readers grasp core concepts and practical techniques in network security through systematic teaching materials. Moreover, the book emphasizes case-based teaching, with real-world case analysis and operational exercises, enabling readers to apply the knowledge they have learned in real-world environments.

This book has the following features:

Case-driven: Through numerous real-world cases and operational steps, it helps readers understand and master network security technologies in practice.

Comprehensive coverage: It covers everything from basic concepts to advanced applications, providing detailed content to help readers fully understand and master network security technologies.

Cutting-edge technology: It introduces the latest network security technologies and trends, ensuring the content is timely and advanced.

This book is structured as a workbook-style textbook, divided into five major sections, which respectively cover overview of security and information security, operating system and host security, network security foundation, encryption decryption and application, and fundamentals of security operations and analysis. Each section is self-contained and subdivided into several projects and multiple tasks, with content progressing from basic to advanced and tightly integrating theory with practice. This allows readers to deeply understand theoretical principles and effectively master application and problem analysis skills through project-based learning.

The authors of this book have extensive teaching experience and practical knowledge. Centered around the HCIA-Security certification, it focuses on equipping readers with basic knowledge and related technologies for small and medium-sized network information security (Huawei firewall technology, encryption and decryption technology, information security system operation and maintenance, etc.), enabling them to assist in designing, deploying, and maintaining network security architectures for small and medium-sized enterprises, thereby ensuring the security of networks and applications for such enterprises.

In selecting, organizing, and designing the teaching methods for the content, this textbook attempts some innovative reforms. We warmly welcome and hope that readers will provide feedback and suggestions for this textbook.

Liu Cheng, Yang Yang are the chief editors of this book, Guan Jian, Sui Weiyue, Li Su, Zhang Yunheng, Zhang Liang, Yu Yang, Ren Yanhui are the deputy editors of this book, Shen Ting, Han Meiqi, Wang Zhanyun are the editors. Due to the limited knowledge and experience of the editor, there are inevitably some omissions and deficiencies in this book, and we sincerely hope that readers can criticize and correct them.

Author

October, 2024

Contents

Section 1 Overview of security and information security 1

Project 1 Understand the basic concepts of information security 2

Task 1.1 Understand the concepts of information and information security 2

Subtask 1.1.1 Understand the definition of information and information security 2

Subtask 1.1.2 Grasp the attributes of security 3

Subtask 1.1.3 Understand the development of information security 4

Subtask 1.1.4 Case study of information security 5

Subtask 1.1.5 Cause analysis of information security incidents 6

Subtask 1.1.6 Understand the significance of building information security 7

Task 1.2 Understand information security risks and management 7

Subtask 1.2.1 Physical risks analysis 7

Subtask 1.2.2 Information risks analysis 8

Subtask 1.2.3 System risks analysis 9

Subtask 1.2.4 Application risks analysis 9

Subtask 1.2.5 Network risks analysis 9

Subtask 1.2.6 Management risks analysis 10

Subtask 1.2.7 Understand the importance of information security management 10

Subtask 1.2.8 Understand the development status of information security management 10

Project 2 Understand the computer networks 11
Task 2.1 TCP/IP architecture analysis 11
Subtask 2.1.1 Analysis of typical campus network architecture 11
Subtask 2.1.2 Introduction of OSI reference model 12
Subtask 2.1.3 Understand OSI seven-layer model 12
Subtask 2.1.4 Understand TCP/IP model 14
Subtask 2.1.5 Master the functions of each layer of TCP/IP protocol stack 15
Subtask 2.1.6 Analysis of encapsulation and decapsulation process of TCP/IP protocol stack 16
Subtask 2.1.7 Understand the five-tuple 18
Task 2.2 Understand network layer protocols 18
Subtask 2.2.1 Working process analysis of ARP protocol 18
Subtask 2.2.2 Working process analysis of ICMP 21
Subtask 2.2.3 Understand routing protocols 24
Subtask 2.2.4 Working process analysis of OSPF 25
Subtask 2.2.5 Working process analysis of SNMP 26
Subtask 2.2.6 Working process analysis of NTA 27
Subtask 2.2.7 Working process analysis of NetStream 28
Task 2.3 Understand transport layer protocols 29
Subtask 2.3.1 Working process analysis of TCP protocol 29
Subtask 2.3.2 Working process analysis of UDP protocol 31
Task 2.4 Understand application layer protocols 32
Subtask 2.4.1 Working process analysis of DNS protocol 32
Subtask 2.4.2 Working process analysis of FTP protocol 33
Subtask 2.4.3 Working process analysis of HTTP protocol 35
Subtask 2.4.4 The analysis of SMTP/POP3/IMAP-mail sending and receiving mechanism 36
Project 3 Master the configuration of common network equipment 37
Task 3.1 Understand common network equipment 38
Subtask 3.1.1 Master the working principle of switch 39

Subtask 3.1.2 Master the working principle of router 41
Subtask 3.1.3 Master the routing 42
Subtask 3.1.4 Master the working principle of firewall 43
Task 3.2 Login network equipment 49
Subtask 3.2.1 Understanding VRP 49
Subtask 3.2.2 Master command line configuration mode 49
Subtask 3.2.3 Learn to use command line help 50
Task 3.3 Network device interface configuration 51
Task 3.4 Routing configuration of network equipment 52
Task 3.5 Equipment login management configuration method and process 54
Subtask 3.5.1 Console port login 54
Subtask 3.5.2 Web login 55
Subtask 3.5.3 Telnet login 57
Subtask 3.5.4 SSH login 59
Task 3.6 Network equipment system settings 61
Subtask 3.6.1 Profile management 61
Subtask 3.6.2 Software upgrade 62
Subtask 3.6.3 License configuration 63
Project 4 Understand common information security threats 64
Task 4.1 Understand the current situation of information security threats 64
Subtask 4.1.1 Understand the evolution of information security attacks 64
Subtask 4.1.2 Master the classification of security threats 64
Task 4.2 Network security threat analysis 65
Subtask 4.2.1 Analysis of attack principle 66
Subtask 4.2.2 Analysis of deception attack principle 66
Subtask 4.2.3 Analysis of DDoS attack principle 67
Subtask 4.2.4 Master the defense means of network attack 68
Task 4.3 Apply security threat analysis 68
Subtask 4.3.1 Analysis of vulnerability principle 68

Subtask 4.3.2 Analysis of fishing attack principle 69
Subtask 4.3.3 Principle analysis of malicious code 70
Subtask 4.3.4 Defense methods for application attacks 70
Task 4.4 Understanding data transmission and terminal security threats 71
Project 5 Understand the development trend of threat prevention and information security 72
Task 5.1 Master security threat prevention 72
Subtask 5.1.1 Understand the key elements of information security 72
Subtask 5.1.2 Master the information security precautions 73
Task 5.2 Cultivate awareness of information security 73
Subtask 5.2.1 Safety awareness test 73
Subtask 5.2.2 Understand social engineering 74
Subtask 5.2.3 Cultivate safety awareness 74
Task 5.3 Analysis of information security development trend 75
Subtask 5.3.1 Analysis of Gartner's top ten security technologies 75
Subtask 5.3.2 Future development trend of security defense 77

Section 2 Operating system and host security 80

Project 6 Operating system 81
Task 6.1 Understand the basic knowledge of operating system 81
Subtask 6.1.1 Understand the definition of operating system 81
Subtask 6.1.2 Master the main functions of the operating system 81
Subtask 6.1.3 Understand operating system classification 83
Task 6.2 Master the Windows operating system 83
Subtask 6.2.1 Understand the origin of Windows 83
Subtask 6.2.2 Meet Windows 84
Subtask 6.2.3 Windows system architecture analysis 84
Subtask 6.2.4 Understanding the Windows family 86

Subtask 6.2.5 Master Windows features 87
Task 6.3 Master Linux operating system 88
Subtask 6.3.1 Understand the history of Linux 88
Subtask 6.3.2 Analysis of Linux system architecture 89
Subtask 6.3.3 Understand the basic idea of Linux 89
Subtask 6.3.4 Understand common Linux operating systems 89
Subtask 6.3.5 Master the main features of Linux 90
Subtask 6.3.6 Understand the application fields of Linux 91
Subtask 6.3.7 Comparison between Windows and Linux 92
Project 7 Common server types and threats 92
Task 7.1 Understanding servers 92
Subtask 7.1.1 Understand server definitions 92
Subtask 7.1.2 Master the characteristics of the server 93
Subtask 7.1.3 Understand server classification 94
Task 7.2 Understand common server software 96
Subtask 7.2.1 Analysis of file working process 96
Subtask 7.2.2 Analysis of working process of database 97
Subtask 7.2.3 Analysis of mail working process 97
Subtask 7.2.4 Analysis of working process of web page 98
Subtask 7.2.5 Working process analysis of FTP server 98
Subtask 7.2.6 Analysis of working process of domain name server 99
Subtask 7.2.7 Analysis of working process of time synchronization server 99
Subtask 7.2.8 Analysis of proxy working process 99
Task 7.3 Understand server security threats 100
Subtask 7.3.1 Malicious program attack 100
Subtask 7.3.2 Hacker brute force cracking 101
Subtask 7.3.3 SQL injection attack 101
Subtask 7.3.4 DDoS attack 102
Subtask 7.3.5 Master security threat defense methods 102

Task 7.4 Understand vulnerabilities and patches ········ 103
Subtask 7.4.1 Understand the definition of vulnerabilities ········ 103
Subtask 7.4.2 Understand the hazards of loopholes ········ 104
Subtask 7.4.3 Understand the classification of vulnerabilities ········ 104
Subtask 7.4.4 Analyze the causes of system vulnerabilities ········ 105
Subtask 7.4.5 Master vulnerability scanning technology ········ 106
Subtask 7.4.6 Understand the definition of patch ········ 107
Subtask 7.4.7 Analysis of typical vulnerability attack cases ········ 108
Subtask 7.4.8 Master vulnerability defense methods ········ 108
Project 8 Host firewall and antivirus software ········ 109
Task 8.1 Know firewall ········ 109
Subtask 8.1.1 Understand firewall definition ········ 109
Subtask 8.1.2 Master firewall classification ········ 110
Task 8.2 Windows firewall settings ········ 110
Task 8.3 Linux firewall settings ········ 115
Subtask 8.3.1 Meet iptables ········ 115
Subtask 8.3.2 Structural analysis of iptables ········ 116
Subtask 8.3.3 Understand iptables rules ········ 116
Subtask 8.3.4 Understand iptables chain ········ 116
Subtask 8.3.5 Understand iptables table ········ 117
Subtask 8.3.6 Analysis of the process of iptables transmitting data packets ········ 117
Subtask 8.3.7 Application of iptables commands ········ 118
Task 8.4 Use of host antivirus software ········ 119
Subtask 8.4.1 Understand the definition of antivirus software ········ 119
Subtask 8.4.2 Understand the basic functions of antivirus software ········ 119
Subtask 8.4.3 Understand the composition of antivirus software ········ 119
Subtask 8.4.4 Master the key technologies of antivirus software ········ 120
Subtask 8.4.5 Understand common knowledge of antivirus software ········ 121

Section 3 Network security foundation 123

Project 9 Working principle and configuration of hardware firewall 124

Task 9.1 Understand hardware firewalls 124

Subtask 9.1.1 Understand hardware firewall characteristics 124

Subtask 9.1.2 Understand hardware firewall classification 125

Subtask 9.1.3 Master firewall networking mode 128

Task 9.2 Analysis of firewall forwarding principle 129

Subtask 9.2.1 Understand packet filtering technology 129

Subtask 9.2.2 Master firewall security policy 129

Subtask 9.2.3 Understanding firewall inter-domain forwarding 130

Subtask 9.2.4 Create and query session entries 131

Subtask 9.2.5 Understand the state detection mechanism 131

Subtask 9.2.6 Understand conversation entries 133

Task 9.3 Firewall security policy and its application 134

Subtask 9.3.1 Understand the matching principles of security policies 134

Subtask 9.3.2 Analysis of security policy matching process 135

Subtask 9.3.3 Application of security policy 136

Subtask 9.3.4 Thinking of configuring security policy 137

Subtask 9.3.5 Configure security policy in Web mode 138

Task 9.4 Firewall security policy configuration case 143

Task 9.5 ASPF technology 147

Subtask 9.5.1 Understand multichannel protocols 147

Subtask 9.5.2 Understand ASPF technology 147

Subtask 9.5.3 ASPF's support for multichannel protocols 148

Subtask 9.5.4 Generation process analysis of Server-map 149

Subtask 9.5.5 Port identification support for multichannel protocols 150

Subtask 9.5.6 Understand fragmented caching 151

Subtask 9.5.7 Understand the background and configuration of long connections 152

Project 10 Network address translation technology 153
Task 10.1 Analysis of network address translation principle 153
Subtask 10.1.1 Understand the background of NAT 153
Subtask 10.1.2 Analysis of the basic principle of NAT technology 154
Subtask 10.1.3 Distinguish the advantages and disadvantages of NAT 156
Subtask 10.1.4 Master NAT classification 157
Task 10.2 Principle analysis of source NAT technology 157
Subtask 10.2.1 Implementation of source NAT without port translation 157
Subtask 10.2.2 Implementation of source NAT with port translation 158
Subtask 10.2.3 Internship of Easy IP 159
Subtask 10.2.4 Understand NAT ALG 160
Subtask 10.2.5 Implementation process analysis of NAT ALG 161
Task 10.3 Implementation of server mapping 162
Subtask Understand the relationship between NAT server and Server-map table 163
Task 10.4 NAT application scenario and configuration 164
Subtask 10.4.1 Source NAT configuration of firewall 165
Subtask 10.4.2 Firewall NAT server configuration 167
Subtask 10.4.3 Two-way NAT technology 168
Project 11 Dual-system hot backup technology of firewall 171
Task 11.1 Principle analysis of dual-system hot standby technology 171
Subtask 11.1.1 Understand the causes of hot standby technology 171
Subtask 11.1.2 Understand router redundancy deployment scheme 171
Subtask 11.1.3 The application of VRRP in multi-zone firewall networking 172
Subtask 11.1.4 Understand the defects of VRRP in firewall application 173
Task 11.2 Working principle analysis of VGMP 174
Subtask 11.2.1 Master the basic principles of VGMP 174
Subtask 11.2.2 Understand VGMP group management 175

Task 11.3 Analysis of HRP working principle..176
Subtask 11.3.1 Understand the basic concepts of HRP..........................176
Subtask 11.3.2 Master the working principle of HRP..............................177
Subtask 11.3.3 Understand the status of the heartbeat interface..................177
Subtask 11.3.4 Master the backup mode of dual-system hot standby 179
Task 11.4 Basic networking and configuration of dual-system hot standby.....181
Subtask 11.4.1 Dual system hot standby configuration command-CLI.......181
Subtask 11.4.2 Example of firewall dual-system hot standby configuration-CLI..183
Subtask 11.4.3 Example of firewall dual-system hot standby configuration-Web 185
Project 12 Firewall user management.. 187
Task 12.1 Principle analysis of user authentication and AAA technology.......187
Subtask 12.1.1 Understand the background of user authentication.............187
Subtask 12.1.2 Know AAA..188
Subtask 12.1.3 Understand the working mechanism of RADIUS protocol.....191
Subtask 12.1.4 Understand the working mechanism of HWTACACS protocol 193
Subtask 12.1.5 Understand how LDAP works..194
Subtask 12.1.6 Master user authentication classification..........................196
Task 12.2 User management and application 197
Subtask 12.2.1 Understand the organizational structure and classification of users...197
Subtask 12.2.2 Administrator certification process and configuration 199
Subtask 12.2.3 Authentication process of Internet users and access users 203
Subtask 12.2.4 Single sign-on...205
Subtask 12.2.5 User Portal authentication...208
Subtask 12.2.6 Access user authentication 209

Task 12.3 Formulation of user authentication strategy........210
Task 12.4 User authentication configuration case........212
Subtask 12.4.1 Configure servers........214
Subtask 12.4.2 Users and user groups........215
Subtask 12.4.3 Configure authentication options........217
Subtask 12.4.4 Configure authentication policies........219
Subtask 12.4.5 Import users through local and server........219
Subtask 12.4.6 Online user management........220
Project 13 Intrusion prevention........222
Task 13.1 Overview of invasion prevention........222
Subtask 13.1.1 Understand the current situation of cyber threats........222
Subtask 13.1.2 Understand hacking........223
Subtask 13.1.3 Understand the threat of denial of service attacks........223
Subtask 13.1.4 Understand virus and malware security threats........224
Subtask 13.1.5 Understand the definition of intrusion........225
Task 13.2 Understanding intrusion prevention system........225
Subtask 13.2.1 Understand the position of safety equipment in the safety system........225
Subtask 13.2.2 Master the intrusion prevention system........227
Subtask 13.2.3 Check intrusion and defense behavior........228
Task 13.3 Understanding network antivirus........229
Subtask 13.3.1 Understand the basic concepts of computer viruses........229
Subtask 13.3.2 Distinguish the relationship between viruses, worms and Trojans........230
Subtask 13.3.3 Analysis of antivirus technology........230
Subtask 13.3.4 Implementation of gateway antivirus........231
Subtask 13.3.5 Analysis of working principles of firewall antivirus........232
Subtask 13.3.6 Antivirus response action........234
Subtask 13.3.7 View firewall antivirus results........234

Section 4 Encryption decryption and application 236

Project 14 Understand the principles of encryption and decryption 236
Task 14.1 Understand the development of encryption technology 237
Subtask 14.1.1 Understand the encryption technology 237
Subtask 14.1.2 Understand the function of encryption technology 237
Subtask 14.1.3 Understand the history of encryption technology 238
Task 14.2 Master the principle of encryption and decryption technology 239
Subtask 14.2.1 Master the classification of encryption technology 239
Subtask 14.2.2 Data encryption application scenario — digital envelope 241
Subtask 14.2.3 Data verification application scenario — digital signature 242
Task 14.3 Understand common encryption and decryption algorithms 243
Subtask 14.3.1 Understand symmetric encryption algorithms 243
Subtask 14.3.2 Understand asymmetric encryption algorithms 245
Subtask 14.3.3 Understand hashing algorithms 245
Project 15 PKI certificate system 246
Task 15.1 Understand digital certificates 246
Subtask 15.1.1 Understand the definition of digital certificate 246
Subtask 15.1.2 Analysis of digital certificate structure 248
Subtask 15.1.3 Analysis of digital certificate file format 249
Task 15.2 Understand PKI architecture 249
Subtask 15.2.1 Understand the basic concepts of PKI 250
Subtask 15.2.2 Analysis of PKI architecture 250
Subtask 15.2.3 Understand PKI life cycle 252
Task 15.3 Analysis of PKI working mechanism 254
Subtask Certificate application scenario 257
Project 16 Encryption technology application 259
Task 16.1 Application of cryptography 259
Task 16.2 Master the classification and configuration of VPN 260

Subtask 16.2.1 Understand the definition of VPN 260
Subtask 16.2.2 Master VPN classification 261
Subtask 16.2.3 Implementation of L2TP VPN 263
Subtask 16.2.4 Implementation of GRE VPN 265
Subtask 16.2.5 Implementation of IPSec VPN 266
Subtask 16.2.6 Implementation of SSL VPN 270
Task 16.3 L2TP VPN application case configuration 272
Task 16.4 GRE VPN application case configuration 275
Task 16.5 Point-to-point IPSec VPN application case configuration 277

Section 5 Fundamentals of security operations and analysis 281

Project 17 Overview and core elements of cybersecurity operation 282
Task 17.1 Understand safe operation 282
Subtask 17.1.1 Understand the concept of safe operation 282
Subtask 17.1.2 Master the basic requirements for safe operation 282
Task 17.2 Understand the content of safe operation 284
Subtask 17.2.1 Understand business continuity plans 284
Subtask 17.2.2 Understanding incident response management 287
Subtask 17.2.3 Understand disaster recovery plans 288
Subtask 17.2.4 Understand investigation and evidence collection 289
Project 18 Data monitoring and analysis 290
Task 18.1 Active analysis method and its implementation 290
Subtask 18.1.1 Security scan 291
Subtask 18.1.2 Manual audit 295
Subtask 18.1.3 Penetration test 296
Subtask 18.1.4 Questionnaire 297
Subtask 18.1.5 Interview survey 297
Task 18.2 Passive acquisition method and its implementation 298
Subtask 18.2.1 Passive acquisition 298

Subtask 18.2.2 Grab a bag 298
Subtask 18.2.3 Port mirroring 300
Subtask 18.2.4 Network log 300
Subtask 18.2.5 Operating system log 304
Task 18.3 Method and implementation of data analysis 307
Subtask 18.3.1 Log analysis 307
Subtask 18.3.2 Network log analysis 308
Subtask 18.3.3 Operating system log analysis 308
Subtask 18.3.4 Introduction to analysis tools 310
Project 19 Electronic evidence collection 311
Task 19.1 Understanding electronic evidence collection 311
Subtask 19.1.1 Understand computer crime 311
Subtask 19.1.2 Understanding electronic evidence 312
Task 19.2 Electronic forensics process and implementation 315
Subtask 19.2.1 Principle of obtaining evidence 315
Subtask 19.2.2 Electronic forensics process 315
Subtask 19.2.3 Protect the scene 316
Subtask 19.2.4 Obtain evidence 316
Subtask 19.2.5 Preserve evidence 318
Subtask 19.2.6 Identification evidence 320
Subtask 19.2.7 Analyze evidence 321
Subtask 19.2.8 Track it 321
Subtask 19.2.9 Show evidence 323
Project 20 Case study of information security 323
Task 20.1 Discussion on operation steps of information security deployment 323
Task 20.2 Network security case discussion 324

Section 1

Overview of security and information security

Xiaoyi had the honor to be an intern in a large enterprise, and his position was the operation and maintenance of the company's computer network. During the work and study, he found that the enterprise's business needs more and more information networking and cloudization, which doubled the work efficiency. However, at the same time, events endangering information security, such as webpage tampering, viruses, Trojans, illegal intrusion, data leakage, website deception, DOS attacks and vulnerability attacks, emerged one after another, which brought serious threats to information security.

This section breaks down the security and information security into five projects:

Understand the basic concepts of information security: This project describes the basic concept of information security and how to ensure it, and lists information security risks and how to assess and avoid these risks.

Understand the computer networks: Understand the basic concept of network, such as the basic communication principle of network, the composition of network and common network protocols. On the basis of understanding the network protocols, we can understand the network security threats, and then deploy security defense strategies.

Master the configuration of common network equipment: Know the working principle and configuration of common network equipment, such as switches, routers, firewalls, etc., and make relevant configuration on the network equipment to connect the network or meet the network security requirements.

Understand common information security threats: Introduce the threats that information systems may suffer.

Understand the development trend of threat prevention and information security: Specific preventive measures of security threats, preventive awareness and information security development trend.

Through these five projects, readers can clearly understand the concepts, norms and precautions of information security, and master the basic configuration methods of common network equipment in common network equipment projects.

Project 1 Understand the basic concepts of information security

In the process of data communication, various unsafe factors will cause information leakage, incompleteness, unavailability and other problems, so information security must be ensured in the communication process.

This project describes the basic concepts of information security and how to ensure information security, lists information security risks and how to assess and avoid these risks.

Task 1.1 Understand the concepts of information and information security

Subtask 1.1.1 Understand the definition of information and information security

Information is given a specific meaning by imposing certain agreed methods on the data. In a general sense, information can be understood as news, signals, data, intelligence or knowledge. Information exists in many forms. It can be data and programs stored and processed in information facilities; it can be printed or written papers, emails, design drawings, and business plans; it can also be displayed on films such as media or expressed in messages in the conversation.

In the 20th century, the greatest achievement of mankind in the field of science and technology was the invention and manufacture of electronic computers. In order to continuously improve its performance and expand the range of computer functions and applications, scientists and technicians around the world have been tirelessly experimenting and improving. In the process of computer upgrading, electronic technology, digital technology, communication technology and network technology have been continuously integrated and widely used, so that the Internet technology with computers as the main load can break through the limitations of time and space and spread globally, and thus create an information age marked by electronic information exchange. With the rapid development of my country's science and technology, especially information technology and network technology, as well as the continuous advancement of the informatization process, various information systems have become the country's key infrastructure. They support network communications, e-commerce, e-government, and e-finance, e-taxation, online education, public security, medical care, social welfare and other applications. Compared with traditional systems, the characteristics of digital networks make the operation of information systems fundamentally different in terms of information collection, storage, data exchange, data processing, and information transmission. Therefore, whether it is in the storage, processing and application on the computer, or in the exchange and transmission of the communication network, the information may be leaked due to unauthorized access, incomplete due to tampering and destruction, or unrecognized due to being replaced by impersonation, it is more likely to be inaccessible due to blocking and interception. The above-mentioned problems are the Achilles heel of network information security.

Information security refers to the use of computer software and hardware technology, network technology, key technology and other security technologies and various organizational and management measures to protect the confidentiality of information in each link of its life cycle: generation, transmission, exchange, processing and storage, integrity and availability are not destroyed. If information assets are damaged, it will not only affect personal privacy and property, but also affect the normal operation and sustainable development of organizational systems, and even directly affect national security. Therefore, the task of information security is to take measures (technical means and effective management) to protect information assets from threats. Its purpose is to protect the hardware and software of the information network and the data in the system from being damaged, altered, or leaked due to accidental or malicious reasons, so that the system can operate continuously and reliably, and the information service is not interrupted.

Subtask 1.1.2 Grasp the attributes of security

Information security involves the confidentiality, integrity, availability, controllability and non-repudiation of information. Taken together, it is to ensure the effectiveness of electronic information. Confidentiality is to counter the passive attack of opponents and ensure that information is not leaked to unauthorized persons. Integrity is to resist the adversary's active attack and prevent the information from being tampered with by unauthorized persons. Availability is to ensure that information and information systems are actually used by authorized users. Controllability is the implementation of security monitoring of information and information systems.

Confidentiality: Confidentiality, also known as security, is a feature that ensures that information can only be obtained by those authorized to use it. The information that all personnel can access is public information, and the information that needs to be restricted to access is generally sensitive or secret. Secrets can be classified into different levels of confidentiality according to the importance of the information or confidentiality requirements. For example, the state divides state secrets into three levels: A (secret level) , B (confidential level) , and C (top secret level) according to the impact (consequence) of the disclosure of secrets on the national economy and security interests. Secrets cannot be leaked to unauthorized users and cannot be used illegally. Even if the unauthorized users obtain the information, they cannot know the content of the information. Confidentiality usually uses access control to prevent unauthorized users from obtaining confidential information, and encryption technology to prevent unauthorized users from obtaining information.

Integrity: Integrity refers to the characteristics of information that is not modified, damaged, inserted, delayed, out of order, and not lost during the process of storage, transmission, and extraction. Generally, access control is used to prevent tampering, and information digest algorithms are used to check whether information has been tampered with. Integrity is the characteristic that data cannot be changed without authorization. Its purpose is to ensure that the data on the information system is in a complete and undamaged state.

Usability: Usability refers to the characteristic that information can be accessed by legitimate users and used in the required order, that is, the required information can be accessed when needed. Availability is a measure of the service function and performance reliability of information resources, and it is a requirement for the overall reliability of information systems. At present, to ensure that the system and network can provide normal services, there is no particularly effective method other than backup and redundant configuration.

Controllability: Controllability refers to the ability to control the flow of information and behavior within the scope of authorization, and has the ability to control the dissemination and content of information. In order to ensure controllability, users are usually authenticated through handshake protocols and authentication, access control lists are used to control user access methods, and all user activities are monitored, inquired, and audited through logging.

Non-repudiation: Non-repudiation refers to ensuring that users cannot deny afterwards the behaviors of generating, issuing, and receiving information. It is a security requirement for the authenticity and identity of information in all aspects of communication. Generally, digital signatures and notarization mechanisms are used to ensure non-repudiation.

Subtask 1.1.3 Understand the development of information security

Since the development of information security, it has gone through three stages: the communication confidentiality stage, the information security stage, and the information assurance stage, as shown in Figure 1-1.

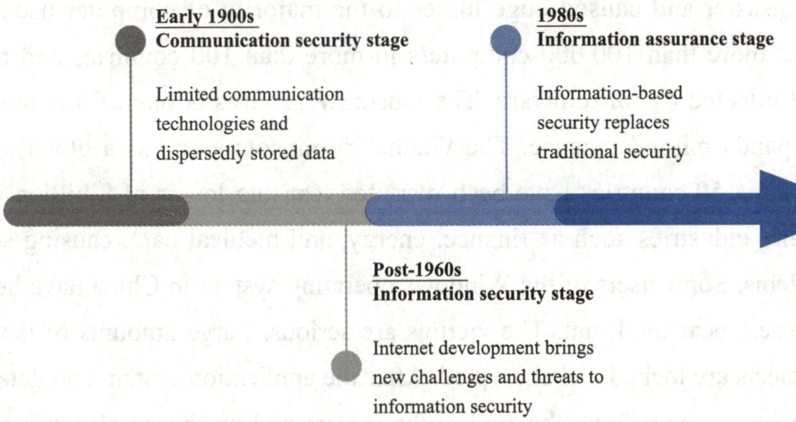

Figure 1-1 Information security development history

Communication security stage: In the early twentieth century, communication technology was not developed in the communication security stage. Data was only scattered in different locations. The security of the information system was limited to ensure the physical security of information and solving through passwords (mainly serial passwords). Confidentiality of communication security. Placing the information in a relatively safe place and not allowing unauthorized users to approach it can basically guarantee the security of the data.

Information security stage: Since the 1960s, due to the rapid development of Internet technology, both internal and external information of the enterprise has been greatly opened up. The resulting information security issues have spanned time and space. Information security the focus has been derived from the traditional three principles of confidentiality, integrity and availability into other principles and goals such as controllability and non-repudiation.

Information assurance stage: In recent years, information security has entered the business-oriented security assurance stage, and information security issues are considered from multiple angles. In terms of business, different business flows have different risk points and defense methods; In terms of security systems, more technical means are used to link security management and technical protection, and actively defend against attacks rather than passive protection; In terms of management, cultivate safety management personnel establish a safety management system.

Subtask 1.1.4 Case study of information security

In 2017, WannaCry (also known as Wanna Decryptor) , a ransomware virus software, began to spread on the Internet. It uses the loopholes in port 445 of the Windows operating system to spread, and has the characteristics of self-replication and active transmission. The size of this "worm-like" ransomware is 3.3 MB, and it was spread by criminals using the dangerous vulnerability

"EternalBlue" leaked by the NSA (National Security Agency). The ravages of ransomware caused a global Internet disaster and caused huge losses to the majority of computer users. The latest statistics show that more than 100,000 computers in more than 100 countries and regions have been attacked and infected by ransomware. The ransomware virus is one of the most influential viruses since the panda burned incense. The WannaCry ransomware has a global outbreak. At least 300,000 users in 150 countries have been recruited, causing losses of 8 billion U.S. dollars. It has affected many industries such as finance, energy, and medical care, causing serious crisis management problems. Some users of the Windows operating system in China have been infected. Campus network users bear the brunt. The victims are serious. Large amounts of laboratory data and graduation projects are locked and encrypted. After the application system and database files of some large enterprises are encrypted, the application system and database will not work normally, causing a huge impact.

Subtask 1.1.5 Cause analysis of information security incidents

Through in-depth analysis of the above-mentioned information security cases, it is found that the reason for the information security incidents is that viruses, vulnerabilities, trojan horses, backdoor programs, and DDOS attack capabilities exist in the network on the surface, but the deeper root causes are:

(1) Information system complexity: In the design and work of the information system, it may be attacked due to its own vulnerabilities and defects, which mainly include the following three types of problems:

① Complicated process: From a design point of view, the consideration of safety in the priority during design is placed in a secondary position relative to factors such as ease of use and implementation. Due to the weakness of human nature and the imperfect design methodology, There will always be loopholes in the information system.

② Complex structure: The information system may have multiple terminals and data services. There may be multiple terminals in a network, such as employee terminals, remote users, mobile terminals, routing equipment, servers, etc. At the same time, there may be multiple types of data in a network, such as business data, management data, voice data, etc. All terminals and data types must be considered when managing network security.

③ Complex application: Priority is given to network redundancy and network stability when designing the network topology. Redundant links and backup devices may be added. The complex application of the network also makes it impossible to quickly locate and resolve security issues.

(2) Man-made and environmental: Mainly includes environmental threats and man-made destruction.

Subtask 1.1.6 Understand the significance of building information security

The significance of building information security is reflected in two aspects: importance and applicability.

The importance is reflected in the fact that the information network has become the foundation of economic prosperity, social stability and national development. It profoundly affects the adjustment of the national strategy for the integration of the global economy and the change of security concepts. Therefore, information security has changed from a purely technical issue to a global issue related to national security.

Applicability is reflected in the application of information security to many technical fields, such as military computer communications command and control and intelligence (C4I) systems, e-commerce systems, biomedical systems, intelligent transportation systems (ITS) , etc.

Task 1.2 Understand information security risks and management

The risks involved in information security are shown in Figure 1-2.

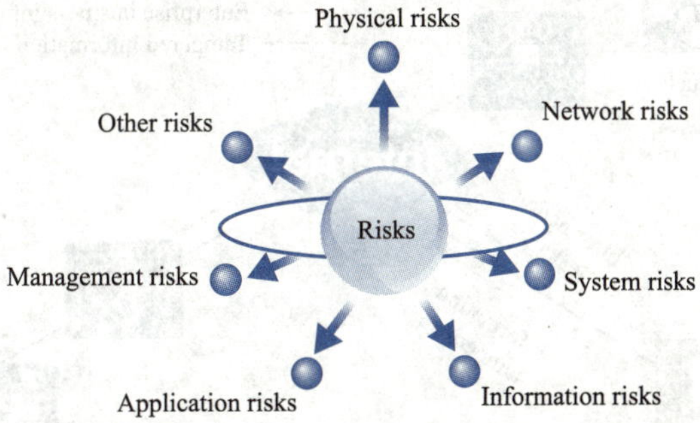

Figure 1-2 Risks involved in information security

Subtask 1.2.1 Physical risks analysis

Physical risks refers to factors that are tangible and can directly affect the physical functions of things, such as earthquakes, severe weather, collapse of houses, mass deaths of people due to disease transmission, etc., which causes or increases the chance of personal or property loss and the extent of loss. Physical risks are reflected in:

(1) Equipment anti-theft and anti-destroy.

(2) The link is aging, man-made damage, bitten by animals, etc.

(3) Failure of the network equipment itself.

(4) The network equipment cannot work due to power outages.

(5) Electromagnetic radiation in the computer room.

Subtask 1.2.2 Information risks analysis

Information risks mainly refers to the fact that in the process of information transmission and communication, due to the inaccuracy of information and the inability to reach the recipient in time, it may lead to managerial personnel's decision-making errors. Information risk is reflected in:

(1) Information storage security: Information storage security includes the protection of server disks, and the encryption of stored information to prevent theft.

(2) Information transmission security: As shown in Figure 1-3, the information transmission security may be stolen by an attacker when transmitting corporate business information between the headquarters and subordinate organizations. The attacker may tamper with the information and send it out as an error message.

(3) Information access security: Information access security is shown in Figure 1-4. When a remote user accesses the company's internal network resources, illegal users appear to imitate legitimate users.

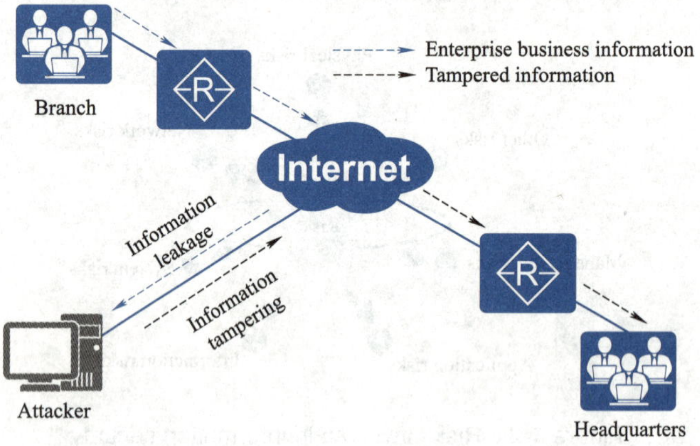

Figure 1-3 Information transmission security

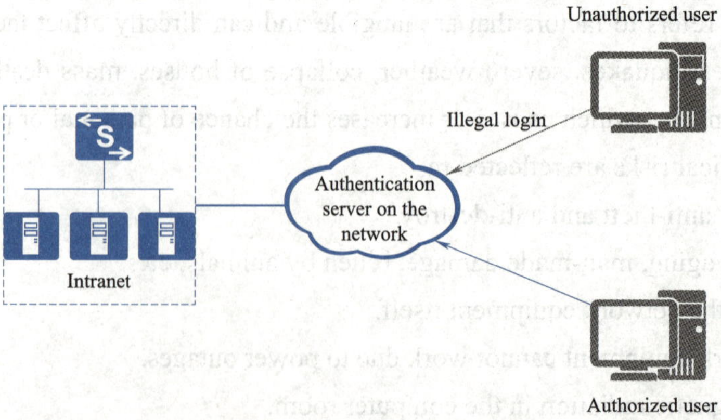

Figure 1-4 Information access security

Subtask 1.2.3 System risks analysis

The system risks is mainly caused by the reasonable degree of the software system architecture and its adaptability to external changes. The system risk is reflected in:

(1) Database system configuration security.

(2) Security database.

(3) Security of services running in the system.

Subtask 1.2.4 Application risks analysis

Application risks mainly refers to the risk arising from network connection or operation in the application system or software process, especially in the network environment. Application risk is reflected in:

(1) Network virus.

(2) Operating system security.

(3) E-mail application security.

(4) Web service security.

(5) FTP service security.

(6) DNS service security.

(7) Business application software security.

Subtask 1.2.5 Network risks analysis

Network security includes the security of network structure and network system, and is the foundation and technical support of the entire network security. The secure network structure adopts a layered system structure, which is convenient for maintenance and management, security control and function expansion, and is equipped with redundant links and firewalls, intrusion detection and other equipment; Network system security mainly involves access control and effective isolation of internal and external networks, isolation and access control of different areas of the internal network, network security testing, auditing and monitoring (recording the activity process of users using computer network systems) , network antivirus and network backup, etc. The content is shown in Figure 1-5.

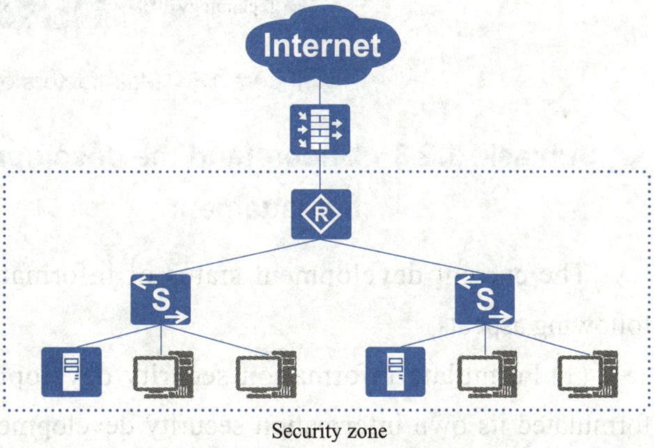

Figure 1-5 Network security

Subtask 1.2.6 Management risks analysis

Whether there are management risks in the information system can be discussed from the following aspects:

(1) National policy: Has the country formulated sound information security regulations, and has the country established a special agency to manage information security.

(2) Enterprise system: The company formulates safety management rules and a computer room management system with clear responsibilities and powers. The company may consider establishing its own safety management organization.

(3) Management system: Clear and effective safety strategies, high-quality safety management personnel, and effective supervision and inspection system to ensure the smooth implementation of rules and regulations.

Subtask 1.2.7 Understand the importance of information security management

Security technology is only a means of information security control. For security technology to play its due role, it must be supported by appropriate management procedures. According to statistics, most of the company's information loss is caused by the negligence or deliberate leaks of internal employees, as shown in Figure 1-6. This shows that paying attention to the key role of management for information security is particularly important for realizing the goal of information security.

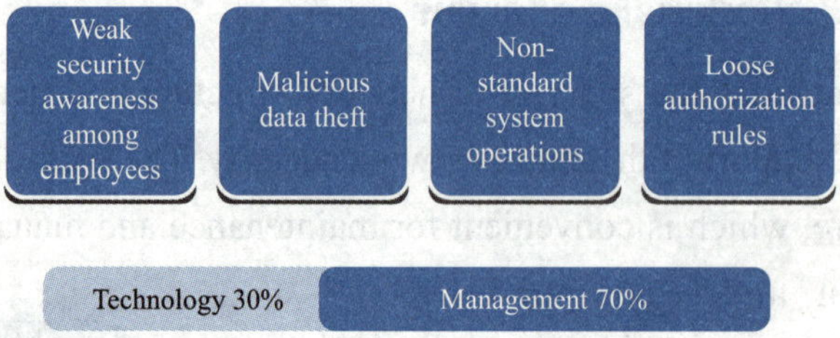

Figure 1-6 Main factors of information leakage

Subtask 1.2.8 Understand the development status of information security management

The current development status of information security management is reflected in the following aspects:

(1) Formulate information security development strategies and plans: Each country has formulated its own information security development strategies and development plans to ensure that information security develops in the right direction.

(2) Strengthen information security legislation to achieve unified and standardized management: To stipulate and standardize information security work in the form of law is the most

beneficial guarantee for effective implementation of security measures.

(3) Stepping into the era of standardization and systematic management: Information security management entered the era of standardization and systematic management in the 1990s. Among them, the 27000 standard system formulated by ISO/IEC is the most well-known.

Project 2 Understand the computer networks

Before learning security technology, we should first understand the basic concepts of the network, such as the basic communication principles of the network, the composition of the network and common network protocols. On the basis of understanding network protocols, we can understand network security threats and deploy security defense strategies.

Task 2.1 TCP/IP architecture analysis

Subtask 2.1.1 Analysis of typical campus network architecture

Before learning the TCP/IP architecture, first understand the typical campus network architecture, as shown in Figure 2-1. In large and medium-sized networks, the network function structure is usually decomposed in a modular manner. However, within each module, there are still problems of structural expansion and flexibility. For example, a campus network needs to access a large number of users. This problem is generally solved through the hierarchical network. Traditional networks use three levels: core, convergence, and access. Each performs its duties. The

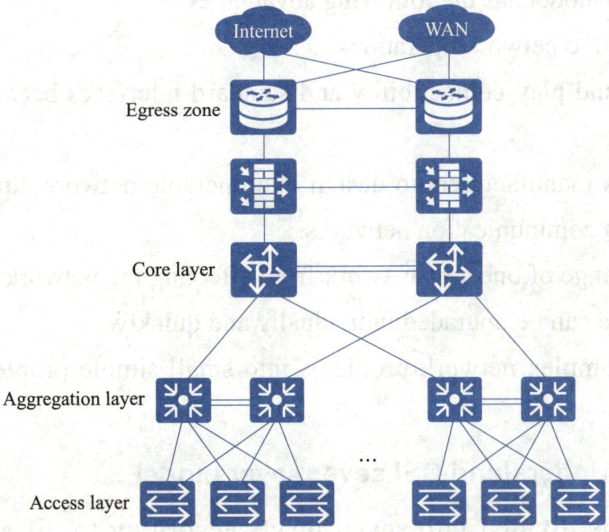

Figure 2-1 Typical campus network architecture

core layer provides high-speed data access, the aggregation layer performs traffic convergence and control strategies, and the access layer provides multiple access methods for terminals.

If we only focus on physical interconnection here, we can think of a computer network as a super large network composed of several network devices and transmission media interconnected. It is a huge and complicated project to finally realize the communication between tens of billions of terminals on this network.

Subtask 2.1.2 Introduction of OSI reference model

In order to achieve efficient mutual visits between tens of billions of terminals around the world, ISO (International Organization for Standardization) proposed the famous OSI (Open System Interconnect) model, The design purpose of the OSI model is to become an open network interconnection model to overcome the interconnection difficulties and inefficiencies caused by the use of many network models.

The OSI reference model soon became the basic model of computer network communication. The following principles were followed in the design:

(1) There are clear boundaries between each layer for easy understanding.

(2) Each layer implements specific functions and does not affect each other.

(3) Each layer is both a server and a served, and each layer is served by the next layer while providing services for the upper layer.

(4) The division of levels is conducive to the formulation of international standard agreements.

(5) The number of levels should be large enough to avoid duplication of functions of each level.

The OSI reference model has the following advantages:

(1) Simplify related to network operations.

(2) Provide plug-and-play compatibility and standard interfaces between different manufacturers.

(3) Enable various manufacturers to design interoperable network equipment and accelerate the development of data communication networks.

(4) Prevent the change of one area network from affecting the network of another area, so that the network of each area can be upgraded individually and quickly.

(5) Decompose complex network problems into small simple problems, easy to learn and operate.

Subtask 2.1.3 Understand OSI seven-layer model

The OSI model is divided into seven levels according to different functions in the communication process, from bottom to top as the physical layer, data link layer, network layer, transport layer, session layer, presentation layer and application layer. As shown in Figure 2-2.

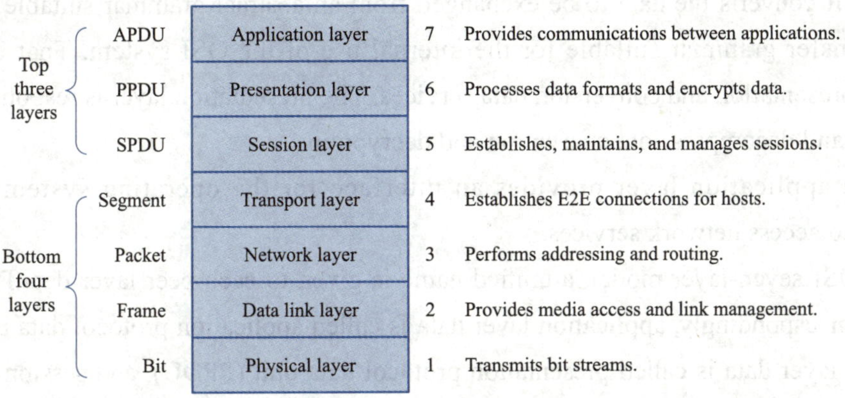

Figure 2-2 OSI seven-layer model

(1) The physical layer involves the original bit stream transmitted on the communication channel, and is the basis of the OSI reference model, which implements the mechanical and electrical functional characteristics required for data transmission. It does not care about the meaning represented by each bit stream (0,1) (for example, it represents an address or application data) , but only cares about how to transmit the bit stream to the opposite end through different physical links. Typical devices such as repeaters and hubs are physical layer devices.

(2) The main task of the link layer is to provide control over the physical layer, detect and correct possible errors, make it appear an error-free line to the network layer, and perform flow control.

(3) The network layer checks the network topology to determine the best route to transmit packets, and forwards the data packets. The key issue is to determine how the data packet is routed from the source to the destination. The network layer device calculates the best route to the destination by running the routing protocol, finds the next network device to which the data packet should be forwarded, then uses the network layer protocol to encapsulate the data packet, and uses the services provided by the lower layer to send the data to the next network device.

(4) The transport layer is located at the fourth layer of the OSI reference model. The ultimate goal is to provide users (generally referring to the application layer process) with effective and reliable services.

(5) In the high level of the conversation layer and above, the unit of data transmission is no longer named separately, and is collectively referred to as a message. The session layer does not participate in specific transmissions. It provides mechanisms for establishing and maintaining communication between applications, including access verification and session management. For example, the server verifies that the user logs in, and that is done by the session layer.

(6) The presentation layer mainly solves the problem of grammatical presentation of user information. It converts the data to be exchanged from an abstract grammar suitable for a certain user to a transfer grammar suitable for the internal use of the OSI system. That is to provide formatted representation and conversion data services. The presentation layer is responsible for data compression and decompression, encryption and decryption.

(7) The application layer provides an interface for the operating system or network applications to access network services.

In the OSI seven-layer model, a unified name is given to each peer layer data:PDU(protocol data unit). Correspondingly, application layer data is called application protocol data unit (APDU), presentation layer data is called presentation protocol data unit (PPDU), and session layer data is called session protocol data unit (SPDU). Usually, we call the transport layer data a segment, the network layer data is called a data packet, the data link layer is called a frame, and the physical layer data is called a bit stream.

Note:

Encapsulation means that the network node packs the data to be transmitted with a specific protocol header to transmit the data. Similarly, when performing data processing at certain layers, a message will be added to the end of the data, which is also called encapsulation at this time. Each layer of the OSI seven-layer model encapsulates the data to ensure that the data can reach the destination correctly and be accepted and executed by the terminal host.

Subtask 2.1.4 Understand TCP/IP model

Although the OSI reference model is very good, many companies have already launched their own computer network products and protocols at that time. If all the newly defined protocols in the OSI reference model are adopted, it will bring a lot of economic losses to these companies and their customers. And the OSI reference model cannot be released to the public for a long time in the laboratory.

Therefore, after organizations and teams organized and sorted out the existing protocols, they launched the TCP/IP (transfer control protocol/internet protocol) model. This is a very snobbery model. It mainly studies some problems in network interconnection. When there is a problem in the network connection process, it is considered to solve it. Make use of reality to make corrections. Over the course of a few years, manufacturers have found that TCP/IP is also very useful, so they have occupied the entire market. When OSI came out of the laboratory, it was discovered that the real world was already occupied by TCP/IP.

The difference between the TCP/IP model and the OSI reference model is that TCP/IP puts the presentation layer and the session layer into the application layer, so the TCP/IP model is divided into four layers from bottom to top: network interface layer, network layer, transport layer and application layer. In some documents, it is divided into five layers for better understanding, and the physical layer is listed separately. Such a model is called a TCP/IP peer-to-peer model, as shown in Figure 2-3.

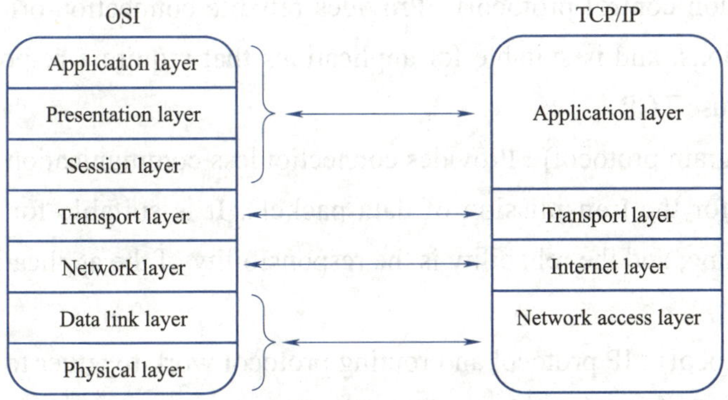

Figure 2-3 TCP/IP model

Subtask 2.1.5 Master the functions of each layer of TCP/IP protocol stack

Each layer of the TCP/IP protocol has a corresponding related protocol, and it is produced by reaching a certain network application. For some protocols, it cannot be strictly defined in terms of layers. For example, ICMP, IGMP, ARP, RARP protocols we put them on the same layer as the network layer IP protocol. But in some occasions, we may put ICMP and IGMP on the upper layer of the IP protocol, and put ARP and RARP on the lower layer of the IP protocol, as shown in Figure 2-4.

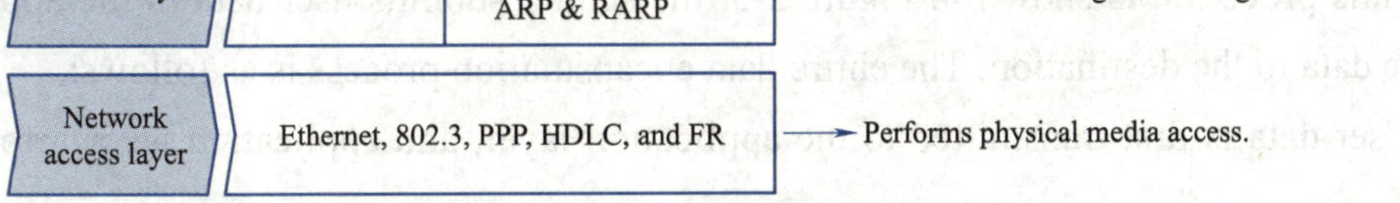

Figure 2-4 The role of each layer of the TCP/IP protocol stack

(1) Application layer:

HTTP (hypertext transfer protocol) : Used to access various pages on the WWW server.

FTP (file transfer protocol) : Provides a way for file transfer, which allows data to be transferred from one host to another.

DNS (domain name service system) : Used to realize the conversion from host domain name to IP address.

(2) Transport layer:

TCP (transmission control protocol) : Provides reliable connection-oriented communication services for applications, and is suitable for applications that require a response. Currently, many popular applications use TCP.

UDP (user datagram protocol) : Provides connectionless communication and does not provide reliable guarantees for the transmission of data packets. It is suitable for transmitting a small amount of data at a time, and the reliability is the responsibility of the application layer.

(3) Internet layer:

IP (Internet protocol) : IP protocol and routing protocol work together to find the optimal path that can transmit data packets to the destination. The IP protocol does not care about the content of the data message and provides a connectionless and unreliable service.

ARP (address resolution protocol) : Resolve a known IP address into a MAC address.

RARP (reverse address resolution protocol) : Used to resolve the IP address when the data link layer address is known.

ICMP (Internet control message protocol) : Defines the function of network layer control and message delivery.

IGMP (Internet group management protocol) : Used for multicast group member management.

(4) Network access layer:

The network access layer is divided into two sublayers: LLC (logic link control sublayer) and MAC (media access control sublayer).

Subtask 2.1.6 Analysis of encapsulation and decapsulation process of TCP/IP protocol stack

When a host needs to access an application of another host, how does the computer network complete this process? As shown in Figure 2-5, the sender submits user data to the application to deliver the data to the destination. The entire data encapsulation process is as follows:

(1) User data is first transmitted to the application layer, and application layer information is added.

(2) After the application layer processing is completed, the data will continue to be transmitted to the lower transport layer, adding transport layer information (such as TCP or UDP, the application layer protocol has specified whether it is TCP or UDP).

(3) After the transport layer processing is completed, the data will continue to be transmitted to the lower network layer, adding network layer information (such as IP).

(4) After the network layer processing is completed, the data will continue to be transmitted to the lower data link layer, adding data link layer information (such as Ethernet, 802.3, PPP, HDLC, etc.), and then transmitted to the opposite end in a bit stream (depending on different types of equipment in the middle, choose a different processing method. The switch generally only processes the data link layer information, while the router performs higher-level network layer processing. Only when it reaches the final destination can the original user data be restored).

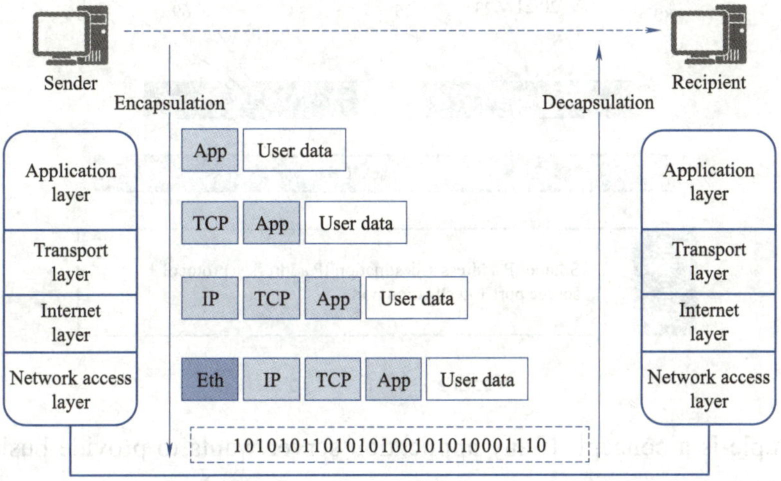

Figure 2-5 TCP/IP protocol stack encapsulation and decapsulation process

After the user data arrives at the destination, the decapsulation process will be completed:

(1) The data packet is first transmitted to the data link layer, after parsing, the data link layer information is stripped, and the network layer information is known according to the analysis information, such as IP.

(2) After the network layer receives the data packet, the network layer information is stripped after parsing, and the upper layer processing protocol, such as TCP, is known based on the parsing information.

(3) After the transport layer (TCP) receives the data packet, the transport layer information is stripped after parsing, and the upper layer processing protocol, such as HTTP, is known based on the parsing information.

(4) After the application layer receives the data packet, the application layer information is stripped after parsing, and the final user data displayed is exactly the same as the data sent by the sender host.

In the data communication process, the application layer and the transport layer provide end-to-end services, and the network layer and data link layer provide segment-to-segment services.

Subtask 2.1.7 Understand the five-tuple

In the process of data communication and when the terminal generates and processes data, how to uniquely identify a piece of traffic? It can be identified by the five-tuple. As shown in Figure 2-6, the five-tuple composition: source IP address, destination IP address, protocol, source port, and destination port. The protocol information, for example, TCP is 6 and UDP is 17.

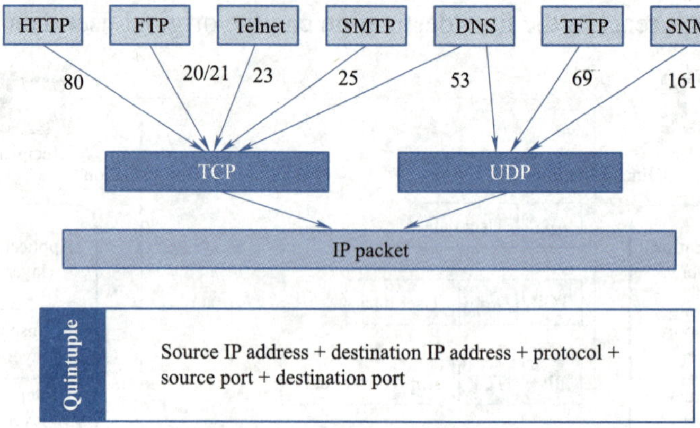

Figure 2-6 Five-tuple

The five-tuple is a concept. If any application server wants to provide business services to the outside world, it needs to register on TCP/UDP port during service startup in order to respond to business service requests. Through the five-tuple, the application server can respond to any concurrent service request, and can ensure that each link is unique in the system.

Destination port: Generally, well-known application services have standard ports, such as HTTP, FTP, Telnet, etc. Some of these applications are generally defined by developers themselves due to non-popular reasons, but they must be registered on the same server. The service port is unique.

Source port: Generally, ports of 1024 and above are used for incremental allocation, but some operating systems may use higher port numbers as their initial ports for incremental allocation. Because the source port is unpredictable, it is rarely involved in ACL policies.

Task 2.2 Understand network layer protocols

Subtask 2.2.1 Working process analysis of ARP protocol

In a local area network, the network structure is often connected to multiple computers by a switch, as shown in Figure 2-7.

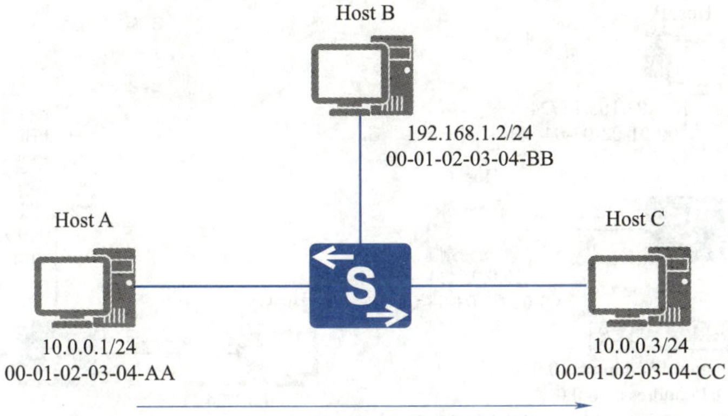

Figure 2-7 ARP protocol

For example, the host with the IP address of 10.0.0.1 wants to send data to the host with the IP address of 10.0.0.3. For the application layer of the 10.0.0.1 host, the data is passed to the transport layer of the host, and then to the network layer of the host. For the network layer, it is known that the data is to be sent to the 10.0.0.3 host, and then the data is handed over to the link layer (Ethernet). The problem is that the network layer knows that the data should be sent to the host with the IP address 10.0.0.3, and the identifier used by the link layer is the MAC address. The network layer passes the data to the link layer, so the link layer must know 10.0.0.3 MAC address of this host, only then can the data be sent to the physical layer. So how can we know the MAC address of the host 10.0.0.3? There are two ways. One is that the administrator tells the host how much the MAC address of the host 10.0.0.3 is, and the other is to resolve the MAC address of the host 10.0.0.3 through the ARP protocol.

ARP is used when packets are forwarded to a host or gateway on the same network segment, the destination address is known, the MAC address corresponding to the destination address is obtained, and the MAC address is used for communication within the same network segment.

The principle of the ARP protocol is relatively simple. As shown in Figure 2-8, when the 10.0.0.1 host needs to know the MAC address of the 10.0.0.3 host, the 10.0.0.1 host will inquire on the network (send the data to all hosts, this method is called broadcast) "10.0.0.3, what is your MAC address", the switch guarantees that this message (data message/frame) can be received by all hosts in the network. For 192.168.1.2 this host will not care about this information, because 10.0.0.1 is not ask for its own MAC address. The host 10.0.0.3 receives the message and processes it, and caches the correspondence between the IP address and MAC address of the 10.0.0.1 host (the inquiry message just now contains the sender's MAC address and IP address information), and writes it to the computer in the ARP cache table in memory, as shown in Figure 2-9.

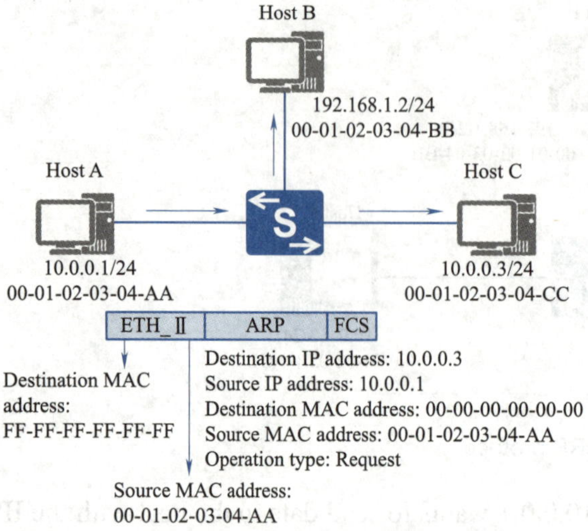

Figure 2-8 ARP request

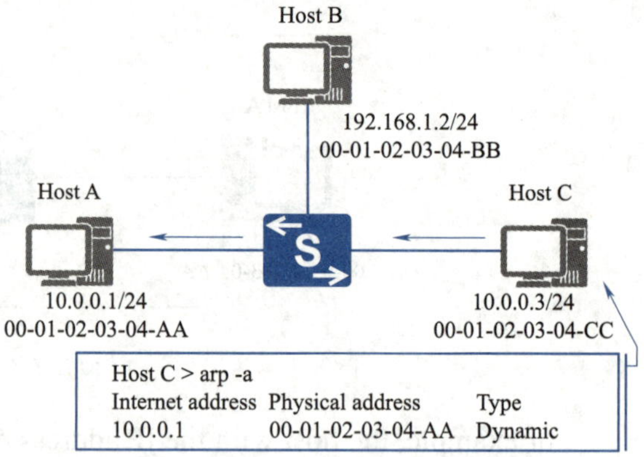

Figure 2-9 ARP response (Ⅰ)

At the same time, the 10.0.0.3 host will respond to the message, he will respond with a message "My MAC address is XX-XX-XX-XX-XX-XX", this message is only sent to the 10.0.0.1 host (data send to a certain host, this method is called unicast), as shown in Figure 2-10.

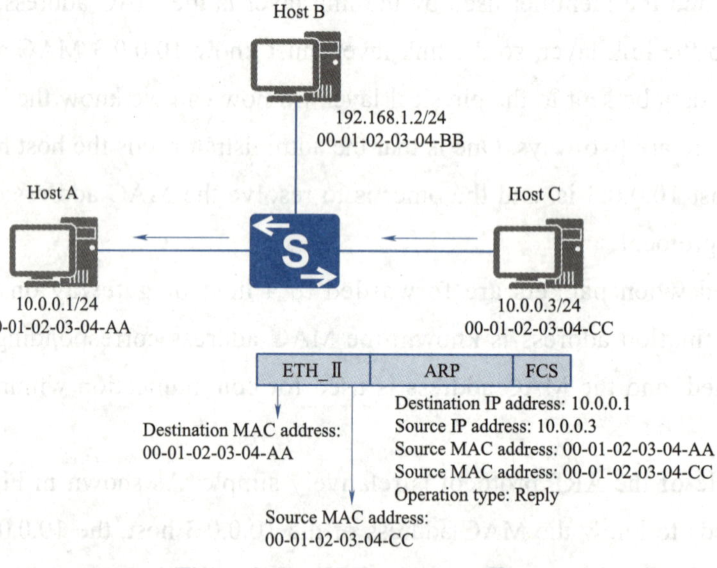

Figure 2-10 ARP response (Ⅱ)

When the 10.0.0.1 host receives the response message from 10.0.0.3, it knows the MAC address of 10.0.0.3, and it will also be cached in the ARP cache table in the computer's memory.

In the above network environment, if the 10.0.0.1 host sends data to 10.0.0.3 or other hosts, the host 10.0.0.1 first looked up the ARP cache table to see if there is a 10.0.0.3 or other host's MAC address, if there is one, the host 10.0.0.1 will send the data directly to the MAC address in the cache table, if not, the ARP protocol will be used to resolve the other party's MAC address.

We can see another ARP feature called gratuitous ARP. It means that the host sends ARP to find its own IP address. Usually, it occurs when the interface is configured during system boot. After the host is assigned an IP address or the IP address is changed, it must be immediately checked whether the assigned IP address is unique on the network to avoid address conflicts. The host performs address conflict detection by sending an ARP Request message.

As shown in Figure 2-10, host A sets the destination IP address field in the ARP Request broadcast message to its own IP address, and all hosts on the network, including gateways, will receive this message. When the destination IP address has been used by a certain host or gateway, the host or gateway will respond with an ARP Reply packet. In this way, host A can detect the IP address conflict.

Subtask 2.2.2 Working process analysis of ICMP

ICMP is used to convey information such as errors, control, and query. ICMP is often considered an integral part of the IP layer. It transmits error messages and other information that needs attention. For example, when the TTL of an IP packet is 0, the router will discard the packet, which means that an error occurred during the transmission of the IP packet. At this time, the router that discards the packet must notify the sender of the packet. Use the ICMP protocol to tell the sender "The message you sent to the target was discarded by me. The reason for discarding is because the TTL is 0." ICMP messages are usually used by the IP layer or higher protocol (TCP or UDP) to return error messages to the user process. At the same time, ICMP is also used to transmit control information. If A wants to test whether it can reach C correctly, A will use ICMP information to send a control message (requires the other party to respond/echo), if A receives C's response message, it means that A can reach C correctly.

A typical application of ICMP is Ping. Ping is a common tool for detecting network connectivity, and it can also collect other relevant information. The user can specify different parameters in the Ping command, such as the length of the ICMP message, the number of ICMP messages sent, the timeout time for waiting for a reply, etc. The device constructs and sends ICMP messages according to the configured parameters to perform a Ping test, as shown in Figure 2-11.

```
<Router A>ping ?
STRING<1-255> IP address or hostname of a remote system
-a            Select source IP address, the default is the IP address of the
              output interface
-c            Specify the number of echo requests to be sent, the default is
              5
-d            Specify the SO_DEBUG option on the socket being used
-f            Set Don't Fragment flag in packet (IPv4-only)
-h            Specify TTL value for echo requests to be sent, the default is
              255
-i            Select the interface sending packets
```

Figure 2-11 Ping command

The common configuration parameters of Ping are described as follows:

(1) -a source-ip-address specifies the source IP address for sending ICMP ECHO-REQUEST packets. If the source IP address is not specified, the IP address of the outgoing interface will be used as the source address of the ICMP ECHO-REQUEST message.

(2) -c count specifies the number of times to send ICMP ECHO-REQUEST packets. By default, 5 ICMP ECHO-REQUEST messages are sent.

(3) -h ttl-value specifies the value of TTL. The default value is 255.

(4) -t timeout specifies the timeout period to wait for ICMP ECHO-REPLY after sending ICMP ECHO-REQUEST.

The output information of the Ping command includes the destination address, ICMP packet length, sequence number, TTL value, and round-trip time. The sequence number is a variable parameter field included in the Echo reply message (Type=0), and the TTL and round trip time are included in the IP header of the message, as shown in Figure 2-12.

```
[Router A]ping 192.168.1.2
  PING 192.168.1.2 : 56  data bytes, press CTRL_C to break
    Reply from 192.168.1.2 : bytes=56 Sequence=1 ttl=255 time=340 ms
    Reply from 192.168.1.2 : bytes=56 Sequence=2 ttl=255 time=10 ms
    Reply from 192.168.1.2 : bytes=56 Sequence=3 ttl=255 time=30 ms
    Reply from 192.168.1.2 : bytes=56 Sequence=4 ttl=255 time=30 ms
    Reply from 192.168.1.2 : bytes=56 Sequence=5 ttl=255 time=30 ms

  --- 192.168.1.2 ping statistics ---
    5 packet(s) transmitted
    5 packet(s) received
    0.00% packet loss
    round-trip min/avg/max = 10/88/340 ms
```

Figure 2-12 Ping command output result

Another typical application of ICMP is Tracert. Tracert traces the forwarding path of the message hop by hop based on the TTL value in the message header. In order to track the path to a specific destination address, the source first sets the TTL value of the message to 1. After the message arrives at the first node, the TTL timeout, so the node sends a TTL timeout message to the source, which carries a timestamp. Then the source sets the TTL value of the message to 2, and the message times out when it reaches the second node. The node also returns a TTL timeout message, and so on, until the message reaches the destination. In this way, the source can track every node the message passes through based on the information in the returned message, and calculate the round trip time based on the timestamp information. Tracert is an effective means to detect network packet loss and delay, and it can also help administrators find routing loops in the network, as shown in Figure 2-13.

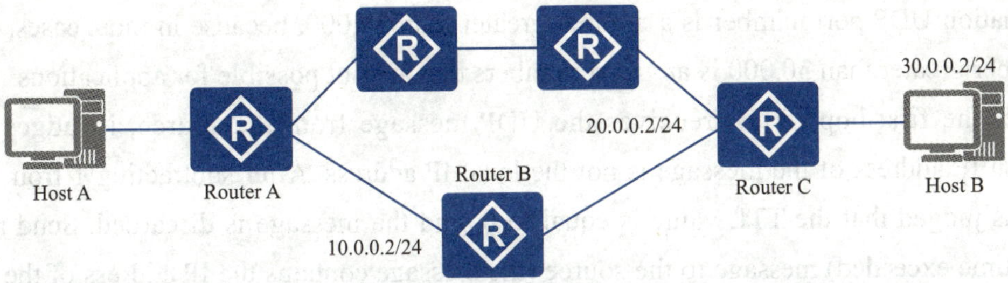


```
<Router A>tracert ?
STRING<1-255> IP address or hostname of a remote system
-a            Set source IP address, the default is the IP address of the
              output interface
-f            First time to live, the default is1
-m            Max time to live, the default is 1
-name         Display the host name of the router on each hop
-p            Destination UDP port number, the default is 33434
-q            Number of probe packets, the default is 3
-s            Specify the length of the packets to be sent. The default
              length is 12 bytes
…
```

Figure 2-13 Tracert command

Tracert commonly used configuration parameters are described as follows:

(1) -a source-ip-address specifies the source address of the tracert packet.

(2) -f first-ttl specifies the initial TTL. The default value is 1.

(3) -m max-ttl specifies the maximum TTL. The default value is 30.

(4) -name enables to display the host name of each hop.

(5) -p port specifies the UDP port number of the destination host.

Tracert displays each hop that a data packet passes through during network transmission, as shown in Figure 2-14.

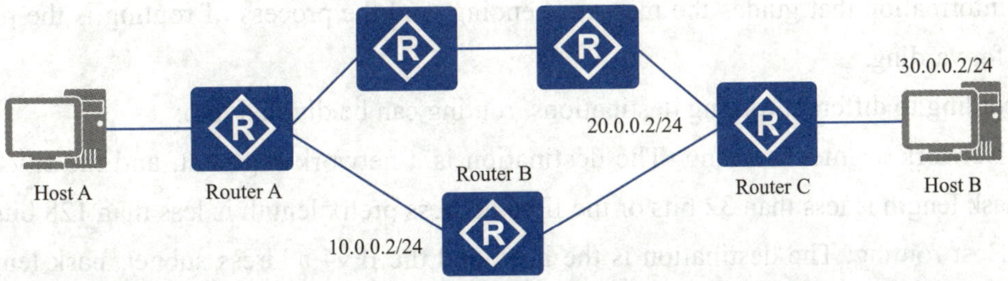


```
<Router A>tracert 30.0.0.2
Tracert to 30.0.0.2(30.0.0.2), max hops:30, packet length:40, press CTRL_C to
break
 1 10.0.0.2 130 ms  50 ms  40 ms
 2 20.0.0.2 80 ms  60 ms  80 ms
 3 30.0.0.2 80 ms  60 ms  70 ms
```

Figure 2-14 Tracert command output result

The source (RTA) sends a UDP message to the destination (host B) with a TTL value of 1, and the destination UDP port number is a number greater than 30,000, because in most cases, the UDP port number greater than 30,000 is any port numbers that are not possible for applications.

After the first hop (RTB) receives the UDP message from the source, it judges that the destination IP address of the message is not the local IP address. After subtracting 1 from the TTL value, it is judged that the TTL value is equal to 0, and the message is discarded. Send an ICMP timeout (time exceeded) message to the source (the message contains the IP address of the first hop 10.0.0.2), so that the source gets the RTB address.

After receiving the RTB ICMP timeout message, the source sends a UDP message to the destination again with a TTL value of 2.

After receiving the UDP message from the source, the second hop (RTC) responds with an ICMP timeout message, so that the source gets the RTC address (20.0.0.2).

The above process continues until the destination end receives the UDP packet sent by the source end and determines that the destination IP address is the local IP address, and then processes the packet. According to the destination UDP port number in the message, search for the upper-layer protocol occupying this port number. Because no application at the destination end uses the UDP port number, an ICMP port unreachable (Destination Unreachable) message is returned to the source end.

After the source receives the ICMP port unreachable message, it judges that the UDP message has reached the destination, it stops the Tracert program, and obtains the path that the data message travels from the source to the destination (10.0.0.2; 20.0.0.2; 30.0.0.2).

Subtask 2.2.3 Understand routing protocols

Routing is the most basic element in a data communication network. Routing information is the path information that guides the message sending, and the process of routing is the process of message forwarding.

According to different routing destinations, routing can be divided into:

(1) Network segment routing: The destination is a network segment, and the IPv4 address subnet mask length is less than 32 bits or the IPv6 address prefix length is less than 128 bits.

(2) Host routing: The destination is the host, and the IPv4 address subnet mask length is 32 bits or the IPv6 address prefix length is 128 bits.

According to whether the destination is directly connected to the router, the route can be divided into:

(1) Direct routing: The network where the destination is located is directly connected to the router.

(2) Indirect routing: The destination network is not directly connected to the router.

According to the different destination address types, routes can also be divided into:

(1) Unicast routing: Indicates that the destination address to which the message is forwarded is a unicast address.

(2) Multicast routing: Indicates that the destination address of the forwarded packet is a multicast address.

According to different configuration methods, routing can also be divided into:

(1) Static routing is easy to configure, has low system requirements, and is suitable for small-scale networks with simple and stable topologies. The disadvantage is that it cannot automatically adapt to changes in the network topology and requires manual intervention.

(2) The dynamic routing protocol has its own routing algorithm, which can automatically adapt to changes in the network topology, and is suitable for networks with a certain number of Layer 3 devices. The disadvantage is that the configuration requires higher requirements for users, higher requirements for the system than static routing, and will occupy a certain amount of network resources and system resources.

According to different scopes of dynamic routing, routing protocols can be divided into:

(1) IGP (interior gateway protocol) : Run within an autonomous system. Common IGP protocols include RIP, OSPF and IS-IS.

(2) EGP (exterior gateway protocol) : Run between different autonomous systems. BGP is currently the most commonly used EGP protocol.

According to different algorithms used, dynamic routing can be divided into routing protocols:

(1) DVP (distance-vector protocol) : Including RIP and BGP. Among them, BGP is also called path-vector protocol.

(2) LSP (link-state protocol) : Including OSPF and IS-IS.

Subtask 2.2.4 Working process analysis of OSPF

As shown in Figure 2-15, OSPF is a routing protocol based on link state. It is designed to ensure that there is no routing loop. OSPF supports the division of areas. The routers in the area use the SPF shortest path algorithm to ensure that there is no loop inside the area. OSPF also uses inter-area connection rules to ensure that there is no routing loop between areas.

OSPF supports triggered updates, can quickly detect and notify topology changes in autonomous systems.

OSPF can solve the problems caused by network expansion. When there are more and more routers on the network and the flow of routing information increases sharply, OSPF can divide each autonomous system into multiple areas and limit the scope of each area. The sub-area feature of

OSPF makes OSPF particularly suitable for large and medium-sized networks. OSPF can provide authentication functions. The packets between OSPF routers can be configured to be authenticated before they can be exchanged.

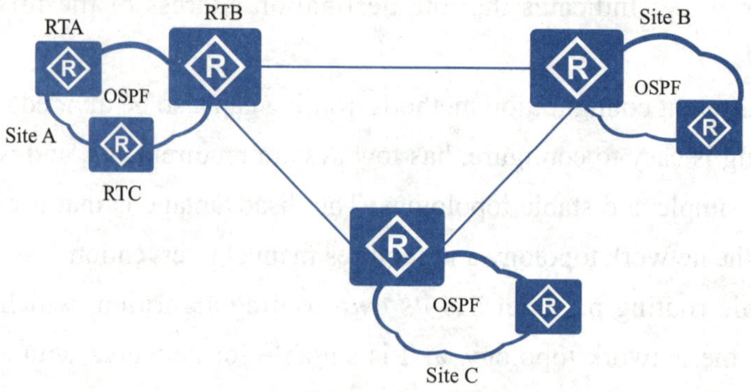

Figure 2-15 OSPF protocol

Subtask 2.2.5 Working process analysis of SNMP

With the rapid development of network technology, the number of network devices in enterprises has increased exponentially, and the types of network devices have also increased, which makes the management of enterprise networks very complicated. Simple network management protocol SNMP (simple network management protocol) can realize unified management of network equipment of different types and different manufacturers, which greatly improves the efficiency of network management.

SNMP is a network management protocol widely used in TCP/IP networks. SNMP provides a way to manage network devices through a network management workstation running the network management software NMS (network management system), as shown in Figure 2-16.

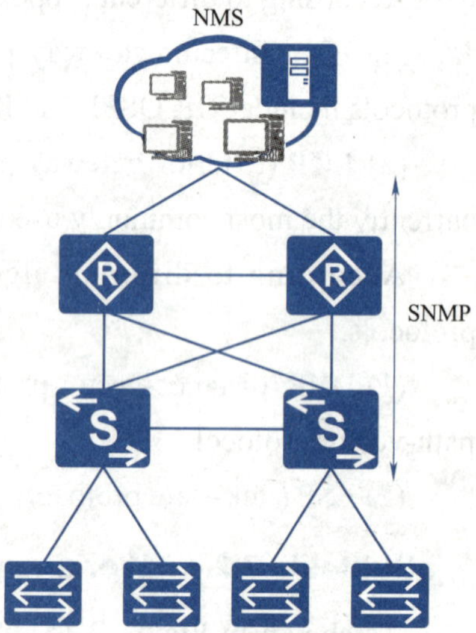

Figure 2-16 SNMP protocol

SNMP supports the following operations:

(1) NMS sends configuration information to network devices through the SNMP protocol.

(2) NMS uses SNMP to query and obtain resource information in the network.

(3) The network equipment proactively reports alarm messages to the NMS, so that the network administrator can deal with various network problems in a timely manner.

As shown in Figure 2-17, SNMP includes NMS, Agent, and MIB.

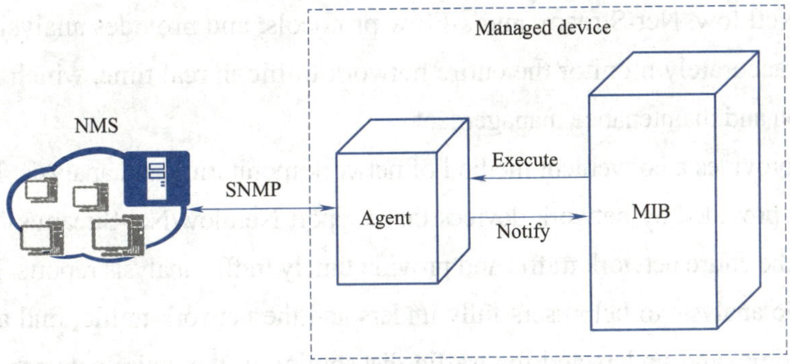

Figure 2-17 SNMP architecture

NMS is a network management software running on a network management host. The network administrator sends a request to the managed device by operating the NMS, so that the network device can be monitored and configured.

Agent is an agent process running on the managed device. After the managed device receives the request sent by the NMS, the Agent responds. The main functions of the Agent include: collecting equipment status information, realizing the remote operation of the equipment by the NMS, and sending alarm messages to the NMS.

MIB (management information base) is a virtual database, a collection of device status information maintained on the managed device. Agent collects device status information by looking up MIB.

The versions of SNMP are as follows:

(1) SNMPv1: The NMS on the NMS workstation and the Agent on the managed device can exchange SNMPv1 messages to realize the management of the managed device by the NMS. SNMPv1 basically has no security at all.

(2) SNMPv2: On the basis of inheriting SNMPv1, SNMPv2c has greatly improved its performance, security, and confidentiality.

(3) SNMPv3: SNMPv3 adds and improves the security and management mechanism on the basis of SNMPv2. The SNMPv3 system structure embodies the modular design idea, enabling the administrator to easily and flexibly realize the addition and modification of functions. The main feature of SNMPv3 is that it is highly adaptable and can be applied to a variety of operating environments. It can not only manage the simplest network and realize basic management functions, but also provide powerful network management functions to meet the management needs of complex networks.

Subtask 2.2.6 Working process analysis of NTA

eSight NTA (network traffic analyzer) is a pure software solution that does not require hardware probes and does not increase user investment; it collects and analyzes common IP packets through NetFlow, Net-Stream, and s-Flow protocols, and provides analysis reports close to customers. It can accurately monitor the entire network traffic in real time, which is a great tool for enterprise operation and maintenance management.

eSight NTA provides a convenient method of network monitoring and analysis. It uses IP network traffic information provided by network devices that support NetFlow/Net-Stream/s-Flow technologies to deeply analyze the entire network traffic and provide timely traffic analysis reports. Rich graphs show the results of traffic analysis to help users fully understand the network traffic, find abnormal network traffic at the first time, and understand the traffic distribution in the entire network. The eSight NTA network flow analysis component provides users with reliable and convenient traffic analysis solutions, which can monitor the entire network traffic in real time and provide multi-dimensional traffic analysis reports; Help users discover abnormal traffic in the network in time, understand the usage of network bandwidth and traffic distribution, and help enterprises to truly realize traffic visibility, trouble-checking, and planning. NTA has the following features:

(1) Traffic visibility: It provides real-time IP traffic monitoring, displays network traffic trends, and facilitates managers to discover abnormalities in time, and then deal with them.

(2) Checkable exceptions: Through the NTA network flow analysis component, the original IP traffic can be analyzed and audited to find the source of the abnormal traffic.

(3) Well-founded planning: The traffic trend and custom report functions provided by the NTA network flow analysis component provide a reference for managers when planning network capacity.

Subtask 2.2.7 Working process analysis of NetStream

NetStream is a patented technology of Huawei Technologies Co., Ltd. It is a technology for statistics and release of network stream information. The NDE sends the statistical information obtained to the NSC on a regular basis, and the NSC performs further processing, and then passes it to the subsequent NDA for data analysis. The result of the analysis provides a basis for network billing and planning, as shown in Figure 2-18.

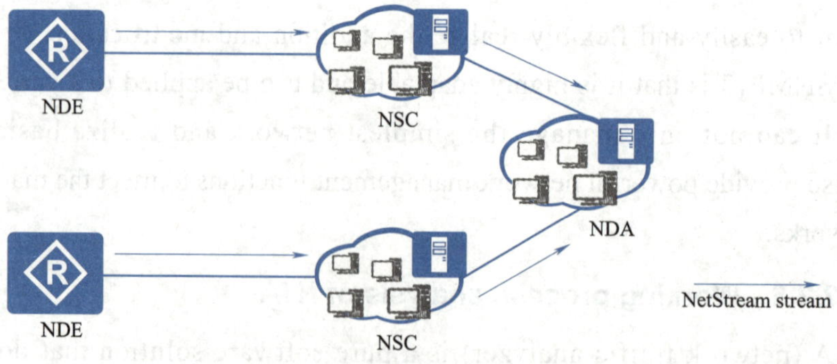

Figure 2-18 NetStream architecture

The data output through NetStream technology can be used for many purposes, including:

(1) Network management and planning.

(2) Corporate accounting and departmental billing.

(3) ISP prepares bills.

(4) Data reserve.

(5) Commercial data collection.

The realization of the NetStream function mainly involves three types of devices:

(1) NDE (NetStream data exporter) : NDE is responsible for the collection and transmission of traffic.

(2) NSC (NetStream collector) : NSC is responsible for parsing messages from NDE and collecting statistical information in the database for NDA to analyze.

(3) NDA (NetStream data analyzer) : NDA is responsible for analyzing statistical information, further processing, generating reports, and providing basis for various businesses (such as traffic accounting, network planning, and monitoring).

The working process of the NetStream system is as follows:

(1) Devices configured with the NetStream function (i. e. NDE) periodically send the collected detailed statistical information about the stream to the NSC.

(2) The information is initially processed by the NSC and then sent to the NDA.

(3) NDA analyzes data for billing, network planning and other applications.

Task 2.3 Understand transport layer protocols

The transport layer defines the end-to-end connectivity between host applications. The two most common protocols in the transport layer are TCP (transmission control protocol) and UDP (user datagram protocol).

Subtask 2.3.1 Working process analysis of TCP protocol

TCP is a connection-oriented transport layer protocol that can provide reliable transmission services. TCP allows a host to run multiple application processes at the same time. Each host can have multiple application ports, and each pair of port number, source and destination IP address combination uniquely identifies a session. Ports are divided into well-known ports and dynamic ports. Some network services use fixed ports. Such ports are called well-known ports, and the port number ranges from 0 to 1023. For example, FTP, HTTP, Telnet, and SNMP services all use well-known ports. Dynamic port numbers range from 1024 to 65535. These port numbers are generally not fixedly assigned to a service, which means that many services can use these ports. As long as the running program requests the system to access the network, the system can allocate one of these port numbers for the program to use.

TCP connection establishment is a three-way handshake process, the purpose of which is to confirm the starting sequence number for both parties in communication, so that subsequent communication can proceed in an orderly manner. The main steps are shown in Figure 2-19.

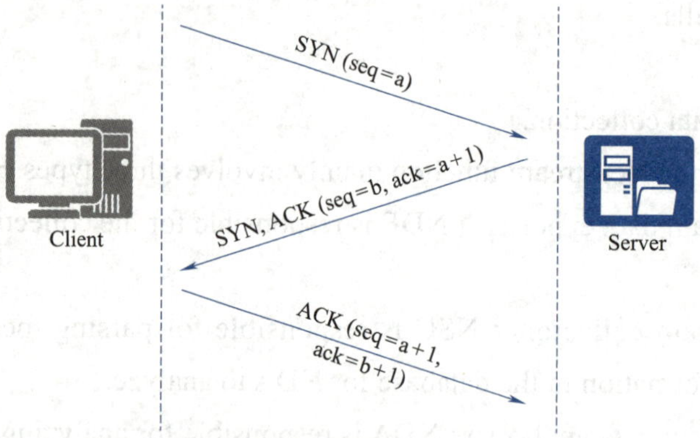

Figure 2-19 TCP three-layer handshake

(1) At the beginning of the connection, the connection establishment party (client) sends a SYN packet, including its own initial sequence number a.

(2) When the connection recipient (server) receives the SYN packet, it will reply with a SYN+ACK packet, which contains the response message ACK to the previous packet a, and the response sequence number is the sequence number of the next desired packet, that is, a+1. Then it also contains its own initial serial number b.

(3) After the connection establishment party (client) receives the response SYN packet, it responds with an ACK packet, which contains the sequence number of the next desired packet, namely b+1.

When the data transmission is completed, TCP needs to terminate the connection to release resources. The four waved processes of TCP termination of the connection are shown in Figure 2-20.

(1) The client process sends a connection release message and stops sending data. The client releases the header of the data message, FIN=1, and its sequence number is seq=a (equal to the sequence number of the last byte of the previously transmitted data plus 1). At this time, the client enters the FIN-WAIT-1 (termination waiting 1) state.

(2) The server receives the connection release message and sends an acknowledgment message, ACK=1, ack=a+1, and brings its own serial number seq=b. At this time, the server enters CLOSE-WAIT (close waiting) state.

(3) After the client receives the server's confirmation request, at this time, the client enters the FIN-WAIT-2 (terminated waiting 2) state, waiting for the server to send a connection release message.

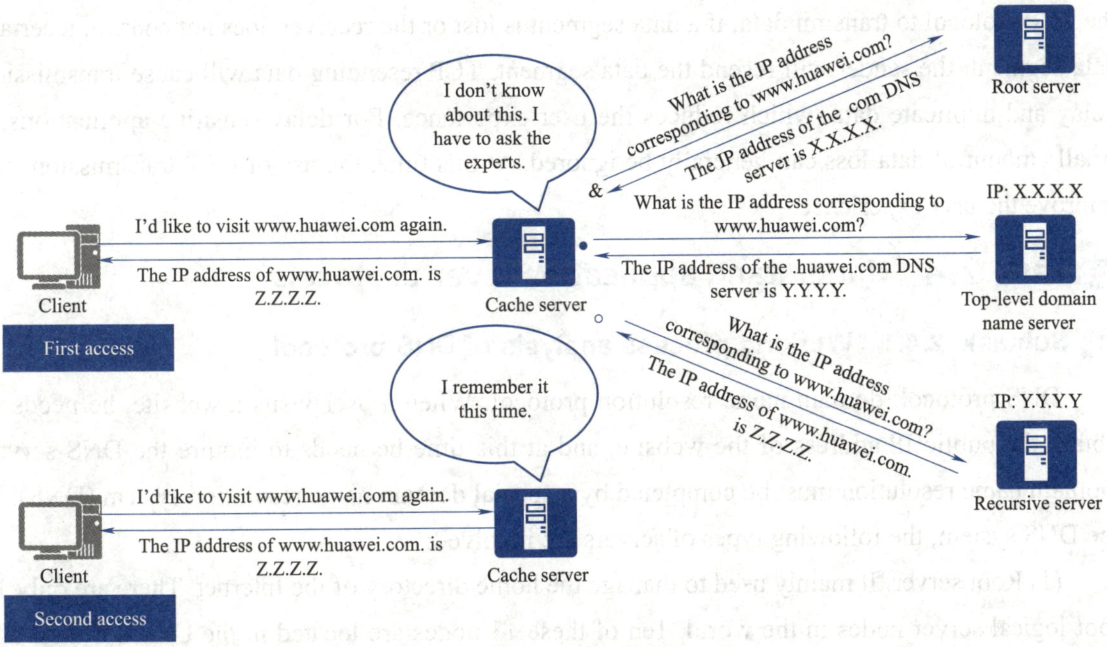

Figure 2-20 DNS working principle

(4) After the server sends the final data, it sends a connection release message to the client, FIN=1, ack=a+1. Because it is in the semi-closed state, the server is likely to send some more data, assuming the serial number at this time is seq=c, at this time, the server enters the LAST-ACK (last confirmation) state, waiting for the client's confirmation.

(5) After the client receives the server's connection release message, it must send an acknowledgment, ACK=1, ack=c+1, and its own serial number is seq=a+1, at this time, the client enters TIME-WAIT (time waiting) status.

(6) As long as the server receives the confirmation from the client, it immediately enters the CLOSED state and ends the current TCP connection.

Subtask 2.3.2 Working process analysis of UDP protocol

UDP is a connectionless-oriented transport layer protocol, and transmission reliability is not guaranteed. When the application program does not require high transmission reliability, but requires high transmission speed and delay, UDP protocol can be used instead of TCP protocol to control data forwarding at the transport layer. When UDP sends data from the source to the destination, there is no need to establish a connection in advance. UDP uses a simple and easy-to-operate mechanism to transfer data between applications. It does not use the confirmation technology or sliding window mechanism in TCP. Therefore, UDP cannot guarantee the reliability of data transmission and cannot avoid receiving duplicate data.

UDP is suitable for transmitting delay-sensitive traffic, such as voice and video. When using the TCP protocol to transmit data, if a data segment is lost or the receiver does not confirm a certain data segment, the sender will resend the data segment. TCP resending data will cause transmission delay and duplicate data, which reduces the user experience. For delay-sensitive applications, a small amount of data loss can generally be ignored. At this time, the use of UDP transmission will improve the user experience.

Task 2.4 Understand application layer protocols

Subtask 2.4.1 Working process analysis of DNS protocol

DNS protocol, domain name resolution protocol. When a user visits a website, he needs to obtain the public IP address of the website, and at this time he needs to inquire the DNS server. Domain name resolution must be completed by a special domain name resolution system (DNS). In the DNS system, the following types of servers are involved:

(1) Root server: It mainly used to manage the home directory of the Internet. There are only 13 root logical server nodes in the world. Ten of these 13 nodes are located in the United States, and the other three nodes are located in the United Kingdom, Sweden, and Japan. Although the network is borderless, the server still has borders. All root servers are managed uniformly by ICANN, an Internet domain name and number allocation agency authorized by the US government.

(2) Top-level domain name server: generally used to store top-level domain names such as .com, .edu, and .cn.

(3) Recursive server: It can also be understood as an authorization server that stores official domain name resolution authorizations. Generally, it stores the resolution relationship between domain names and IP addresses in this network, which is the role of DNS Pod. Imagine that if every online user sends a request to the authorization server while surfing the Internet, the authorization server will inevitably not be able to bear it, so there is a need for a cache server.

(4) Cache server: It is equivalent to a proxy of the authorization server, which can relieve the pressure of the authorization server. Every time we go online, the request for domain name resolution is sent to the cache server. When the cache server receives a user request for the first time, it will request the domain name and IP address resolution table from the authorized server, and then store it locally, etc. When another user requests the same domain name in the future, it will reply directly and not request it again. After all, the IP address of a website does not change frequently. Of course, this analytic table has a certain validity period. When the validity period expires, the analytic table will automatically age, and it will ask the authorization server again when there is a user request next time. This regular aging mechanism can ensure the regular update of the domain name resolution on the cache server.

The DNS resolution process is shown in Figure 2-20.

(1) DNS client queries usually use recursive methods, and the cache server first determines whether there is a local resolution cache for this domain name.

(2) If there is no local cache, the domain name will be sent to the root server. After the root server receives the www.vmall.com request, it will determine who is authorized to manage .com and return the IP address of the top-level domain name server where .com is located.

(3) The cache server continues to send www.vmall.com resolution requests to the top-level domain name server. After receiving the request, the top-level domain name server will return the IP address of the recursive server of the next level .vmall.com.

(4) The cache server continues to send www.vmall.com resolution requests to the recursive server, and after receiving the request, the recursive server returns the resolution address of www.vmall.com. If there are many levels of domain names, there will also be multiple levels of recursive servers.

(5) After the cache server obtains the resolved IP address of www.vmall.com, it sends the IP address to the client and caches it locally.

(6) In the subsequent period of time, when a client requests the resolution of the domain name www.vmall.com again, the cache server directly responds to the resolved IP address without repeating the query.

Subtask 2.4.2 Working process analysis of FTP protocol

FTP (file transfer protocol) provides an effective way to upload and download files between the server and the client, as shown in Figure 2-21. When using FTP to transfer data, it is necessary to establish a control connection and a data connection between the server and the client.

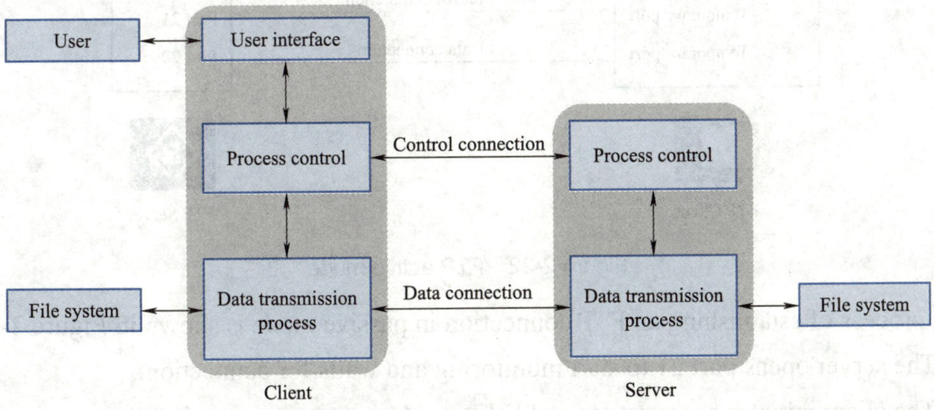

Figure 2-21 FTP architecture

When using FTP for file transfer, two TCP connections are used. The first connection is the control connection between the FTP client and the FTP server. The FTP server opens port 21 and waits for the FTP client to send a connection request. The FTP client randomly opens the port and sends a request to establish a connection to the server. The control connection is used to transmit control commands between the server and the client. The second connection is the data connection between the FTP client and the FTP server. The server uses TCP port 20 to establish a data connection with the client. Normally, the server actively establishes or interrupts the data connection.

FTP connection establishment is divided into active mode and passive mode. The difference between the two is whether the data connection is initiated by the server or the client. By default, the active mode is adopted, and the user can switch through commands. In the active mode, when the client has a firewall, because the data connection is initiated by the server, problems may occur in the data connection. In passive mode, this problem is solved. The active mode is conducive to the management of the FTP server, but not conducive to the management of the client; the passive mode is the opposite. By default, port 21 of the server is used to transmit control commands, and port 20 is used to transmit data.

The process of establishing an FTP connection in active mode is shown in Figure 2-22.

(1) The server opens port 21 to start monitoring and waits for connection. Start listening and wait for connection.

(2) The client initiates a request to establish a control connection, and the server responds.

(3) The client sends the PORT command through the control connection, telling the server the temporary slogan of the client's data connection.

(4) The server's port 20 establishes a data connection with the client.

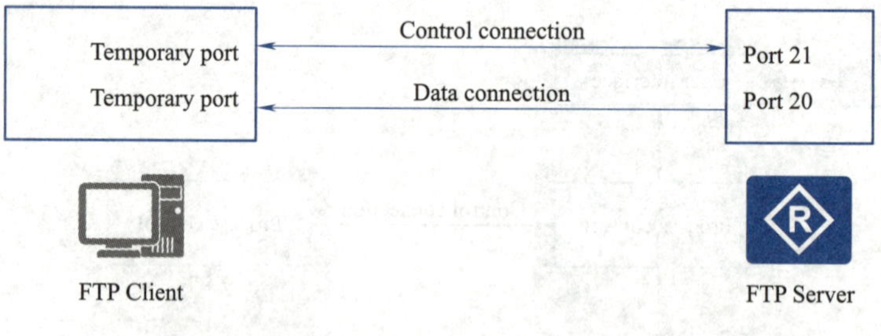

Figure 2-22 FTP active mode

The process of establishing an FTP connection in passive mode is shown in Figure 2-23.

(1) The server opens port 21 to start monitoring and waits for connection.

(2) The client initiates a request to establish a control connection, and the server responds.

(3) The client sends the command word PASV through the control connection to inform the server that it is in passive mode.

(4) The server responds, telling the client the temporary port number of the data connection.

(5) The temporary port of the client and the server establishes a data connection.

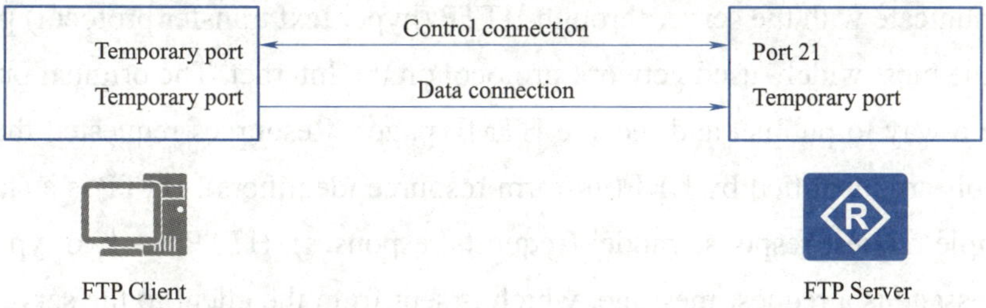

Figure 2-23 FTP passive mode

Subtask 2.4.3 Working process analysis of HTTP protocol

WWW is the abbreviation of world wide web, also known as 3W or Web. As a new generation of user interface on the Internet, it abandons the past pure text information interaction means and adopts hypertext. Hypertext is a global information organization. It establishes links between different parts of a document through keywords, so that information can be searched interactively. When hypertext and multimedia are combined in an information browsing environment, it is called hypermedia.

The Internet uses a combination of hypertext and hypermedia to extend information links to the entire Internet. The web is a kind of hypertext information system, which makes the text no longer fixed in a certain position, but can jump from one position to another. It is this kind of multi-linkage that calls it the web.

HTTP is a connection-oriented protocol and does not have an encryption authentication mechanism, which leads to insufficient security; and due to the encryption function of the HTTPS protocol, certain virus attacks may be hidden and cannot be detected by security devices, which has also become a major hidden danger of network security.

As shown in Figure 2-24, the web is implemented based on the Client/Server architecture and consists of three parts:

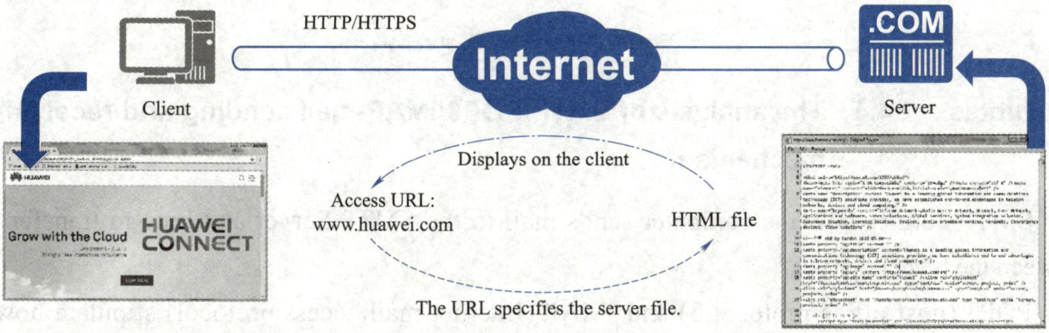

Figure 2-24 Web architecture

(1) Use HTML (hyper text mark-up language) description file.

(2) Use URL (uniform resource locator) to specify the location of the file.

(3) Communicate with the server through HTTP (hyper text transfer protocol) protocol.

HTTP is the most widely used network protocol on the Internet. The original purpose of HTTP was to provide a way to publish and receive HTML pages. Resources requested through HTTP or HTTPS protocols are identified by URI (uniform resource identifiers). HTTP is a stateless protocol based on a simple request-response model (requests/responses). HTTP has two types of messages: One type of message is a request message, which is sent from the client to the server; Another type of message is a response message, which is sent from the server to the client.

The HTTP workflow is shown in Figure 2-25:

(1) The client (browser) actively sends a connection request to the web server.

(2) The server accepts the connection request and establishes a connection. (Steps 1 and Steps 2 are the known TCP three-way handshake).

(3) The client sends GET and other http commands (HTTP request messages) to the server through this connection.

(4) The server receives the command and transmits the corresponding data (HTTP response message) to the client according to the command.

(5) The client receives the data sent from the server.

(6) After the server finishes sending the data, it actively closes the connection (TCP waved four times).

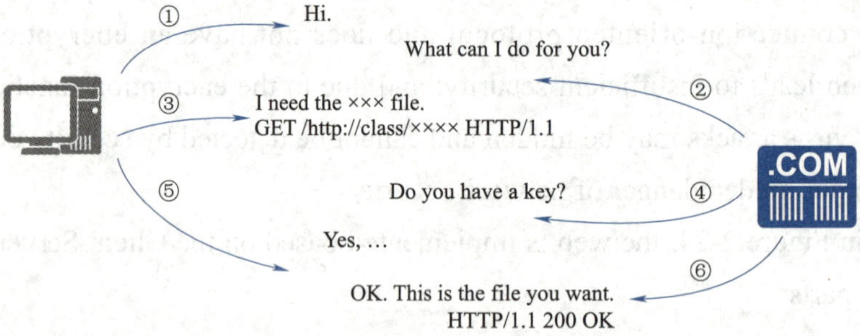

Figure 2-25 HTTP workflow

Subtask 2.4.4 The analysis of SMTP/POP3/IMAP-mail sending and receiving mechanism

SMTP defines how the computer sends mail to the SMTP Server and how to transfer mail between the SMTP Server.

POP3 (post office protocol 3) and IMAP (Internet mail access protocol) stipulate how the computer manages and downloads e-mail on the mail server through the client software.

Based on this method, network administrators need to deploy SMTP service, POP3 service (or IMAP service) on the mail server; end users need to install mail client software (such as Microsoft OutLook, FoxMail and other mail management software) on the PC.

A complete e-mail sending process consists of 4 steps, as shown in Figure 2-26:

(1) The PC encapsulates the e-mail content in an SMTP message and sends it to the sender SMTP Server.

(2) The sender SMTP Server encapsulates the mail in an SMTP message and sends it to the receiver SMTP Server, and then the receiver SMTP Server stores it.

(3) After receiving the user's request, the POP3 Server reads the mail stored in the SMTP Server.

(4) The POP3 Server encapsulates the mail into a POP3 message and sends it to the PC.

Note:

SMTP Server, POP3 Server, and IMAP are management software that provide services to users and need to be deployed on hardware servers. The main difference between IMAP and POP3 is that with POP3, the client software will download all unread mail to the computer, and the mail server will delete the mail. With IMAP, users can directly operate on the mail on the server without downloading all the mails locally to perform various operations.

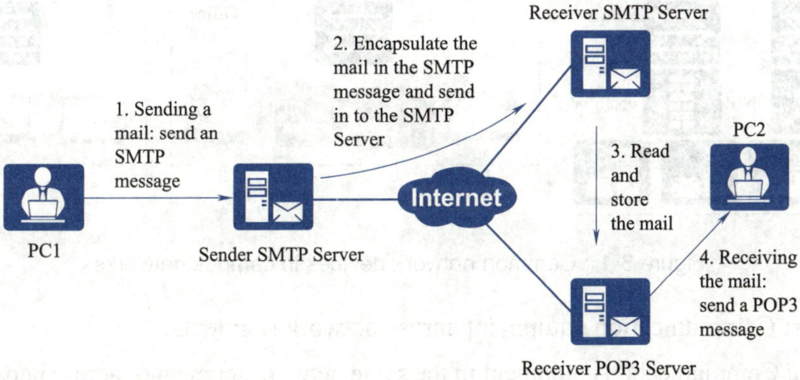

Figure 2-26 Mail sending and receiving mechanism

Project 3 Master the configuration of common network equipment

Network equipment is the basic structure of the network. When we plan and build a network, we often need to deploy network equipment, and we need to configure the network equipment to connect to the network or meet the needs of network security.

Task 3.1 Understand common network equipment

As mentioned in the previous task, computer networks are composed of network devices interconnected through transmission media. What are the common network devices?

Common network devices in campus networks are shown in Figure 3-1, which mainly include:

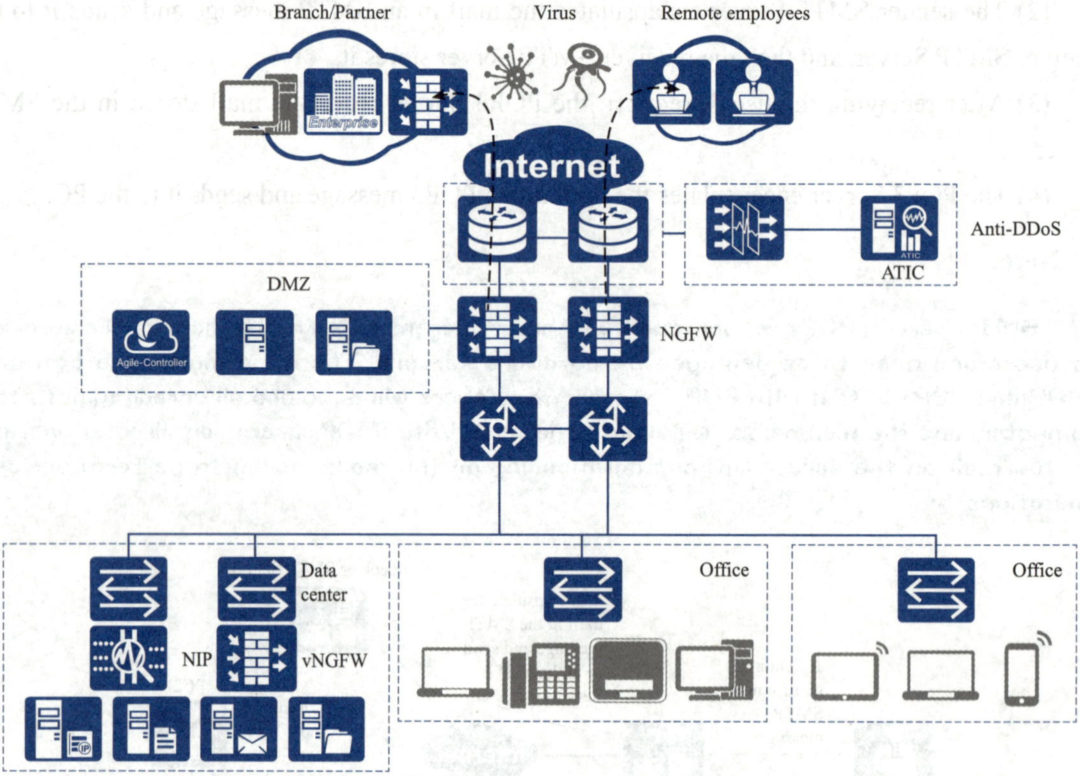

Figure 3-1 Common network devices in campus networks

(1) Router: Communication equipment across network segments.

(2) Switch: Communication equipment in the same network segment or across network segments.

(3) Anti-DDoS: DDoS defense systems are usually deployed at the network exit upstream of the firewall to reduce the burden of firewall packet processing.

(4) NGFW: Next-generation firewalls can be deployed at the egress of the network for preliminary protection, or to protect the data center from being attacked.

(5) vNGFW: The software NGFW is deployed in the VM (virtual machine), which is similar to the hardware firewall function.

(6) NIP: Huawei's next-generation intrusion prevention system, professional intrusion detection equipment/intrusion prevention equipment, and it can be deployed in the data center to

protect the data center from being intruded.

(7) Agile-Controller: An automated control system for network resources based on users and applications is usually deployed in the DMZ service area for admission control.

Subtask 3.1.1 Master the working principle of switch

With the development of enterprise networks, more and more users need to access the network, and the switch provides a large number of access ports to meet this demand. At the same time, the switch completely solves the conflict problem that plagued the early Ethernet, greatly improves the performance of the Ethernet, and also improves the security of the Ethernet. The switch works at the data link layer and operates on data frames. After receiving the data frame, the switch will forward the data frame according to the header information of the data frame.

There is a MAC address table in the switch, which stores the mapping relationship between the MAC address and the switch port. The MAC address table is also called a CAM (content addressable memory) table.

As shown in Figure 3-2, there are three types of switch forwarding operations on frames: flooding, forwarding, and discarding.

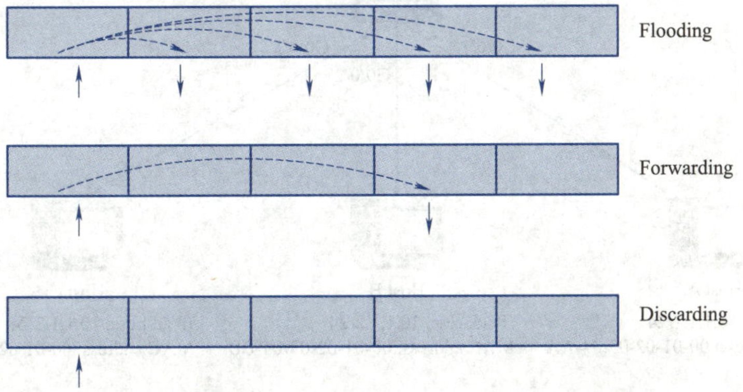

Figure 3-2 Switch forwarding behavior

(1) Flooding: The switch forwards frames coming in from a certain port through all other ports (note that "all other ports" refers to all ports except the port where the frame enters the switch).

(2) Forwarding: The switch forwards the frame coming in from a certain port through another port (note that the "other port" cannot be the port where the frame enters the switch).

(3) Discarding: The switch directly discards the frames coming in from a certain port.

The basic working principle of the switch can be summarized as follows:

If it is a unicast frame that enters the switch, the switch will look up the destination MAC address of the frame in the MAC address table. If the MAC address is not found, the switch performs flooding operation. If the MAC address is found, determine the port corresponding to the MAC address in the MAC address table, and judge whether the port is the port where the frame enters the switch. If it is not, the switch performs a forwarding operation. If it is, the switch performs a discard operation.

If what enters the switch is a broadcast frame or a multicast frame, the switch will not check the MAC address table, but directly perform the flooding operation.

In addition, the switch also has the ability to learn. When a frame enters the switch, the switch checks the source MAC address of the frame, maps the source MAC address and the port where the frame enters the switch, and then stores the mapping relationship in the MAC address table.

Next, we take a small switching network as an example to explain the basic working principle of the switch. As shown in Figure 3-3, the switch SWA connects to host A through G0/0/1, to host B through G0/0/2, and to host C through G0/0/3.

In the initial state, the switch does not know the MAC address of the connected host, so the MAC address table is empty. As shown in Figure 3-3, SWA is the initial state. Before receiving the data frame sent by host A, there is no entry in the MAC address table.

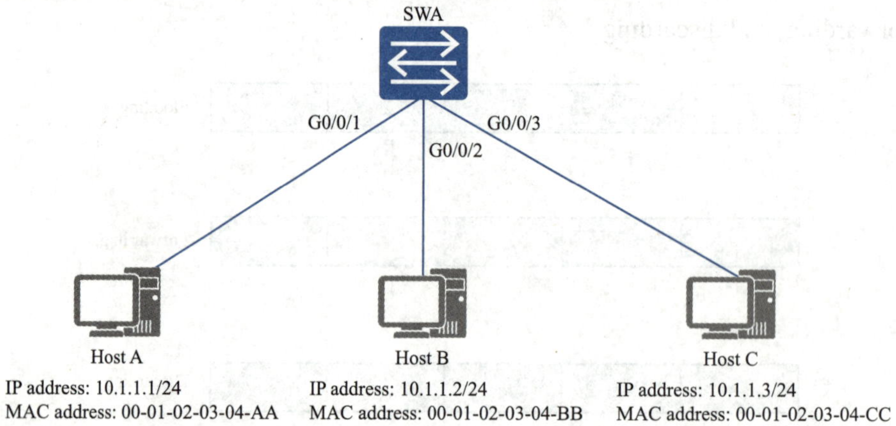

Figure 3-3 Switch initialization

When host A sends data to host C, it usually first sends an ARP request to obtain the MAC address of host C. The destination MAC address in this ARP request frame is the broadcast address, and the source MAC address is its own MAC address. By default, the aging time of MAC address entries learned by X7 series switches is 300 seconds. If the data frame sent by host A is received again within the aging time, the aging time of the MAC address of host A and the mapping of G0/0/1 saved in the SWA will be refreshed. After that, if the switch receives a data frame with a destination MAC address of 00-01-02-03-04-AA, it will all be forwarded through the G0/0/1 port.

As shown in Figure 3-3, the destination MAC address of the data frame sent by host A is the broadcast address, so the switch will broadcast this data frame to host B and host C through ports

G0/0/2 and G0/0/3.

As shown in Figure 3-4, after receiving this data frame, both host B and host C will check the ARP data frame. However, host B will not reply to the frame, and host C will process the frame and send an ARP response. The destination MAC address of the reply data frame is the MAC address of host A, and the source MAC address is the MAC address of host C. As shown in Figure 3-5, when SWA receives a reply data frame, it will add the mapping relationship between the source MAC address of the frame and the interface to the MAC address table. If this mapping relationship already exists in the MAC address table, it will be refreshed. Then SWA queries the MAC address table, finds the corresponding forwarding port according to the destination MAC address of the frame, and forwards the data frame from G0/0/1.

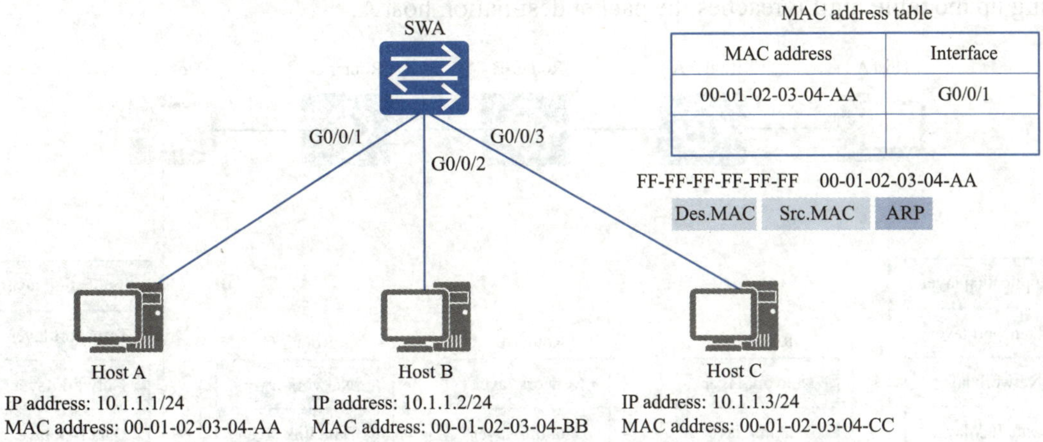

Figure 3-4 The switch floods the broadcast frame

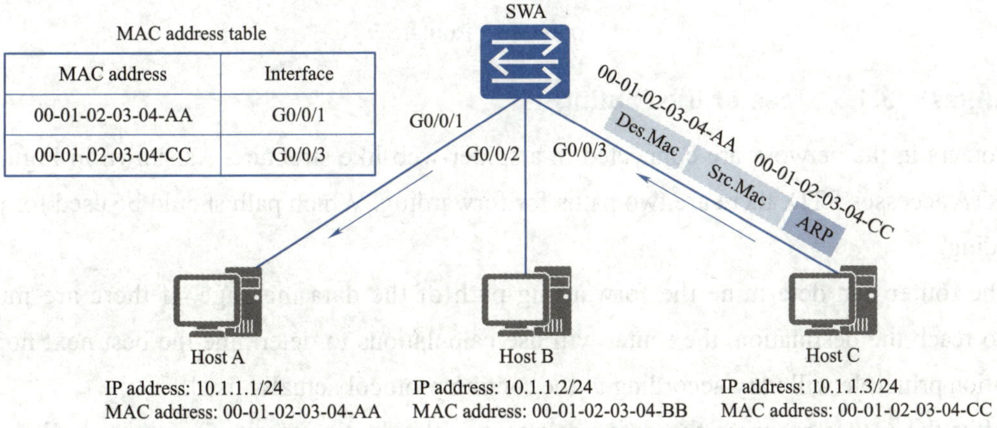

Figure 3-5 The switch forwards unicast frames and performs MAC learning

Subtask 3.1.2 Master the working principle of router

The switch is a data link layer device, and its main function is to realize the communication of data frames in the same broadcast domain. If you want to realize the communication between different broadcast domains or even different types of data link layer protocols, it needs to be realized through a router.

A router is a network-layer device whose main function is to realize the forwarding of packets between different networks. As shown in Figure 3-6, host A and host B on different networks (different links) communicate with each other. The router interface on the same network (same link) as host A receives the data frame sent by host A. After the link layer analysis of the router determines that the frame header is sent to itself, it sends the frame to the network layer for processing. The network layer packet header determines the network segment where the destination address is located, and then forwards the frame to the next hop from the corresponding interface by looking up the table until it reaches the packet destination host A.

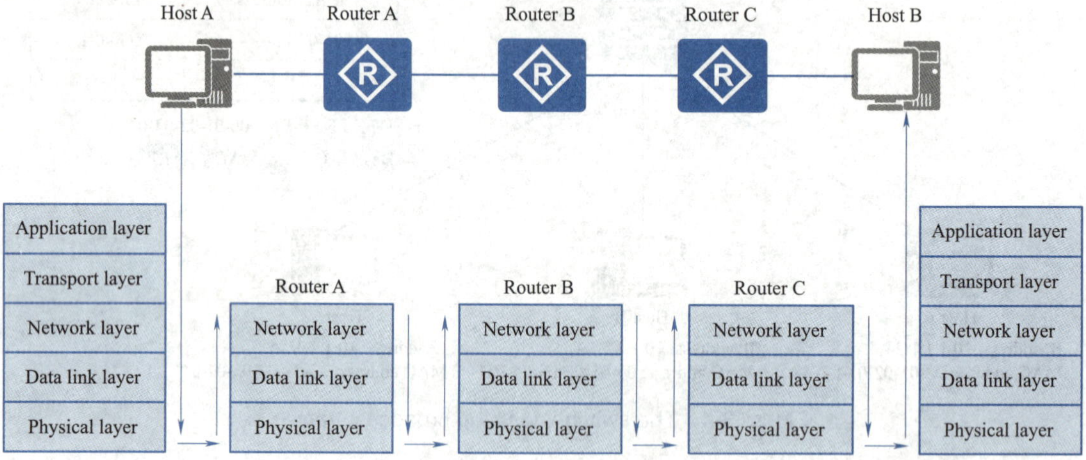

Figure 3-6 Router

Subtask 3.1.3 Master the routing

Routers in the network are connected in a spider-web-like structure. As shown in Figure 3-7, when RTA accesses RTD, there are two paths for forwarding. Which path should be used for packet forwarding?

The router can determine the forwarding path of the data message. If there are multiple paths to reach the destination, the router will use calculations to determine the best next hop. The calculation principle will vary according to the routing protocol actually used.

After the router receives the data packet, it will select an optimal path according to the destination IP address in the data packet and forward the data packet to the next router. The last router on the path is responsible for sending the data packet to the destination host. The transmission of data packets on the network is like a relay race in sports. Each router is responsible for forwarding the data packets to the next hop router according to the optimal path. In the end, the data packet reaches the destination through the optimal path. Of course, sometimes, due to the implementation of some special routing strategies, the path through which the data packet passes may not be the best.

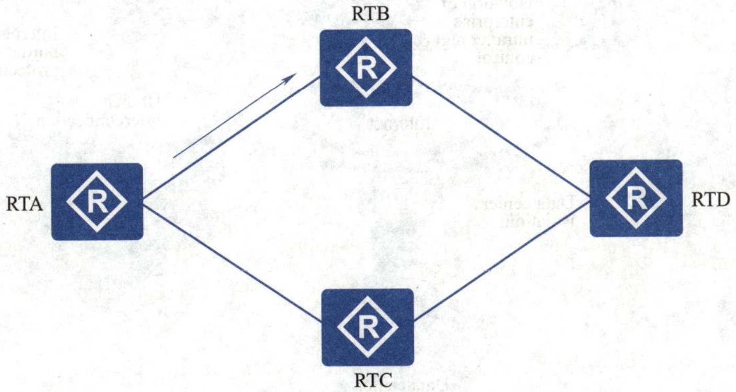

Figure 3-7 Route selection

Subtask 3.1.4 Master the working principle of firewall

In building construction, firewalls are designed to prevent fire from spreading from one part of the building to another. The firewall in the network also has a similar function: to prevent the danger of the Internet from spreading to the private network, or to protect important equipment in the private network of the enterprise.

At the network boundary, on the one hand, the firewall prevents unauthorized or unverified access from the Internet to the protected network, and on the other hand, allows users of the internal network to access the Internet through Web access or send and receive e-mail.

When users on the external network access internal network resources, they must pass through the firewall; When users on the internal network access external network resources, they will also pass through the firewall. In this way, the firewall acts as a "guard" and can drop data packets that need to be forbidden here. As shown in Figure 3-8.

The firewall is not only used to connect the private network to the Internet, but can also be used to protect mainframes and important resources (such as data) within the organization's network. All access to protected data must be filtered by a firewall, even if users inside the network want to access protected resources, they must go through the firewall.

The firewall can also be used as an authority control gateway for accessing the Internet, such as allowing specific people in the organization to access the Internet. Many firewalls now also have other functions, such as identity authentication, information security (encryption) processing, and so on.

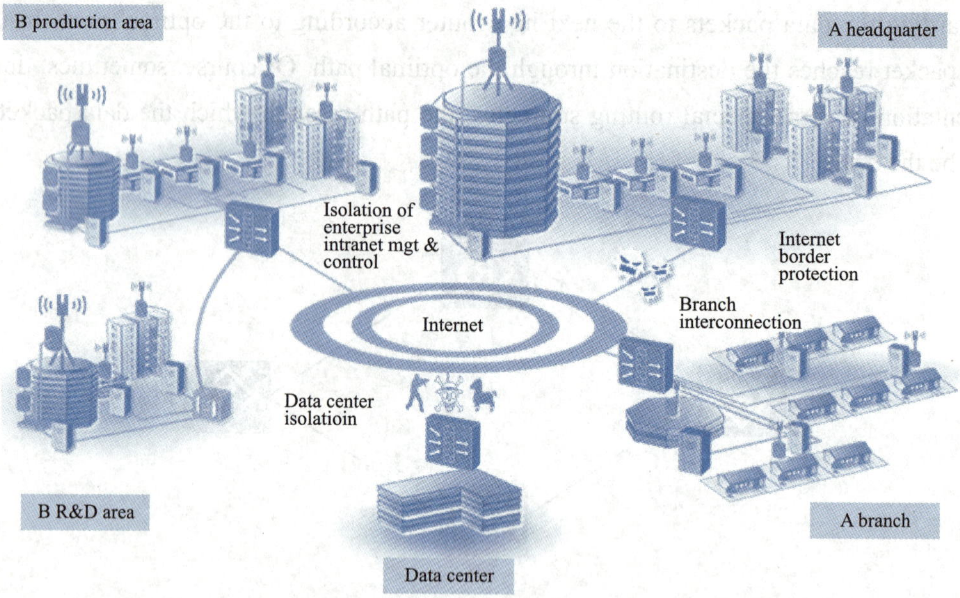

Figure 3-8 Schematic diagram of firewall

There is a difference between a firewall, a router, and a switch. Routers are used to connect different networks, through routing protocols to ensure interconnection and intercommunication, to ensure that messages are forwarded to the destination; switches are usually used to build a local area network, as an important hub for LAN communication, and quickly forward messages through Layer 2/Layer 3 exchanges. The firewall is mainly deployed at the network boundary to control the access behavior in and out of the network, and security protection is its core feature. The essence of routers and switches is forwarding, and the essence of firewalls is control, as shown in Figure 3-9.

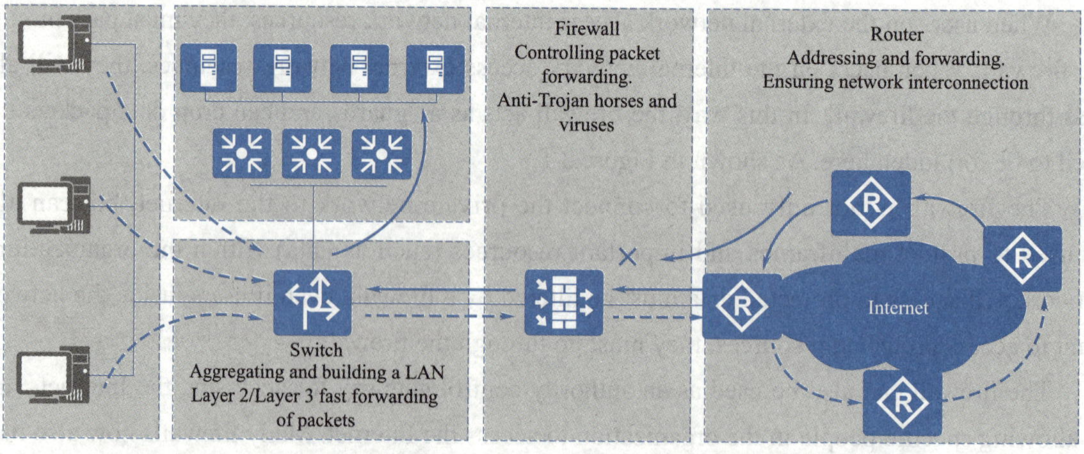

Figure 3-9 The difference between firewall, router and switch

The routers currently on the market basically have a simple firewall function, whether it is consumer-level or enterprise-level, which can realize some functions such as packet filtering and IP filtering. So why do you need a hardware firewall? The differences between the two are as follows:

(1) Background:

The generation of routers is based on the routing of network data packets. What the router needs to accomplish is to carry out effective routing management on the data packets of different network segments. What the router cares about is whether it can route data packets of different network segments for communication.

Firewall is generated from people's demand for security. Whether the data packet can arrive correctly, the time of arrival, the direction, etc. are not the focus of the firewall. The key point is whether the data packet should pass and whether it will cause harm to the network after passing.

(2) Purpose:

The fundamental purpose of the router is to keep the network and data connected.

The fundamental purpose of the firewall: to ensure that any non-allowed data packets are "blocked".

(3) Core technology:

The router core ACL list is based on simple packet filtering, which belongs to the third layer of OSI filtering.

From the perspective of firewall technology implementation, firewall is application-level information flow filtering based on stateful packet filtering.

(4) Security strategy:

The default configuration of the router is not comprehensive enough for security considerations, and advanced configuration is needed to achieve some anti-attack effects. The rules for security are more complicated and the probability of configuration errors is high.

The default configuration of some firewalls can prevent various attacks, and the more user-friendly firewalls are configured using a graphical interface, which is simple to configure and has a low error rate.

(5) Impact on performance:

The original intention of the router is to forward data packets, rather than specifically designed as a full-featured firewall, so the amount of calculation is very large when data is forwarded. If packet filtering is performed again, it will have a great impact on the router's CPU and memory. This is why the data forwarding rate decreases after the router turns on the firewall. Moreover, because the hardware cost of the router is relatively high, the hardware cost of its high-performance configuration is relatively large.

The firewall is not used for data forwarding, as long as it determines whether the packet meets the requirements, if it is, it will pass, otherwise it will not. Its software is also specially optimized for data packet filtering. Its main module runs in the kernel mode of the operating system. In the design, security issues are especially considered, and its performance in packet filtering is very high.

As the router is a simple packet filtering, the increase in the number of packet filtering rules and the increase in the number of NAT rules will increase the impact on the router performance accordingly. The firewall uses stateful packet filtering, the number of rules, and NAT. The number of rules has less impact on performance.

(6) Anti-attack capability:

Under normal circumstances, ordinary routers do not have application layer prevention functions, and do not have functions such as real-time intrusion detection. If you need to have such functions, you need to upgrade the router's system software. At this time, we not only have to bear the software upgrade cost, but also need to upgrade the hardware configuration because these functions require a lot of calculations, which further increases the cost.

The firewall has a variety of attack prevention functions, and even has an application layer defense function.

In summary, a fundamental condition for users to use firewalls is their demand for network security. The simplicity and complexity of the user's network topology and the difficulty of user applications are not the criteria for determining whether a firewall should be used. Even if the user's network topology and applications are very simple, the use of firewalls is still necessary; if the user's environment and applications are more complex, then the firewall will be able to bring more benefits, and the firewall will be an indispensable part of the network construction. For a normal network, the router will be the first pass to protect the internal network, and the firewall will be the second pass and the most stringent pass.

The earliest firewall can be traced back to the late 1980s, with a history of more than 30 years. In the past two decades, the development process of firewalls can be roughly divided into the following three periods, as shown in Figure 3-10.

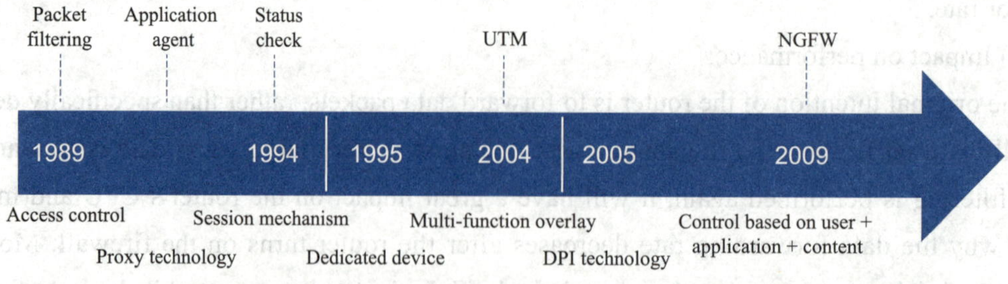

Figure 3-10 The history of firewall development

1. Firewall

A firewall is a network security system designed to monitor and control network traffic, determining whether to allow the transmission of data packets based on predefined security rules. Its main function is to protect the internal network from external threats, prevent unauthorized access, and establish a secure barrier between enterprise networks and the Internet. By filtering traffic, blocking malicious attacks, and recording network activities, firewalls effectively enhance the security and stability of networks. As the first line of defense in network security, firewalls play a crucial role in the network environments of enterprises, institutions, and individual users.

2. UTM

From 1995 to 2004, during this period, stateful inspection of firewalls became a trend. In addition to the access control function, some other functions, such as VPN, have also been added to the firewall.

At the same time, some special equipment also emerged in this period. For example, a WAF (web application firewall) device dedicated to protecting the security of Web servers.

In 2004, the industry put forward the concept of UTM (united threat management), which combines traditional firewall, intrusion detection, antivirus, URL filtering, application control, mail filtering and other functions into a firewall to achieve comprehensive security protection.

3. NGFW

After 2004, the UTM market has developed rapidly, and UTM products have sprung up like mushrooms, but they are facing new problems. The first is that the degree of detection of application layer information is limited, and more advanced detection methods are needed at this time, which makes DPI (deep packet inspection) technology widely used. The second is the performance problem. If multiple functions are running at the same time, the processing performance of the UTM device will be severely reduced.

In 2008, the industry released the next-generation firewall to solve the problem of performance degradation when multiple functions are running at the same time. At the same time, it can also be controlled based on users, applications, and content.

In 2009, the industry defined the next-generation firewall, clarifying the functions and characteristics of the next-generation firewall. Subsequently, various security vendors also launched their own next-generation firewall products, and firewalls have entered a new era.

The firewall is mainly deployed at the network boundary to isolate different networks. How to distinguish different networks on the firewall?

In order to distinguish different networks on the firewall, we introduced an important concept on the firewall: security zone, referred to as zone for short. A security zone is a collection of one or more interfaces, which is the main feature that distinguishes a firewall from a router. The firewall divides the network and identifies the “route” of message flow through security zones. Generally speaking, when messages flow between different security zones, they will be controlled. The role of zone is reflected in:

(1) Security policies are implemented based on security areas;

(2) There is no security risk in the data flow occurring within the same security zone, and no security strategy is required;

(3) Only when data flows between different security areas, the security check of the device will be triggered and the corresponding security strategy will be implemented;

(4) In a firewall, all network devices connected to the same interface must be located in the same security zone, and a security zone can include networks connected to multiple interfaces.

As shown in Figure 3-11, the default security zones of the firewall include: Untrust, DMZ, Trust, and local, which represents the firewall itself.

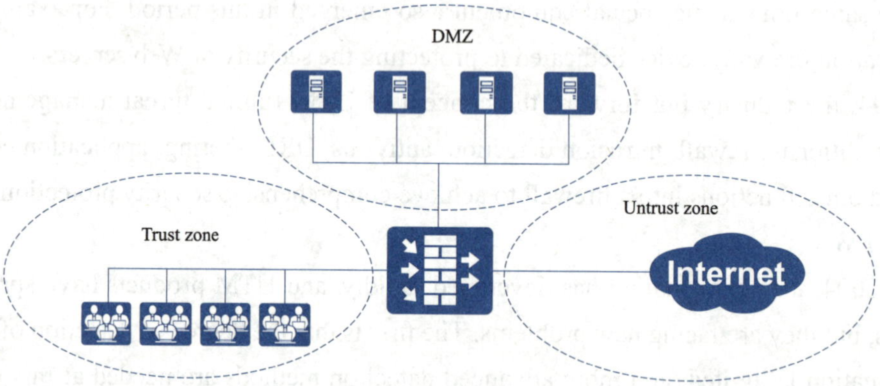

Figure 3-11 Firewall security zone

The firewall is classified in units of interfaces. All network devices connected to the same interface must be located in the same security zone, and a security zone can include networks connected to multiple interfaces. The interface here can be either a physical interface or a logical interface. Therefore, the function of dividing users on different network segments connected to the same physical interface into different security zones can be realized through logical interfaces such as sub-interfaces or VLAN IF. Figure 3-12 shows the relationship between firewall security zones and interfaces.

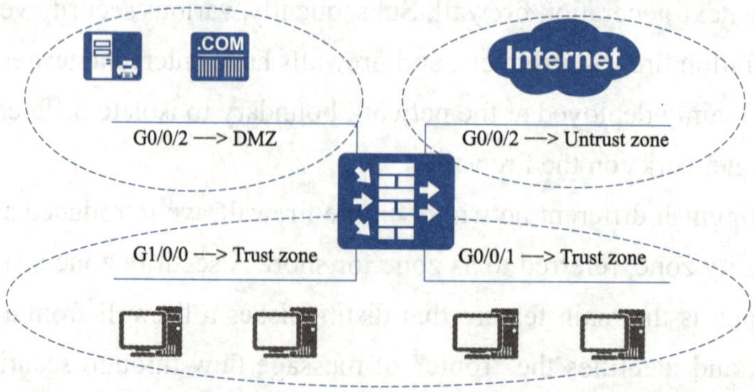

Figure 3-12 The relationship between firewall security zones and interfaces

Task 3.2 Login network equipment

Subtask 3.2.1 Understanding VRP

VRP (versatile routing platform) is a network operating system used by Huawei's data communication products. The network operating system is system software that runs on certain equipment and provides network access and interconnection services.

The VRP general routing platform, as the software core engine for Huawei's full range of routers, Ethernet switches, and service gateways from low-end to core products, realizes a unified user interface and management interface; Realizes control plane functions and defines forwarding plane interface specifications. Realize the interaction between the forwarding plane of each product and the VRP control plane. Realize the network interface layer, shielding the difference between the link layer of each product and the network layer.

Subtask 3.2.2 Master command line configuration mode

The hierarchical command structure of VRP defines many command line views, and each command can only be executed in a specific view. This subtask introduces common command line views. Each command is registered in one or more command views, and the user can only run the corresponding command after entering the view where the command is located. After entering the configuration interface of the VRP system, the first view that appears on the VRP is the user view. In this view, users can view the operating status and statistical information of the device.

To modify the system parameters, the user must enter the system view. Users can also enter other functional configuration views through the system view, such as interface view and protocol view. As shown in Figure 3-13, the current view can be judged through the prompt, for example: "< >" indicates the user view, and "[ ]" indicates other views except the user view.

```
<NGFW>                                       User view
<NGFW>system-view
[NGFW]                                       System view
[NGFW]interface GigabitEthernet 0/0/1
[NGFW-GigabitEthernet0/0/1]quit              Interface view
[NGFW]ospf 1
[NGFW-ospf-1]                                Protocol view
```

Figure 3-13 VRP view

VRP system commands adopt hierarchical protection, and the commands are divided into four levels: visit level, monitoring level, configuration level, and management level.

(1) Visit level: Network diagnostic tool commands (ping, tracert), commands for accessing external devices from the device (including Telnet client, SSH, Rlogin), etc. Commands at this level do not allow configuration file saving operations.

(2) Monitoring level: It is used for system maintenance, service fault diagnosis, etc., including display and debugging commands. Commands of this level do not allow configuration file saving operations.

(3) Configuration level: Business configuration commands, including routing and commands at various network levels, which are used to provide users with direct network services.

(4) Management level: It is related to the basic operation of the system and the commands of the system support modules. These commands provide support for the business, including file system, FTP, TFTP, X-modem download, configuration file switching commands, standby board control commands, user management commands, and command levels setting commands, system internal parameter setting commands, etc.

The system also divides logged-in users into 4 levels, which correspond to the command levels respectively. After users of different levels log in, they can only use commands equal to or lower than their own level. When the user switches from a low-level user to a high-level user, you need to use the following command: super password [level user-level] {simple | cipher} password to switch.

Subtask 3.2.3 Learn to use command line help

VRP supports a command line help system to reduce the difficulty for users to remember commands.

With full help, the system can assist you to give hints of all keywords or parameters when you enter the command line. In any command view, type "?" to get all the commands and their brief descriptions in the command view, as shown in Figure 3-14.

```
<NGFW>?
User view commands:
  anti-ddos                  Defend against DDoS attacks
  arp                        Specify ARP configuration information
```

Figure 3-14 Complete command help (I)

After entering a certain keyword, you can also add a "?" separated by a space for command help, as shown in Figure 3-15.

```
<NGFW>display firewall ?
  blacklist                  Indicate the blacklist command group
  dataplane                  Indicate dataplane to manageplane
  defend                     Indicate attack defense
```

Figure 3-15 Complete command help (II)

With partial help, the system can assist you to give prompts for all keywords or parameters beginning with the string when entering the command line, as shown in Figure 3-16.

```
<NGFW>d?
  debugging                                   delete
  dir                                         display
  download

<NGFW>display b?
  backup-configuration                        bfd
  bgp                                         bridge
  buffer                                      bulk-stat
```

Figure 3-16 Part of the command help

After entering an incomplete keyword, press the Tab key, the system automatically executes some help:

If the matching keyword is unique, the system replaces the original input with the complete keyword and displays it in a new line, with a space between the cursor and the end of the word, as shown in Figure 3-17.

```
[NGFW]info-
[NGFW]info-center
```

Figure 3-17 Command Help Tab (Ⅰ)

For unmatched or matched keywords that are not unique, first display the prefix, and continue to press Tab to cycle through the words. At this time, there is no space between the cursor and the end of the word. Press the space bar to enter the next word, as shown in Figure 3-18.

```
[NGFW]info-center l
[NGFW]info-center lo
[NGFW]info-center loghost
[NGFW]info-center local
[NGFW]info-center logbuffer
[NGFW]info-center logfile
```

Figure 3-18 Command Help Tab (Ⅱ)

If you enter a wrong keyword, press Tab to display in a new line, and the entered keyword remains unchanged.

Task 3.3 Network device interface configuration

When the device performs interface configuration, it supports Web mode and CLI mode.

1. Web mode

(1) Select Network → Interface and select the edit corresponding to the interface.

(2) Configure the interface IP address, switch the interface mode, and join the security zone (required by the firewall device), as shown in Figure 3-19.

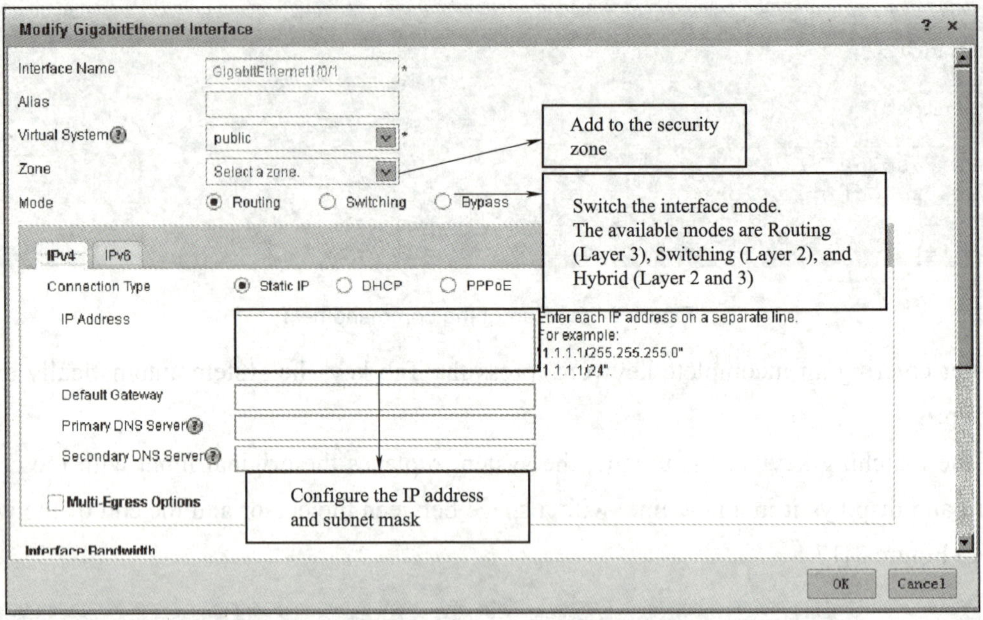

Figure 3-19 Configure interface

2. CLI mode

(1) Enter the interface view.

```
<USG>system-view.
[USG]interface interface-type interface-number.
```

(2) Configure Layer 3 Ethernet interface or Layer 2 Ethernet interface.

```
Configure Layer 3 Ethernet interface.
Ip address Ip-address {mask | mask-length}
Configure Layer 2 Ethernet interface.
Port-switch
```

(3) Add the interface to the security zone.

Execute the `system-view` command to enter the system view.

Execute the command `firewall zone [name] zone-name` to create a security zone and enter the corresponding security zone view.

Execute the `add interface interface-type interface-number` command to add the interface to the security zone.

Task 3.4 Routing configuration of network equipment

When the device is configured for routing, it supports Web mode and CLI mode

1. Web mode

Choose Network → Route → Static Route and click New, as shown in Figure 3-20.

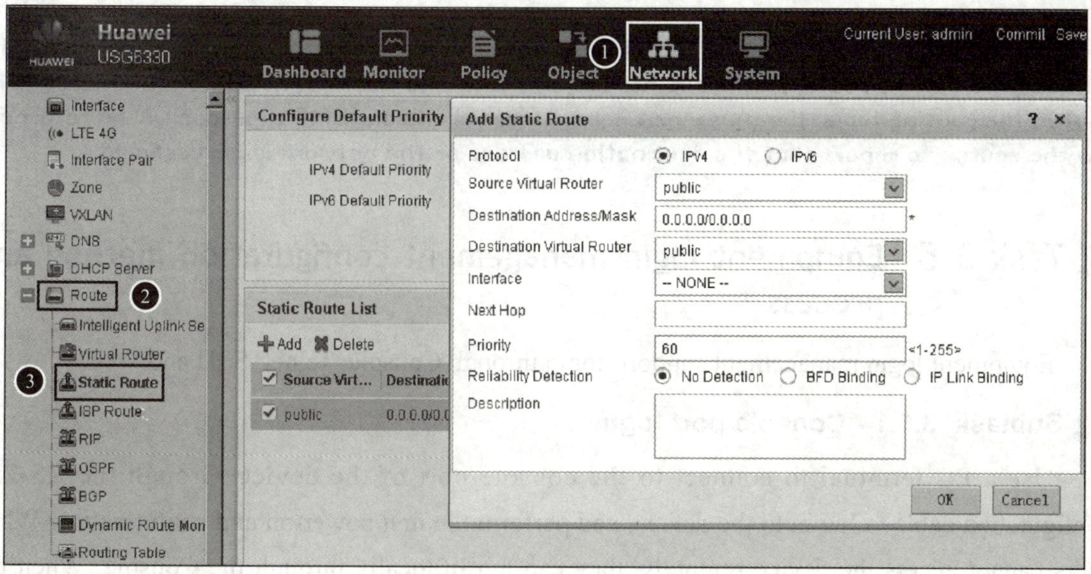

Figure 3-20 Configure routing

2. CLI mode

(1) To configure a static route, you need to perform the following operations:

Execute the `system-view` command to enter the system view.

Run the command `ip route-static Ip-address {mask | mask-length} {interface-type interface-number | next-Ip-address} [preference value] [reject | blackhole]` to add a static route.

Note:

An interoperable network can be established through the configuration of static routing, but the problem with this configuration is that when a network failure occurs, the static routing will not automatically change, and the intervention of the administrator is necessary.

(2) To configure the default route, you need to perform the following operations:

Execute the `system-view` command to enter the system view.

Execute the command `Ip route-static 0.0.0.0 {0.0.0.0 | 0} {interface-type interface-number | next-ip-address} [preference value] [reject | blackhole]` to configure the default route.

Note:

The default route is the route used when no matching routing table entry is found. Only when there is no suitable route, the default route is used. In the routing table, the default route appears as a route to the network 0.0.0.0 (with a mask of 0.0.0.0) . If the destination address of the message cannot match any entry in the routing table, the message will choose the default route. If there is no default route and the destination of the message is not in the routing table, the message will be discarded and an ICMP message will be returned to the source to report that the destination address or the network is unreachable.

Task 3.5 Equipment login management configuration method and process

Equipment login management support: login through Console, Telnet, SSH and Web.

Subtask 3.5.1 Console port login

Use a PC terminal to connect to the console port of the device through the RS-232 configuration cable to log in to the device, and perform the first power-on and configuration. When users cannot access the device remotely, they can log in locally through the Console; When the device system cannot be started, they can perform diagnosis through the console port or enter BootRom to upgrade the system.

Specific steps are as follows:

Step 1: As shown in Figure 3-21, right-click My Computer, select Properties, enter the Device Manager to view the local port parameters, and record which COM port the USB to serial port corresponds to. The Figure 3-22 is COM3.

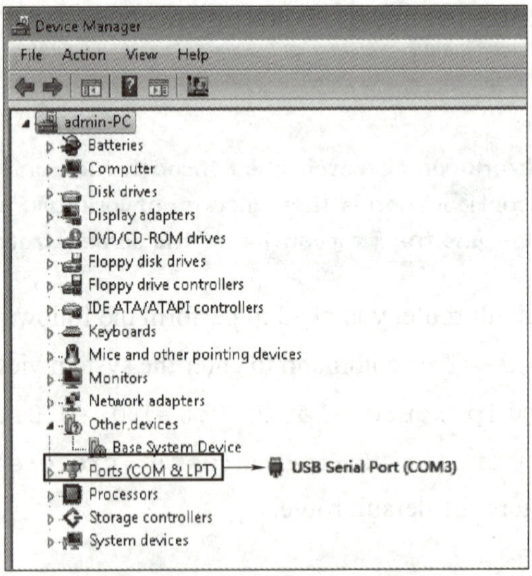

Figure 3-21 Console port settings

Step 2: Open the control software, take PuTTY as an example. After entering the Serial window shown in Figure 3-22, please fill in the space of serial line to connect to according to the port actually used by the PC (or configuration terminal), and fill in the configuration parameters of PuTTY on the left according to the parameter table on the right. Then click the "Open" button.

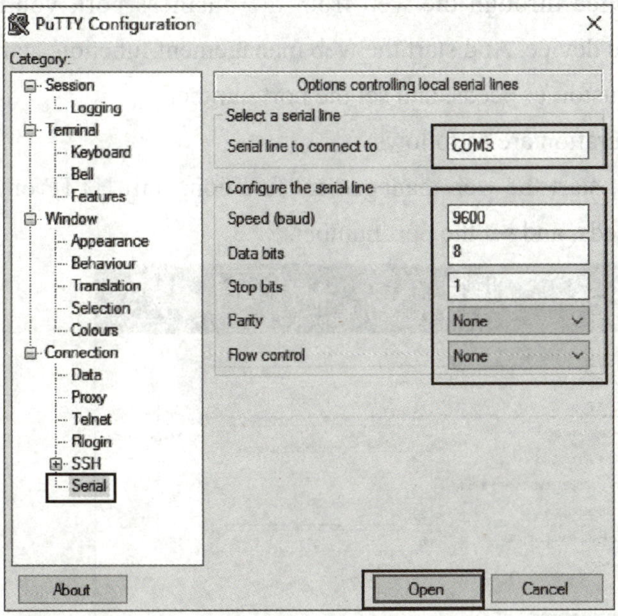

Parameter	Value
Speed/Bd	9 600
Data bits	8
Parity	None
Stop bits	1
Flow control	None

Figure 3-22 PuTTY parameter setting

Step 3: Enter the user name and password to complete the login. The default user name for logging in to the USG configuration port is admin, and the default user password is Admin@123. Among them, the user name is not case sensitive, and the password must be case sensitive.

Subtask 3.5.2 Web login

Access the device through a web browser on the client for control and management. It is suitable for the configuration terminal PC to log in through the Web mode. The HTTP/HTTPS service is enabled by default on the device. At this time, if you log in using HTTP, it will automatically be redirected to HTTPS login. For security reasons, it is recommended not to turn off the HTTPS service.

Specific steps are as follows:

Step 1: The device can log in to the web interface through the GigabitEthernet0/0/0 interface by default.

Step 2: Set the IP address acquisition mode of the network connection of the administrator's PC to "Automatically obtain an IP address".

Step 3: Connect the Ethernet port of the PC to the default management interface of the device directly, or connect through a switch.

Step 4: Visit http://192.168.0.1 in the browser of the PC to enter the login page of the web interface. The default user name is admin and the password is Admin@123. To ensure system security, please change the password after logging in.

If you need to log in to the device through the web from the business port, you need to configure the web login function on the device. And start the web management function, start HTTP or HTTPS management according to customer needs, and set the port number.

The specific steps for web configuration are as follows:

Step 1: As shown in Figure 3-23, start the web management function, start HTTP or HTTPS management according to customer needs, and set the port number.

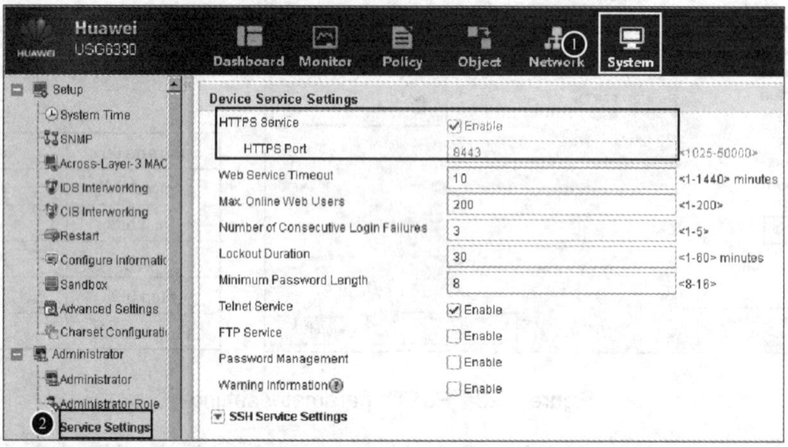

Figure 3-23 Start the web management function

Step 2: As shown in Figure 3-24, configure the web administrator.

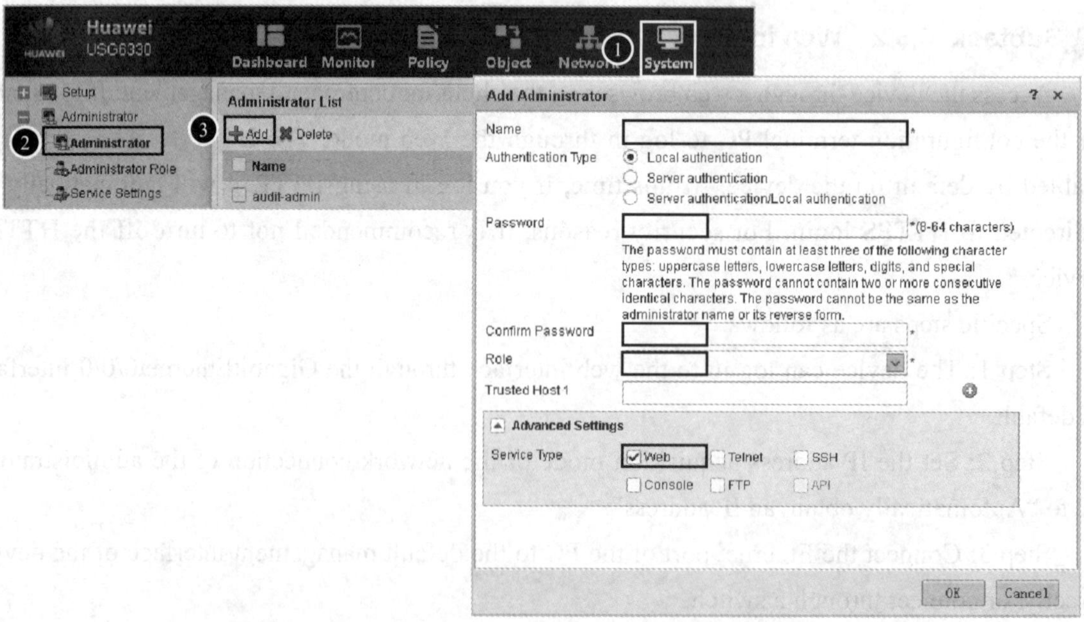

Figure 3-24 Configuring the web administrator

Step 3: As shown in Figure 3-25, choose "Network", configure the login interface.

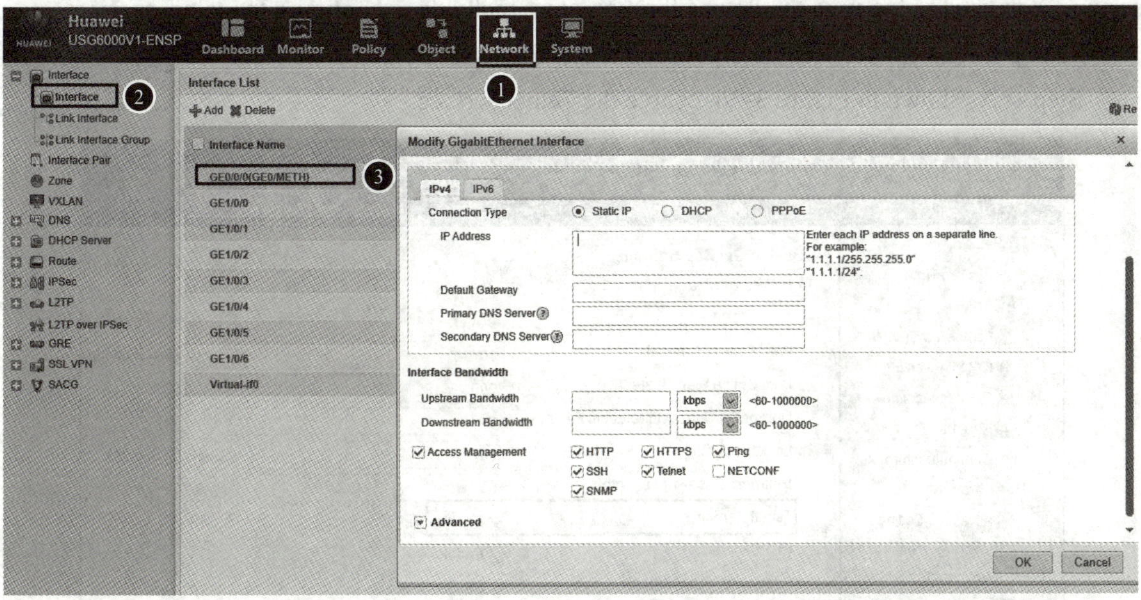

Figure 3-25 Configure login interface

The specific steps of CLI configuration are as follows:

Step 1: Start the web management function.

```
[NGFW] web-manager security enable port 8443
```

Step 2: Configure web users.

```
[NGFW] aaa
[NGFW-aaa] manager-user webuser
[NGFW-aaa-manager-user-webuser] password cipher Admin@123
[NGFW-aaa-manager-user-webuser] service-type web
[NGFW-aaa-manager-user-webuser] level 3
```

Step 3: Configure NGFW interface web device management.

```
[NGFW-GigabitEthernet1/0/1] service-manage enable
[NGFW-GigabitEthernet1/0/1] service-manage https permit
```

Subtask 3.5.3 Telnet login

Connect to the network through a PC terminal, log in to the device using Telnet, and perform local or remote configuration. The target device verifies the user according to the configured login parameters. The Telnet login mode is convenient for remote management and maintenance of the device.

By default, the NGFW does not allow login through telnet. You need to configure the telnet service. You need to log in to the device through other methods to configure the telnet service.

The specific steps for web configuration are as follows:

Step 1: As shown in Figure 3-26, enable the Telnet service.

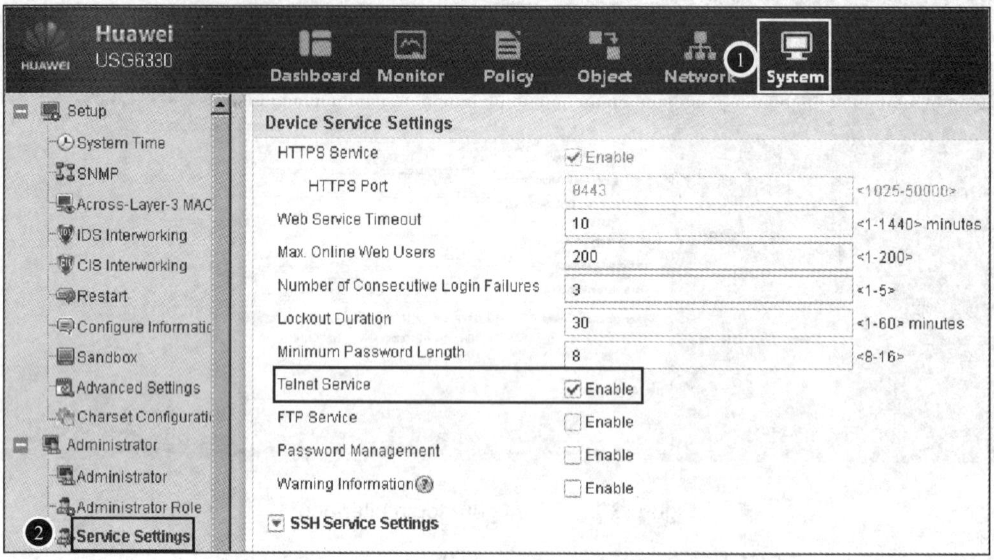

Figure 3-26 Enabling the Telnet service

Step 2: As shown in Figure 3-27, configure the Telnet administrator.

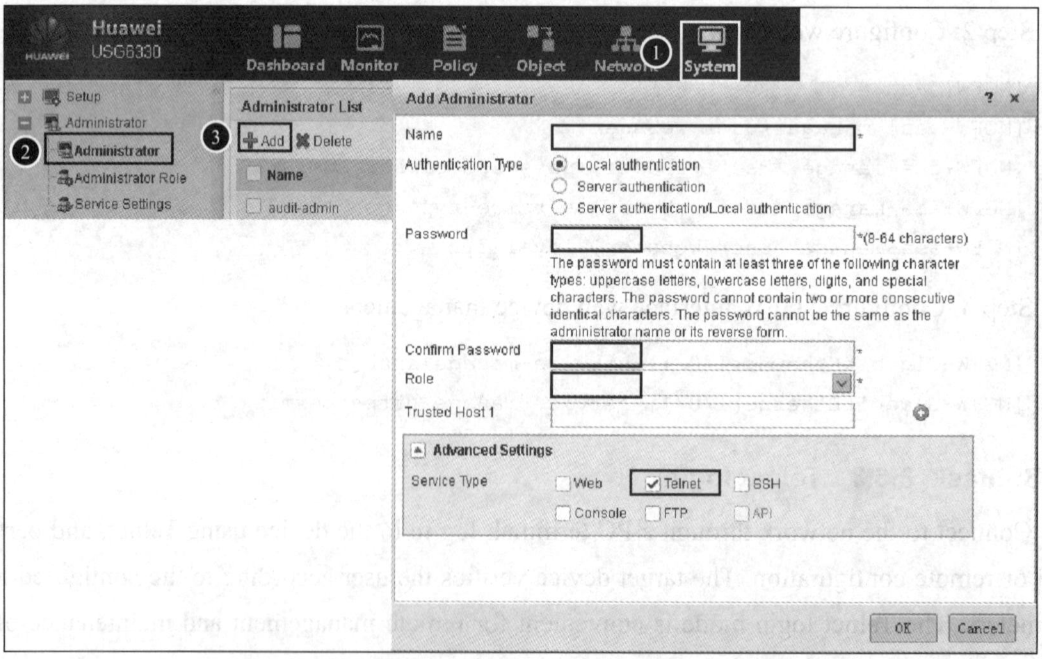

Figure 3-27 Configuring the Telnet administrator

Step 3: As shown in Figure 3-28, configure the login interface.

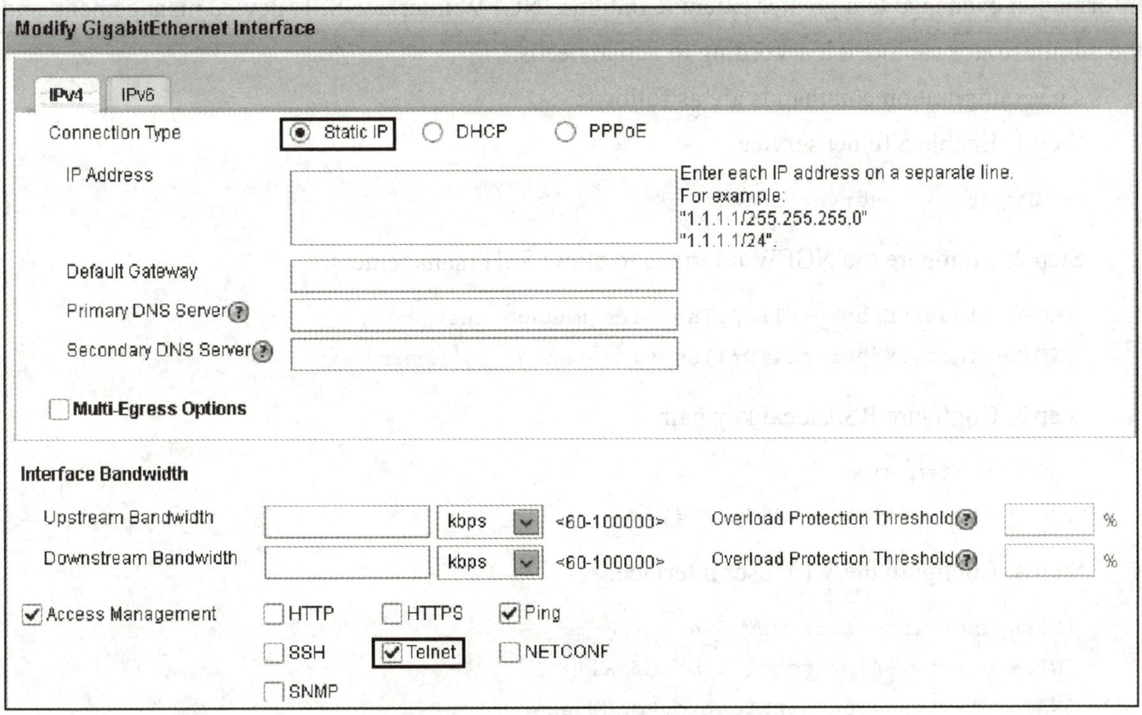

Figure 3-28 Configure login interface

The specific steps of CLI configuration are as follows:

Step 1: Turn on the Telnet service.

```
[NGFW] telnet server enable
```

Step 2: Configure Telnet administrator.

```
[NGFW] user-interface vty 0 4
[NGFW-ui-vty0-4] authentication-mode aaa
[NGFW] aaa
[NGFW-aaa] manager-user vtyadmin
[NGFW-aaa-manager-user-vtyadmin] password
Enter Password
```

Step 3: Configure NGFW interface Telnet device management.

```
[NGFW-GigabitEthernet1/0/1] service-manage enable
[NGFW-GigabitEthernet1/0/1] service-manage telnet permit
```

Subtask 3.5.4 SSH login

SSH provides secure information protection and powerful authentication functions to protect the equipment system from attacks such as IP spoofing. SSH login can ensure the security of data information exchange to a greater extent. Configure NGFW interface SSH device management, and the administrator can open it according to actual needs.

The configuration commands are as follows:

Step 1: Enable STelnet service.

```
[NGFW]stelnet server enable
```

Step 2: Configure the NGFW interface to allow SSH management.

```
[NGFW-GigabitEthernet1/0/1]service-manage enable
[NGFW-GigabitEthernet1/0/1]service-manage ssh permit
```

Step 3: Configure RSA local key pair.

```
<NGFW>system-view
[NGFW]rsa local-key-pair create
```

Step 4: Configure the VTY user interface.

```
[NGFW]user-interface vty 0 4
[NGFW-ui-vty0-4]protocol inbound all
[NGFW-ui-vty0-4]authentication-mode aaa
```

Step 5: Create an SSH administrator account, specify the authentication method to be Password, and the service method to be STelnet.

```
[NGFW]aaa
[NGFW-aaa]manager-user sshadmin
[NGFW-aaa-manager-user-sshadmin]service-type ssh
[NGFW-aaa-manager-user-sshadmin]level 3
[NGFW-aaa-manager-user-sshadmin]password cipher huawei
```

Step 6: Create an SSH user.

```
[NGFW]ssh user sshadmin
[NGFW]ssh user sshadmin authentication-type password
[NGFW]ssh user sshadmin service-type stelnet
```

After the above configuration is completed, run the client software that supports SSH to establish an SSH connection. If the Huawei device is used as an SSH client, you need to enable the SSH login function of the client.

```
[NGFW]ssh client first-time enable
```

Task 3.6 Network equipment system settings

Subtask 3.6.1 Profile management

The configuration file is a configuration item to be loaded when the device starts. The user can save, modify and clear the configuration file, and select the configuration file loaded when the device starts. System files include the software version of the NGFW device, signature database files, etc. General software upgrades needed to manage system files. There are two types of configuration files:

(1) Saved-configuration: The saved-configuration is the configuration file used when the NGFW device is powered on next time. It is stored in the NGFW's Flash or CF card and will not be lost after restarting.

(2) Current-configuration: The current-configuration is the current effective configuration of the NGFW device. Both command line and web operations are used to modify the current-configuration. It is stored in the memory of the NGFW and will be lost after restarting.

The main operations on the configuration file are:

Save the configuration file: In order to make the current configuration the initial configuration when the firewall is powered on next time.

Method 1 (Command line) : In user view, execute the save command.

Method 2 (Web) : Select the "Save" button at the top right of the "Homepage".

(1) Erase configuration file (restore factory configuration) : After the configuration file is erased, the firewall will use the default configuration parameters for initialization when the firewall is powered on next time.

Method 1 (Command) : In the user view, execute the `reset saved-configuration` command.

Method 2 (Web) : Select System → Maintenance → Configuration Management and execute the button to restore factory configuration.

Method 3 (hardware reset button) : If the device is not powered on, first press and hold the reset button, and then turn on the power switch. When the device indicator on the panel flashes at a frequency of 2 times per second at the same time, release the reset button, and the device will start with the default configuration.

Method 4 (hardware reset button) : If the device has started normally, press and hold reset for a long time (more than 10 seconds). The device will restart and start with the default configuration.

(2) Configure the system software at the next launch.

Command line: In the user view, executes the command `startup system-software sysfile`.

Web: Select System → Maintenance → System Update, and select the button to start the system software next time.

(3) Restart the firewall: The function is to restart the firewall and record the restart action in the log.

Method 1 (Command line) : In the user view, execute the reboot command.

Method 2 (Web) : Select System → Maintenance → System Restart.

Subtask 3.6.2 Software upgrade

The system software can be uploaded to the device through TFTP and FTP. Upgrade the system software to configure the software system to be used when the device is started next time. If the storage space of the current device is insufficient, the device will automatically delete the currently running system software. The system software must use ".bin" as the extension and does not support Chinese.

1. Web upgrade method

Use one-click version upgrade, the specific steps are shown in Figure 3-29.

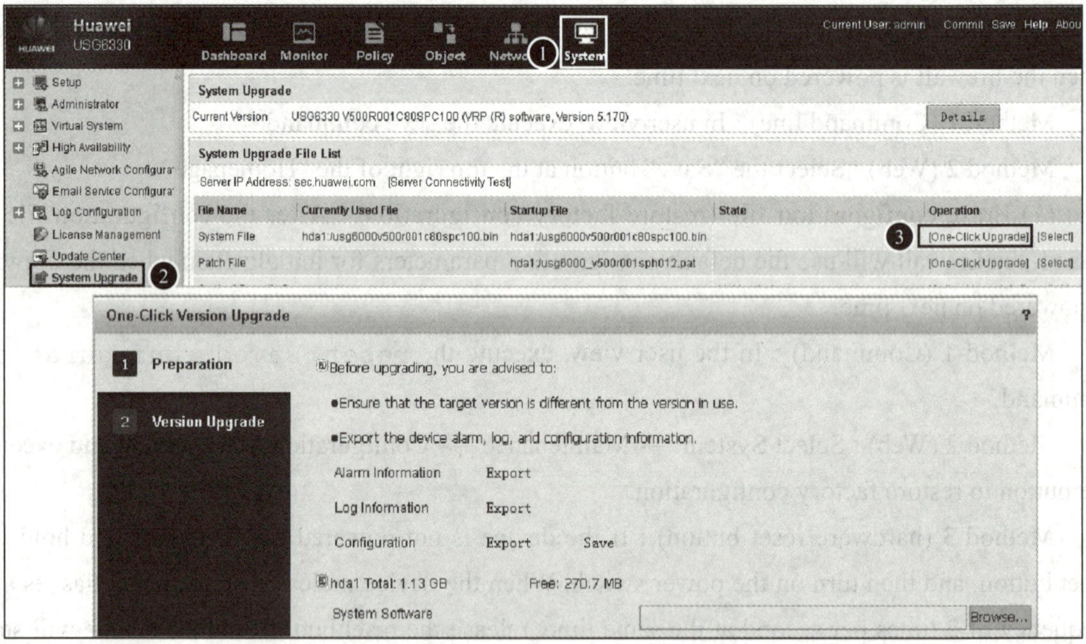

Figure 3-29 One-click device upgrade

Step 1: Choose System → System Upgrade.

Step 2: Click One-click Upgrade to display the one-click system software upgrade wizard interface. Optional: Click "Export" in turn to export the alarm information, log information, and configuration information on the device to the terminal. It is recommended to save the configuration information to the terminal.

Step 3: Click "Browse" and select the system software to be uploaded.

Step 4: After judging whether the current network allows the device to be upgraded, restart it immediately, and select the radio button before "Set to start the system software next time and restart the system" or "Set to start the system software next time without restarting the system".

Step 5: After restarting the device, the upgraded system software can be used.

2. CLI upgrade method

Step 1: Use FTP to download files, configure the command (FW as FTP client) :

Run the `ftp ftp-server [port-number] [Vpn-instance vpn-instance-name]` command to establish a control connection with the FTP server and enter the FTP client view.

Execute the command `get remote-filename [local-filename]` to download the file from the FTP server and save it locally.

Step 2: Configure the system software to be used when the system is started next time.

Execute `startup system-software sys-filename`.

Subtask 3.6.3 License configuration

License is a form of contract for equipment suppliers to authorize the scope and duration of product features. The license can dynamically control whether certain features of the product are available.

License supports online automatic activation and local activation. When you can connect to the Huawei Security Center sdplsp.huawei.com, you can perform online automatic activation. When you cannot connect to the Security Center, you can select local manual activation. The license file must use ".dat" as the extension, and does not support Chinese.

1. Web Configuration method

As shown in Figure 3-30, you can use the Web method to manage licenses.

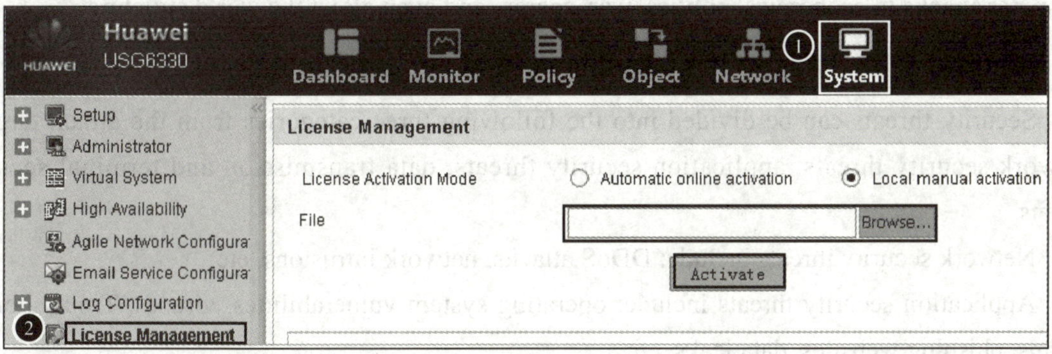

Figure 3-30 License management

2. CLI configuration method

Configuration command:

Run the `system-view` command to enter the system view.

Run the `license active license-file` command to activate the specified license file.

You can use the `display license` command to view the license information.

Project 4 Understand common information security threats

Because information and information systems have various characteristics such as vulnerability, sensitivity, confidentiality, and dissemination, they are also threatened by various scenarios and methods.

Based on some attack cases, this project will introduce the possible threats to the information system.

Task 4.1 Understand the current situation of information security threats

Subtask 4.1.1 Understand the evolution of information security attacks

The evolution characteristics of information security attack events:

(1) The attack method has little change. The attack method is still the viruses, loopholes, phishing, etc. that we can see. It seems that the form has not changed much.

(2) The means of attack have changed from single to complex. A major attack often requires sophisticated deployment, long-term latency, and a combination of multiple attack methods to achieve the ultimate goal.

(3) The purpose of the attack is diversified. The targets of the attack range from personal computer attacks to economics, politics, war, energy, and even affect the world structure.

Subtask 4.1.2 Master the classification of security threats

Security threats can be divided into the following three categories from the attack targets: network security threats, application security threats, data transmission and terminal security threats.

Network security threats include: DDoS attacks, network intrusions, etc.

Application security threats include: operating system vulnerabilities, viruses, Trojan horses, worms, phishing websites, data leaks, etc.

Data transmission and terminal security threats include: communication traffic hijacking, man-in-the-middle attacks, unauthorized personnel logging into the system, and weak wireless network security.

Task 4.2 Network security threat analysis

Let's take the DDoS attack incident on the Dyn DNS service in the United States as an example to help you to analyze the network security threats. On October 21, 2016, from 11: 00 to 17:00 (UTC time), the Dyn DNS service in the United States suffered a DDoS attack, and nearly half of the United States fell into a network disconnection. Dyn is a DNS SaaS provider whose core business is to manage managed DNS services for its users. The DDoS attack severely affected its DNS business, resulting in the inaccessibility of customer websites. Because of its services to many companies, the damage caused is like a spark, and the impact is extremely bad. More than one hundred websites were inaccessible, severely disconnected for up to 3 hours, and Amazon alone lost more than tens of millions of dollars. This large-scale DDoS attack was launched by a botnet composed of IoT devices, which were infected with Mirai malware. The "zombies" attacked were mainly network cameras, digital hard disk video recorders and smart routers. The Mirai botnet infected millions of devices, and only one in ten devices participated in this attack.

The process of this attack is shown in Figure 4-1.

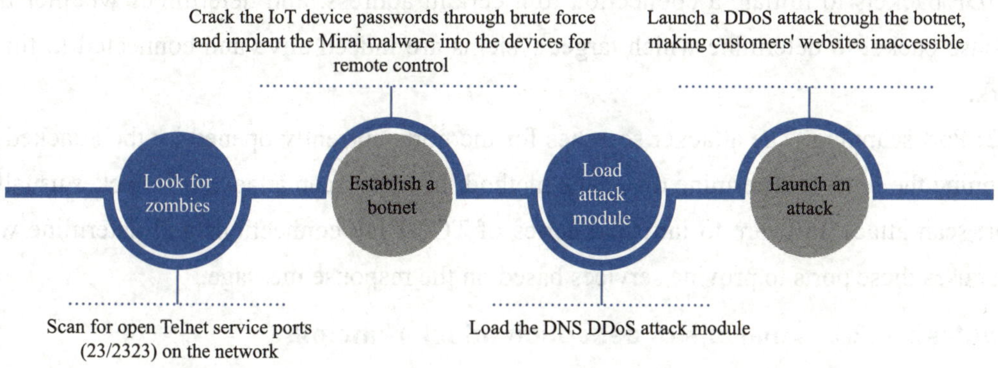

Figure 4-1 The attack process of Mirai virus

(1) Looking for zombies: The Internet of Things devices usually enable the telnet remote login function by default, which is convenient for administrators to remotely manage. Attackers can find surviving IoT devices through IP address scanning, and use port scanning to further determine whether the IoT devices have enabled telnet services.

(2) Establish a botnet: Some IoT device users directly use factory passwords or set simple passwords (simple username/password combinations similar to admin/123456). These passwords can be easily cracked by attackers by brute force. When the attacker successfully cracks the password of the IoT device and successfully logs in via telnet, he then remotely implants the malicious software Mirai on the IoT device to gain absolute control of the device. Malware has absolute control over the infected device. In addition to using the device to initiate DDoS attacks,

it can also cause serious harm to the system, business, and data of the device itself. For example, it can tamper with data, steal privacy, modify configuration, delete files, etc. Use this as a springboard to attack core business systems.

(3) Load attack module: The attacker loads the DNS DDoS attack module on the IoT device.

(4) Launching an attack: The attacker launched a DDoS attack on the US DYN DNS service through a botnet, causing hundreds of websites to become inaccessible.

What methods were used during this DDoS attack? Scanning technology was used in the process of searching for zombies, spoofing attacks were used in the process of forming botnets, and finally, DDoS attacks were launched.

Subtask 4.2.1 Analysis of attack principle

Scanning is a potential attack and does not itself have a direct destructive behavior. It is usually a network detection behavior before the attacker launches a real attack. Scanning can be divided into address scanning and port scanning.

(1) Address scanning: The attacker uses ICMP packets to detect the target address, or uses TCP/UDP packets to initiate a connection to a certain address, and determines whether there is a response packet to determine which target systems are indeed alive and connected to the target network.

(2) Port scanning: The attacker searches for the ports currently opened by the attacked object by scanning the ports to determine the attack method. In a port scan attack, the attacker usually uses the port scan attack software to initiate a series of TCP/UDP connections and determine whether the host uses these ports to provide services based on the response message.

Subtask 4.2.2 Analysis of deception attack principle

Attackers can obtain control authority through brute force password cracking, and can also obtain access and control authority through various spoofing attacks, such as IP spoofing attacks.

IP spoofing is initiated by using the normal trust relationship between hosts. The IP address-based trust relationship between hosts will allow IP address-based authentication, and allow or deny IP address-based access services. Trusted hosts can log in directly without entering a password for verification.

The entire steps of an IP spoofing attack are shown in Figure 4-2.

(1) First, temporarily paralyze the network of the trusted host to avoid interference with the attack.

(2) Then, connect to a port of the target host to guess the serial number and increase the law.

(3) Next, disguise the source address as a trusted host and send a data segment with the SYN

flag to request connection.

(4) After that, wait for the target machine to send a SYN+ACK packet to the paralyzed host.

(5) Finally again disguised as the ACK sent by the trusted host to the target machine, the data segment sent at this time has the predicted sequence number of the target machine +1.

(6) The connection is established and the command request is sent.

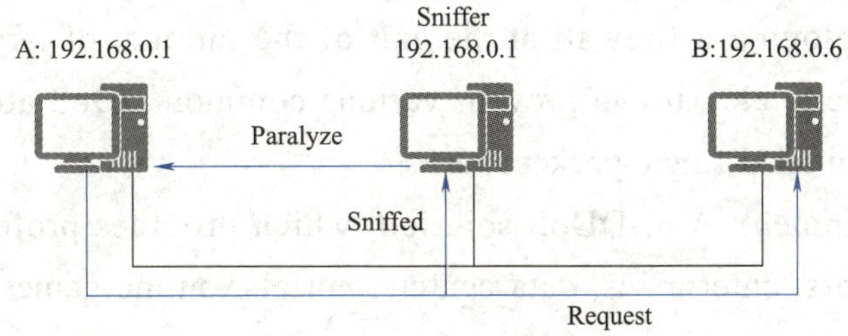

Figure 4-2 IP spoofing attack

Subtask 4.2.3 Analysis of DDoS attack principle

DDos (distributed denial of service) is a distributed denial of service attack, which is a typical traffic attack. As shown in Figure 4-3, a DDoS attack means that an attacker obtains control over a large number of online hosts on the network through various means. These controlled hosts are called zombie hosts, and the network formed by attackers and zombie hosts is called a botnet. When the target is determined, the attacker controls the zombie hosts to send a large number of carefully constructed attack messages to the target. This causes congestion and exhaustion of system resources on the network where the attacked is located, and the attacked has the effect of refusing to provide services to normal users' requests.

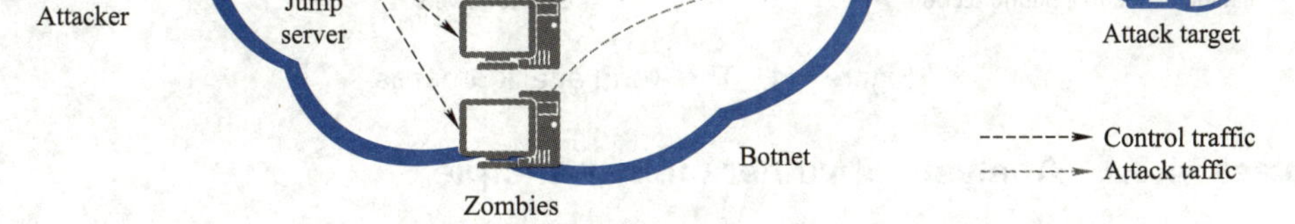

Figure 4-3 DDoS attack

According to the different types of attack messages used, there are many types of DDoS attacks on the network. There are mainly the following common DDoS attacks: SYN Flood, UDP

Flood, ICMP Flood, HTTP Flood, HTTPS Flood, DNS Flood, etc.

Subtask 4.2.4 Master the defense means of network attack

Currently, there are a large number of zombie hosts and botnets on the Internet. Driven by commercial interests, DDoS attacks have become an important security threat to the Internet. Security devices can be deployed in the network to defend against network attacks.

Firewall: By deploying a firewall at the exit of the intranet of large and medium-sized enterprises, data centers, etc., it can prevent various common DDoS attacks, and it can also effectively prevent traditional single-packet attacks.

Anti-DDoS equipment: Anti-DDoS solution, which provides professional DDoS attack protection for operators, enterprises, data centers, portals, online games, online videos, DNS domain name services, etc.

Task 4.3 Apply security threat analysis

Let's take the incident of a worm attacking a micro-blog website as an example to take everyone to analyze application security threats. A Chinese micro-blog website encountered a worm attack. In less than an hour, more than 30,000 users were attacked by the worm. The attack process is shown in Figure 4-4.

What methods were used during this worm attack? This worm attack involves vulnerabilities, phishing attacks, and malicious code.

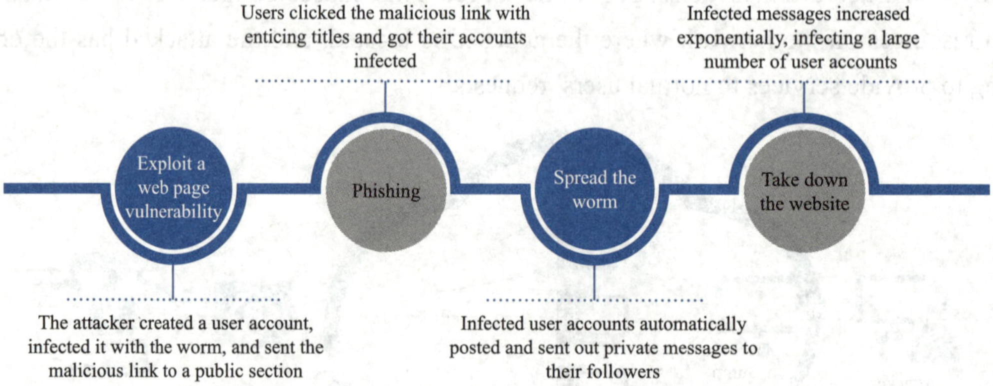

Figure 4-4 The worm attack process

Subtask 4.3.1 Analysis of vulnerability principle

Vulnerabilities are defects in the specific implementation of hardware, software, and protocols or system security policies, which can enable an attacker to access or damage the system without authorization. As the system vulnerabilities are not fixed in time, the following attacks will be brought:

(1) Injection attacks: The main method of SQL injection attacks is to construct clever SQL statements and combine them with the content submitted by the web page to perform injection attacks. The more commonly used techniques include the use of annotation symbols, identities (such as 1=1), the use of union statements for joint queries, and the use of insert or update statements to insert or modify data.

(2) Cross-site scripting attack: Cross-site scripting (usually referred to as XSS or cross-site scripting or cross-site scripting attack) is a security vulnerability attack of website applications and a type of code injection. It allows malicious users to inject code into webpages, and other users will be affected when they watch the webpage. Such attacks usually include HTML and client-side scripting languages.

(3) Malicious code propagation: Malicious code writers generally use three types of methods to spread malicious code—software vulnerabilities, users themselves, or a mixture of the two. Some malicious codes are self-starting worms and embedded scripts, which are software themselves. Such malicious codes have no requirements for human activities. Some malicious codes, such as Trojan horses and e-mail worms, use the psychology of the victims to manipulate them to execute unsafe codes; others trick users into turning off protection measures to install malicious codes.

(4) Data leakage: Data leakage is an incident in which sensitive, protected or confidential data may be plagiarized, stolen or used by an unauthorized organization. Data leakage information may include personal health information (PHI), personal identifiable information (PII), trade secrets or intellectual property.

Subtask 4.3.2 Analysis of fishing attack principle

"Phishing" is a kind of online fraud. It means that criminals use various means to imitate the URL address and page content of the real website, or use the loopholes in the server program of the real website to insert dangerous HTML code into some web pages of the site. This is used to defraud the user's personal information such as bank or credit card account numbers and passwords.

Phishing websites are usually designed in two ways: the first is to trick users into filling in information such as ID numbers, bank accounts, etc. in the name of "winning a lottery"; the second is to imitate bank online payment and electronic transaction websites to defraud the user's bank card information or online payment account password. The whole process is like phishing, and such malicious websites are also called "phishing websites". Such phishing techniques are not highly technical. Some take advantage of people's greed for cheap psychology to be deceived, and some take advantage of the weakness of some netizens in preventing fraud. Once people are fooled, personal privacy information will be leaked and sold, or because the bank account information is filled in on the website, the corresponding assets will be immediately transferred away, and people will regret it.

Subtask 4.3.3 Principle analysis of malicious code

Malicious code refers to computer code that is deliberately programmed or set to cause threats and potential threats to the network or system. The most common malicious codes are viruses, Trojan horses, worms, backdoors, etc. Web browsing and e-mail transmission are the main ways for viruses, Trojan horses, and spyware to enter the intranet. Malicious code is also called malware. These software may also be called adware, spyware, and malicious shareware. It refers to software that is installed and operated on the user's computer or other terminal without explicitly prompting the user or without the user's permission, which infringes the user's legitimate rights and interests.

(1) Virus: An attacker uses the inherent vulnerability of computer software and hardware to compile a set of instruction sets or program codes. By modifying other programs, they can put their own exact copies or possibly evolved forms into other programs, so that other programs can be infect. It can also break computer systems, tamper with, and damage business data.

(2) Trojan horse: A specific program (trojan horse program) is used to control another computer. Trojan horses usually have two executable programs: one is the client, which is the control end; the other is the server, which is the controlled end. The "server" part is implanted into the computer of the planted person, and the so-called "hacker" uses the "controller" to access the computer running the "server". After running the "server" of the Trojan horse program, one or more ports of the victim's computer will be opened, so that the hacker can not only steal important information on the computer, but also damage the intranet computer.

(3) Worm: A malicious program that can use system vulnerabilities to spread itself through the network. It uses the network to copy and spread itself. The way of infection is through the network and e-mail. Main hazard: It consumes host resources, or even damages the host system, causing host denial of service; The traffic caused by the spread of the worm can cause network congestion, and even paralyze and lose control of the entire Internet.

(4) Backdoor: It refers to the secret function hidden in the program, it is usually set by the program designer in order to enter the system at will in the future.

(5) Spyware: A software that can install backdoors and collect user information on the user's computer without the user's knowledge, collect, use, and disseminate sensitive information of the company's employees, seriously interfering with the normal business of the company.

Subtask 4.3.4 Defense methods for application attacks

The defense methods for application attacks mainly include:

(1) We should regularly use professional tools to scan and recover systems or personal

computers for vulnerabilities, and install patches in time to avoid virus invasion and infection to a certain extent.

(2) We should raise our security awareness and be alert to suspicious websites and links.

(3) We should use professional equipment for protection, such as firewalls, WAF, anti-virus software, etc.

Note:

As a border device, the firewall blocks users' online behaviors, web pages, mail viruses, and illegal applications to protect the internal network.

WAF is called web application firewall. It is a protective device that provides protection for Web applications by implementing a series of security policies for HTTP/HTTPS.

Task 4.4 Understanding data transmission and terminal security threats

Let's take the encrypted communication incidents of Google and Yahoo users and the leakage of Tumblr user information monitored by the NSA (National Security Agency) as examples to help you to analyze data transmission and terminal security threats.

The National Security Agency (NSA) monitors encrypted communications between Google (including Gmail) and Yahoo users in the cloud. NSA uses the encryption and decryption features of Google's front-end server to bypass the device and directly monitor the back-end plaintext data.

More than half of the account passwords on the light blog site Tumblr were stolen by hackers. The hacker first invaded Tumblr's server in a certain way and obtained Tumblr user information. Tumblr once claimed that because the database information is encrypted, it will not affect users. However, it turns out that Tumblr uses a weaker algorithm for user information protection. After obtaining the encrypted user information, hackers cracked a large amount of user information in a relatively short period of time.

From the above events, we can see that there are a lot of security risks in the communication process.

(1) Transmission security risks, man-in-the-middle attack, unencrypted data transmission or insufficient encryption.

(2) Terminal security risks, the server has some loopholes, the user uses a weak password, and the user identity is not verified.

1. Man-in-the-middle attack

Man-in-the-middle attack (MITM attack) is an "indirect" intrusion attack. This attack mode is to place a computer controlled by the intruder on the network connection through various technical means. Between the two communicating computers, this computer is called the "middleman", as shown in Figure 4-5. The consequences of man-in-the-middle attacks are information tampering and information theft.

Figure 4-5 Man-in-the-middle attack

2. The information is not encrypted or the encryption strength is not enough

If the information is not encrypted, the security will be problematic, but even if the data is encrypted, the information may be stolen and cracked. Threat prevention recommendations: Information storage must be encrypted, information transmission must be encrypted, and strong encryption algorithms are used.

3. Identity authentication attack

The attacker obtains identity authentication information through certain means, and then steals sensitive information or achieves certain illegal purposes through the identity information. It is a common attack link in the entire attack event. Threat prevention advice: We should install genuine anti-virus software, use strong passwords and reduce the correlation between multiple passwords.

Project 5 Understand the development trend of threat prevention and information security

In the previous project, we introduced some basic concepts and knowledge of information security, and also introduced common types of attacks. So what measures should we take to ensure enterprise information security?

Regarding the prevention of security threats, not only professional tools are required, but also everyone involved in information security construction must have a sense of prevention.

Task 5.1 Master security threat prevention

Subtask 5.1.1 Understand the key elements of information security

The three key elements of information security prevention are: security operation and management, security products and technologies, and personnel. The three are interlinked, and one is indispensable!

(1) Security operation and management: We should ensure the safety compliance in the operation of various processes of the enterprise, and respond, prevent and improve in a timely

manner when threats and accidents occur.

(2) Security products and technologies: Security prevention technologies ensure the operability of prevention behaviors; security prevention technologies have different methods in different technical levels and fields.

(3) Personnel: "In information security, people are the weakest link." It is extremely important to ensure that employees have a good security awareness.

Subtask 5.1.2 Master the information security precautions

Information security prevention methods are shown in Figure 5-1, which include security products and technologies, operation and maintenance management, and security awareness.

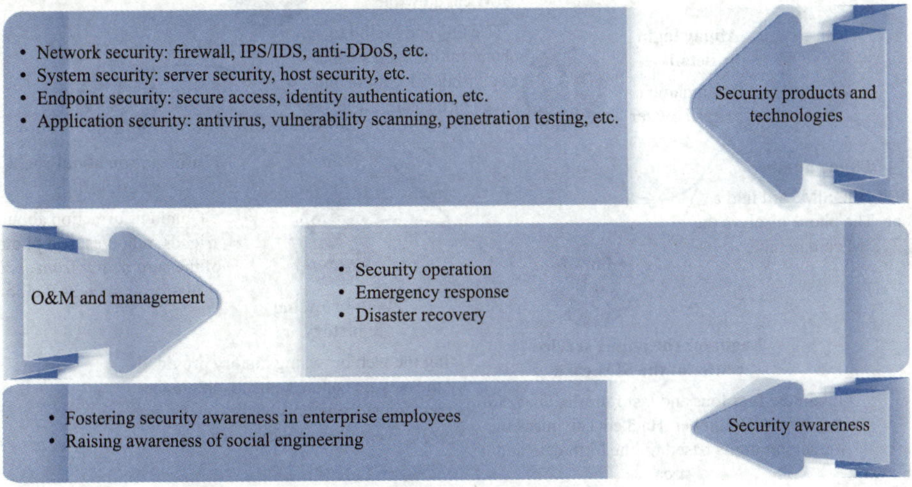

Figure 5-1 Information security prevention methods

Task 5.2 Cultivate awareness of information security

Subtask 5.2.1 Safety awareness test

According to the "Survey Report on Internet Security Awareness of Chinese Netizens" conducted by 360, 24.1% of Internet users have different account passwords; 61.4% of Internet users will distinguish between account passwords; but 13.8% of Internet users still have all accounts use the same password, which is very dangerous.

When users use public Wi-Fi to surf the Internet, the security risks are also different. Statistics show that when connecting to public Wi-Fi, all netizens will browse simple web information or watch videos and listen to music; In addition, 25.1% of netizens will also log in to mailboxes to send e-mails and log in to personal social accounts to chat; 13.6% of netizens also will shop and conduct online banking transactions. If you accidentally connect to the phishing Wi-Fi or hacker Wi-Fi, these operations can easily lead to the theft of the account password and even the loss of property in the financial account. 80% of Wi-Fi can be easily cracked. Every year, net squatting brings people up to 5 billion yuan in economic losses such as the theft of online banking and account stolen.

Subtask 5.2.2 Understand social engineering

Social engineering emerged as a formal discipline around the 1960s. Social engineering was proposed in "The Art of Deception" after hacker Mitnick repented. It is a means of deceiving and harming victims through psychological traps such as psychological weakness, instinctual reaction, curiosity, trust, and greed. As shown in Figure 5-2, if you find a SIM card (phone card), what information can you get from it?

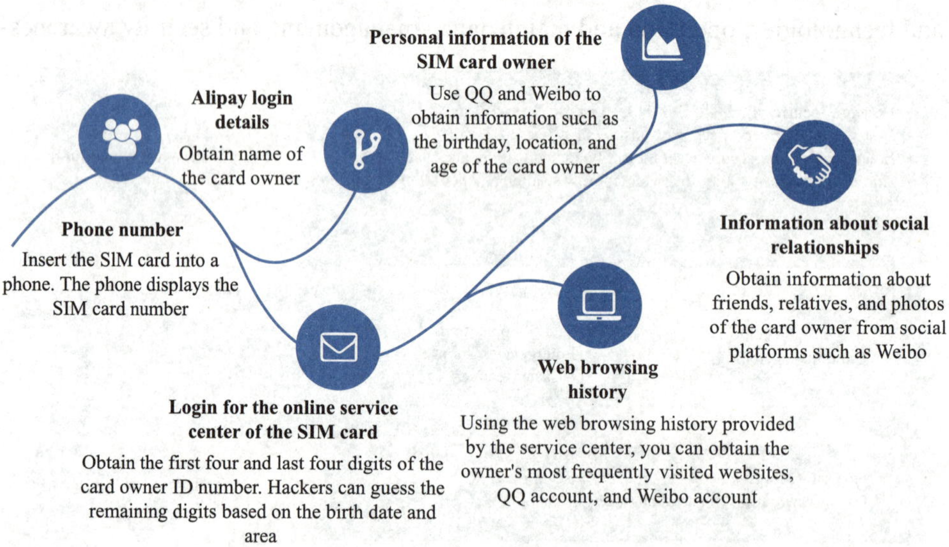

Figure 5-2 Social engineering case

Subtask 5.2.3 Cultivate safety awareness

As ancient saying goes: "Don't harm others, but beware of harm from others". In the process of cultivating and establishing security awareness, developing a keen thinking mode and habit is also an effective way to prevent network security deception, as shown in Figure 5-3.

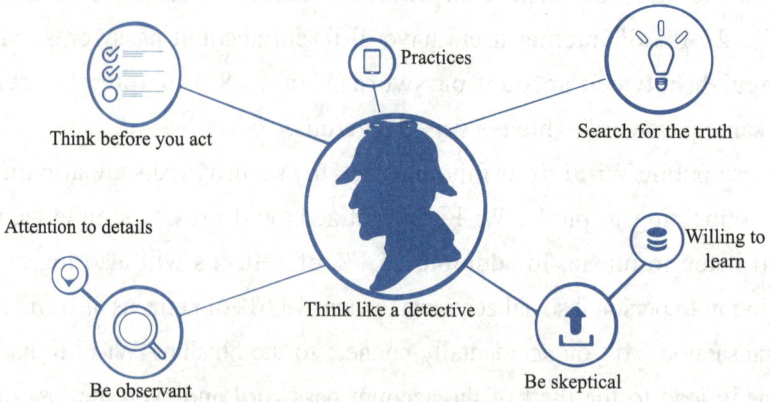

Figure 5-3 Cultivate safety awareness

Task 5.3 Analysis of information security development trend

Subtask 5.3.1 Analysis of Gartner's top ten security technologies

Gartner is the first company of information technology research and analysis. It is also the most authoritative IT research and consulting company in the world. Garter put forward ten security technologies, as shown in Figure 5-4.

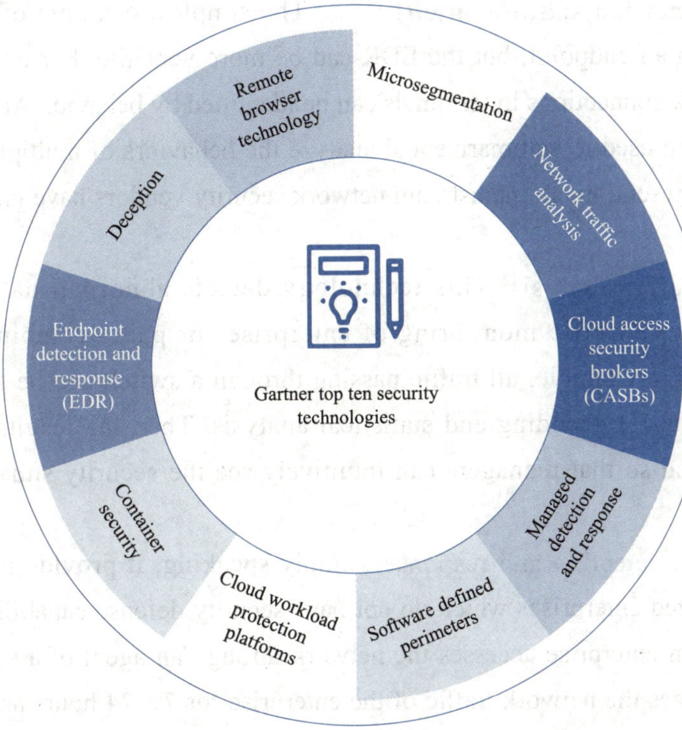

Figure 5-4 Gartner's top ten security technologies

(1) Cloud workload protection platforms: Gartner's definition is abstract, simply speaking it is a platform to protect services running in public cloud and private cloud. The current practice is to deploy an agent at all OS levels where the tasks reside and communicate with the control center. A distributed monitoring and centralized management C/S structure is formed so that operation and maintenance personnel can conveniently monitor the security status of numerous hosts and issue disposal strategies at the same time.

(2) Remote browser technology: The essence of this technology is to isolate the process of browsing the web from the terminal. For example, the simplest way is for everyone to remotely log in graphically to a host for web browsing. Because it is graphical login, even if the browser is attacked, it will not harm the visitors' terminals. After the browsing, the host that performs the browsing task can be reset to the most secure state. This technology can be provided as a service

and must rely on virtualization. The enterprise leases remote browsing server resources, and the provider is responsible for maintaining the status of the browsing server.

(3) Spoofing: A number of fake servers are deployed in the enterprise to trap attackers or confuse them to mislead their judgment of the internal network topology, increasing the cost of invasion. At the same time, once the attacker enters the false server, it will produce early warning. Such fake servers can even be embedded directly into switch devices.

(4) Endpoint detection and response (EDR) : The simplest example of EDR is antivirus software deployed on an endpoint, but the EDR can be more versatile. For example, suspicious processes and network connections in terminals can be identified by behavior. At the same time, big data technology can be used to summarize and analyze the behaviors of multiple terminals to find potential threats. At present, many mainstream network security vendors have gradually introduced EDR solutions.

(5) Network traffic analysis: This technology detects abnormal data in the network through comprehensive traffic monitoring of enterprises or parks combined with big data analysis technology. For example, all traffic passing through a switch can be sent to an analysis device for comprehensive decoding and statistical analysis. Then the results of decoding and analysis are visualized so that managers can intuitively see the security situation of the whole network.

(6) Management detection and response: Simply speaking, it provide intimate service to small and medium-sized enterprises which do not have security defense capabilities. For example, if a small and medium enterprise accesses the network through an agent of a security vendor, the security vendor analyzes the network traffic of the enterprise for 7×24 hours and cleans threats in a timely manner. At the same time, security event warnings are pushed to enterprises, so that SMEs do not even need to purchase and deploy any equipment.

(7) Microsegmentation: The microsegmentation is not the isolation between servers in the traditional machine room, but is refined to the application level. In the era of cloud, the host is weakened, we see a service to provide external applications. So isolation is also evolving from the server level to the application level.

(8) Cloud access security broker: This technology is used to secure the cloud. Many implementations of this technology are to obtain all access traffic to cloud services through reverse proxy, and then conduct security detection and audit to timely find abnormal access that is not compliant, such as infiltration, leakage and so on. This technology is similar to the previous management detection and response, the former protects the service consumer and the latter protects the service provider. Most of these programmes are provided as a service.

(9) Software defined perimeter: This concept emerged entirely to solve the problem of flexible

resource access management in the cloud era. Emphasis is placed on replacing traditional physical devices with software. The same goes for software defined network.

(10) Container security: Traditional security is host-oriented and is based on the host. In the cloud era, where applications are container-level and the concept of host is weakened, container security becomes very important.

Subtask 5.3.2 Future development trend of security defense

The future development trend of security defense is as follows: security servitization; the importance of terminal detection is increasingly prominent; traffic control evolves from IP to applications; software defined security.

1. Security servitization

As shown in Figure 5-5, in the future, security defense solutions may no longer consist of one or more devices. Security defense and analysis are performed remotely. All traffic from users accessing the network is routed to the data center of the security vendor by proxy for analysis, filtering, and cleaning in the data center. Security defense is no longer a device, but a remote service. The customer only needs to configure the address of a secure proxy server. Both "Management detection and response" and "Cloud Access Security Broker" are belong to this type of security services.

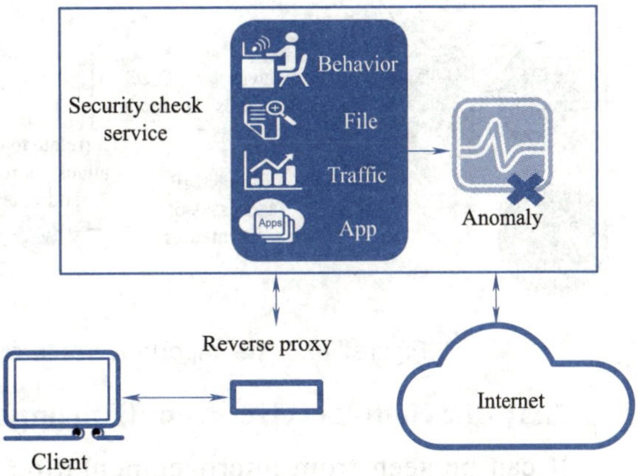

Figure 5-5 Security servitization

2. The importance of terminal detection is increasingly prominent

For the internal of enterprise, the simple terminal antivirus software will eventually evolve into a distributed monitoring and centralized analysis architecture and evolve to the EDR direction, the unified analysis of the behavior and context information of processes in all hosts in the enterprise helps to discover potential threats more efficiently.

The security checking capability of terminal has been paid more and more attention by traditional network security vendors. In the past, terminal security and network security are separated from the two camps, terminal antivirus vendors focus on checking terminal files, network security vendors only focus on network traffic. Now, both functions are converging, as shown in Figure 5-6. In particular, the linkage between terminal security software and network defense devices enables the malicious parts of traffic to be directly associated with processes and files on terminals to accurately trace the source of threats. In the future, security software in terminals will work more closely with devices.

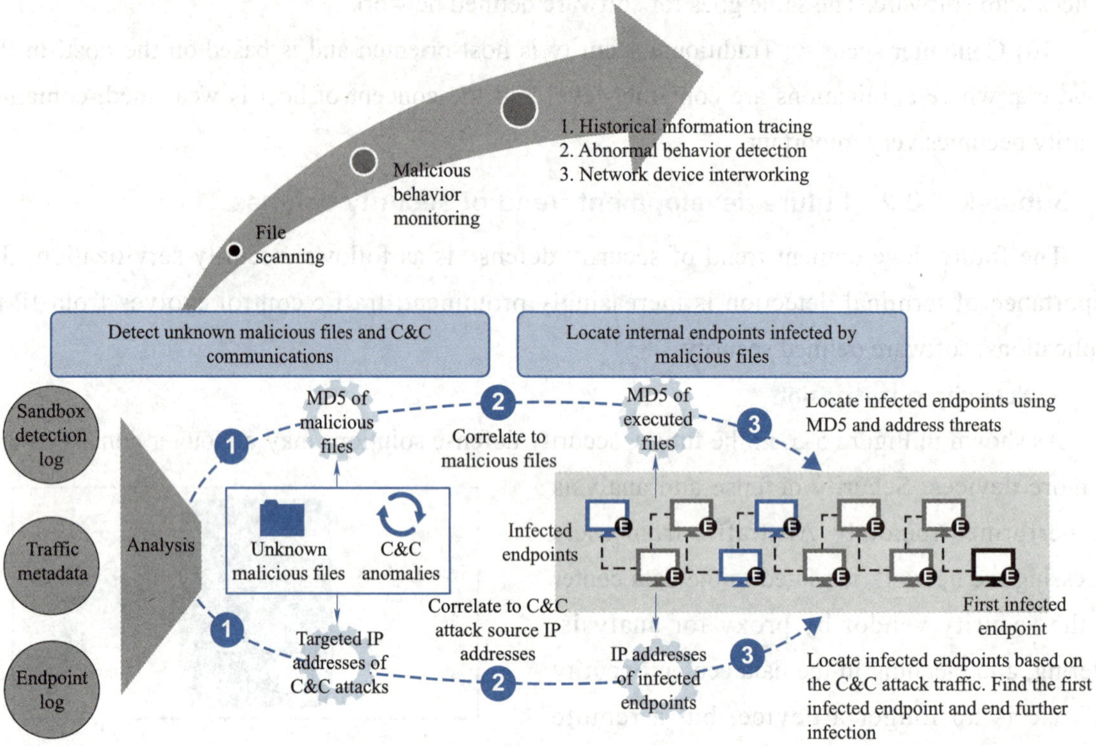

Figure 5-6 The importance of terminal detection is increasingly prominent

3. Traffic control evolves from IP to applications

It can be seen from microsegmentation and container security that the concept of host is weakened and the concept of service is strengthened in the era of cloud. Therefore, traffic management should also be implemented at the application level and container level. The network topology which operation and maintenance personnel see is not between hosts, but between services and clients. At the same time, graph theory can be better applied to security check, timely discovering abnormal communication paths of cloud data centers, and finding potential threats, as shown in Figure 5-7.

4. Software defined security (Huawei SDSec)

Gartner's top security technologies are basically targeted at the cloud, so it is imperative to deploy security defense solutions in the cloud. The future security will also be SDN, that is, software defined security. All inspection devices evolve into software and run in containers or virtual hosts. Operation and maintenance personnel can easily change the check process of different application data flows. For example, some application data flows need to be checked by WAF, and some application data flows need to be checked by virus scan or IPS. Even this change is intelligently implemented based on the analysis of traffic and process behavior as shown in Figure 5-8.

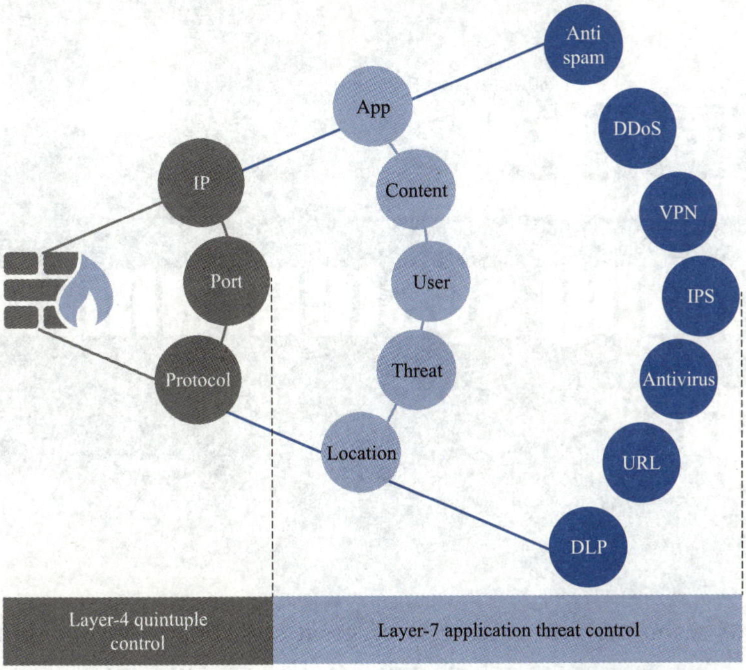

Figure 5-7 Traffic control evolves from IP to applications

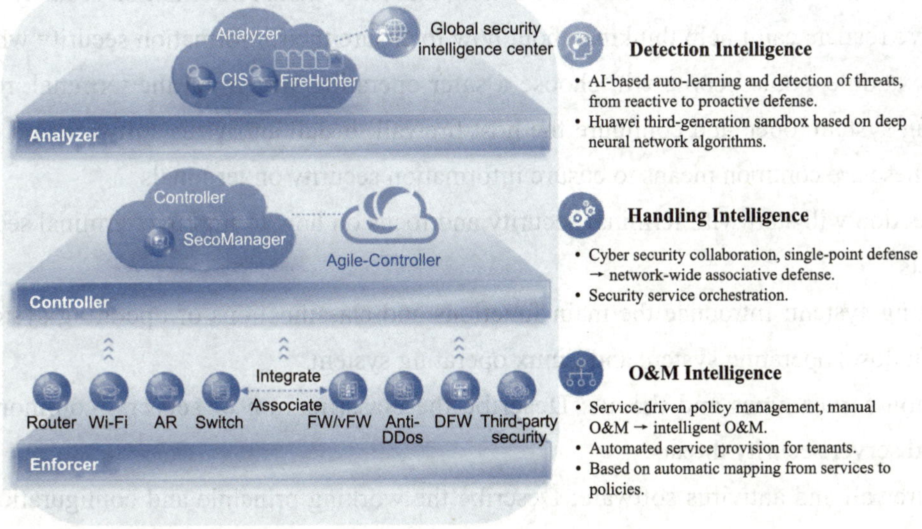

Figure 5-8 Huawei SDSec

Section 2

Operating system and host security

As a kind of resource, information is of great significance to mankind because of its universality, sharing, value-added, handleability and multi-utility. The essence of information security is to protect information resources in information systems or information networks from various types of threats, interference and destruction, that is, to ensure information security.

I believe readers can't help thinking about how to ensure their information security when they see it. For example, some people will choose a safer operating system on the terminal, reinforce the operating system, open and configure the host firewall, install anti-virus software and check it regularly. These are common means to ensure information security on terminals.

This section will start with terminal security and focus on how to improve terminal security in three projects:

Operating system: Introduce the main functions and classifications of operating system, and common Windows operating system and Linux operating system.

Common server types and threats: Describe the classification of servers, common server software and server security threats.

Host firewall and antivirus software: Describe the working principle and configuration mode of Windows firewall, Linux firewall, and the functions and key technologies of host antivirus software.

Through these three projects, readers can understand the technologies of each terminal reinforcement, and master the ability to install and configure firewall and antivirus software of each operating system.

Project 6 Operating system

Operating system is a platform that manages the hardware and software resources of a computer, without it, no computer can run properly. In the history of personal computing, there have been many different operating systems, five of which are most commonly used: DOS, Windows, Linux, Unix(Mac), OS/2.

This project mainly introduces the Windows operating system and Linux operating system.

Task 6.1 Understand the basic knowledge of operating system

Subtask 6.1.1 Understand the definition of operating system

Operating system (OS) is a computer program that manages and controls computer hardware and software resources. It is the most basic system software that runs directly on the "bare computer". Any other software can only run under the support of the operating system. Of course, the operating system itself is also software, different from other software working on the operating system, can be called system software.

The operating system is the interface between the user and the computer, as well as the interface between the computer hardware and other software. The functions of the operating system include managing the hardware, software and data resources of the computer system, controlling the operation of the program, improving the human-computer interface, providing support for other application software, making all the resources of the computer system play a role to the maximum, providing various forms of user interface, so that users have a good working environment. Provide necessary services and corresponding interfaces for other software development. In fact, the user does not need to contact the operating system, the operating system manages the computer hardware resources, according to the resource request of the application program at the same time, allocate resources, such as dividing CPU time, opening up memory space, calling the printer and so on.

Subtask 6.1.2 Master the main functions of the operating system

1. Processor management

The main task of processor management is to allocate processor resources, and control and manage its operation. In the multi-programming environment, the allocation and operation of processor resources are based on the process as the basic unit, so the management of the processor can be attributed to the management of the process. It includes the following aspects:

(1) Process control: Resource control and management of the running process.

(2) Process synchronization: Processes that cooperate with each other share data, concurrent execution of processes occurs here. How to ensure concurrent execution of these processes is called process synchronization.

(3) Process communication: Process communication is the transfer of data between processes.

(4) Process scheduling: Each program requires a processor when multiple programs run, the operating system dynamically assigns the processor to a process for execution.

2. Memory management

The main task of memory management is to provide a good environment for the operation of multiple programs, to facilitate the use of memory, to improve the utilization rate of memory, and to expand memory logically. To this end, memory management should have the following functions:

(1) Memory allocation: The method by which memory is allocated during program execution.

(2) Memory protection: Before memory allocation, it is necessary to protect the operating system from the influence of user processes, while protecting the running user process from other user processes.

(3) Memory expansion: Limited capacity cannot meet the storage requirements of large programs and multiple programs, so the memory management of the operating system needs to support memory expansion.

3. Device management

The main tasks of device management are to complete the I/O requests and classify I/O devices for users. Improve CPU and I/O device utilization; Improve I/O speed; And make it easy for users to use I/O devices. To achieve the above tasks, device management should provide the following functions:

(1) Buffer management: It improves parallelism between CPU and I/O devices to mitigate speed mismatches between CPU and I/O devices. In modern operating systems, almost all I/O devices use buffers when exchanging data with processors. The main responsibility of buffer management is to organize these buffers and provide the means to obtain and release buffers.

(2) Device allocation: Every time a process makes an I/O request to the system, the device allocation program allocates a device to the process according to a certain policy as long as it is secure.

(3) Virtual device: The virtual technology is used to virtualize a physical device into multiple logical devices for use by multiple user processes.

4. File management

In modern computer systems, we store programs and data as files on disks for users to use.

The main task of file management is to manage user files and system files to facilitate user use and ensure file security. Therefore, file management should have the functions of file storage space management, file read and write management, directory management and file sharing and protection.

5. Operation management

The main task of operation management is to provide a good environment for users to use the system, so that users can effectively organize their workflow, and make the whole system run efficiently. Operation management should have the following functions:

(1) Task and interface management.

(2) Human-computer interaction.

(3) Graphic interface.

(4) Language control.

Subtask 6.1.3 Understand operating system classification

There are so many types of current operating systems that it is difficult to use a single standard to classify them uniformly.

According to the application field, it can be divided into desktop operating system, server operating system, and embedded operating system.

According to the number of users supported, it can be divided into single-user operating systems (such as MSDOS, OS/2) and multi-user operating systems (such as UNIX, Linux, Windows).

According to the degree of open source code, it can be divided into open source operating systems (such as Linux, FreeBSD) and closed source operating systems (such as Mac OS X, Windows).

Task 6.2 Master the Windows operating system

Subtask 6.2.1 Understand the origin of Windows

In March 1980, Steve Jobs, founder of Apple, introduced a GUI (graphic user interface) technology that he discovered during a visit to Xerox in Silicon Valley at a conference. Bill Gates, president of Microsoft, realized the potential value of this technology after hearing about it. Then he led Microsoft to start the development of GUI software—Windows.

In 1985, Microsoft officially released the first generation of window type multi-task system-Windows 1.0, because of the limitations of the hardware level at that time, Windows 1.0 did not have the expected social effect, also did not have its advantages. However, the launch of the operating system marked the PC began to enter the era of graphical user interface. In the graphical user interface operating system, most of the operation objects are represented by corresponding icons. This kind of intuitive operation interface makes the computer closer to the psychological characteristics and actual needs of users. Windows uses a graphical GUI, which is more user-friendly than the way DOS used to type commands.

Subtask 6.2.2 Meet Windows

Microsoft Windows is an operating system developed by Microsoft, it was published in 1985. It was originally just a Microsoft-DOS simulation environment, and its later versions gradually developed into an operating system mainly designed for personal computers and server users. And eventually gained the world's monopoly of PC operating systems. This operating system can run on several different types of platforms, such as personal computer(PC), mobile devices, server and embedded systems. Personal computers are the most widely used among these platforms.

Subtask 6.2.3 Windows system architecture analysis

Processors on a computer running Windows have two different modes: "user mode" and "kernel mode". Depending on the type of code running on the processor, the processor switches between the two modes. Applications run in user mode, and core operating system components run in kernel mode. Multiple device drivers run in kernel mode, but some drivers run in user mode.

When you start an application in user mode, Windows creates a "process" for the application. Processes provide dedicated "virtual address spaces" and dedicated "handle tables" for applications. Because an application's virtual address space is a dedicated space, one application cannot change data belonging to another application. Each application runs in isolation, and if an application is corrupted, the damage is limited to that application. Other applications and operating systems are not affected by this damage.

In addition to being a dedicated space, the virtual address space of a user mode application is limited. Processors running in user mode cannot access the virtual address reserved for the operating system. Limiting the virtual address space of a user mode application prevents application changes and potentially corrupts critical operating system data.

All code running in kernel mode shares a single virtual address space. This means that kernel mode drivers are not isolated from other drivers and from the operating system itself. If a kernel mode driver accidentally writes to the wrong virtual address, data belonging to the operating system or other drivers can be corrupted. If the kernel mode driver is corrupted, the entire operating system is corrupted. As shown in Figure 6-1, a simplified version of Windows architecture with four user modes has the following processes:

(1) Fixed system support process, such as login processes and Session managers, which are not Windows services (not started through SCM as a service control manager).

(2) Service process, such as a task scheduler and printer service, typically require users to

log in to the system. Many service applications, such as SQL Server and Exchange Server, run as services.

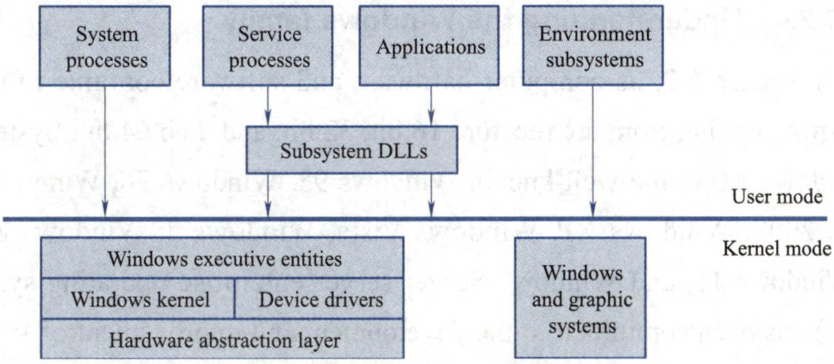

Figure 6-1 Windows architecture

(3) User program, it can be Windows 32-bit or 64-bit, Windows 3.1 16-bit, MS-DOS 16-bit, or POSIX 32-bit or 64-bit. Note that 16-bit programs can only run on 32-bit systems.

(4) Environment subsystem server process, which implements part of the operating system support environment, acts as a personalized interface to the user or developer. Windows NT is originally released with three subsystems, Windows, POSIX, and OS/2. Windows 2000 is the last subsystem with both POSIX and OS/2. The flagship and enterprise versions of Windows also support an enhanced POSIX subsystem. It's called SUA.

Note:

"Subsystem DLL" under server process and user program. Under Windows, user program do not call local Windows services directly, but through subsystem DLL. The role of subsystem DLL is to translate documented functions into non-documented (unexposed) system services for use.

The kernel mode components shown in Figure 6-1 include:

(1) Windows execution entity, includes basic system services, such as memory manager, process and thread manager, security management, I/O management, networking, and interprocess communication.

(2) Windows kernel, including low-level system functions such as thread scheduling, interrupts, exception distribution, and multi-core synchronization. It also provides routine and basic objects for implementing high-level structures.

(3) Device drivers, including hardware device drivers (translate user I/O to hardware I/O), software drivers (such as file and network drivers).

(4) Hardware abstraction layer, a kernel independent layer of code that separates device drivers from platform differences.

(5) Windows and graphics system, which implements GUI functions, handles user interface and graphics.

Subtask 6.2.4 Understanding the Windows family

As shown in Figure 6-2, as computer hardware and software continue to upgrade, so does Microsoft Windows, moving from architecture 16 bit, 32 bit, and then 64 bit. System versions from the original Windows 1.0 to the well-known Windows 95, Windows 98, Windows ME, Windows 2000, Windows 2003, Windows XP, Windows Vista, Windows 7, Windows 8, Windows 8.1, Windows 10, Windows 11, and Windows Server server enterprise operating system, continuous update, Microsoft has been committed to the development and improvement of Windows operating system.

The latest stable version of the Windows operating system is Windows 11, released on October 31, 2023. The latest stable version of Windows Server is Windows Server 2019, released on November 13, 2018.

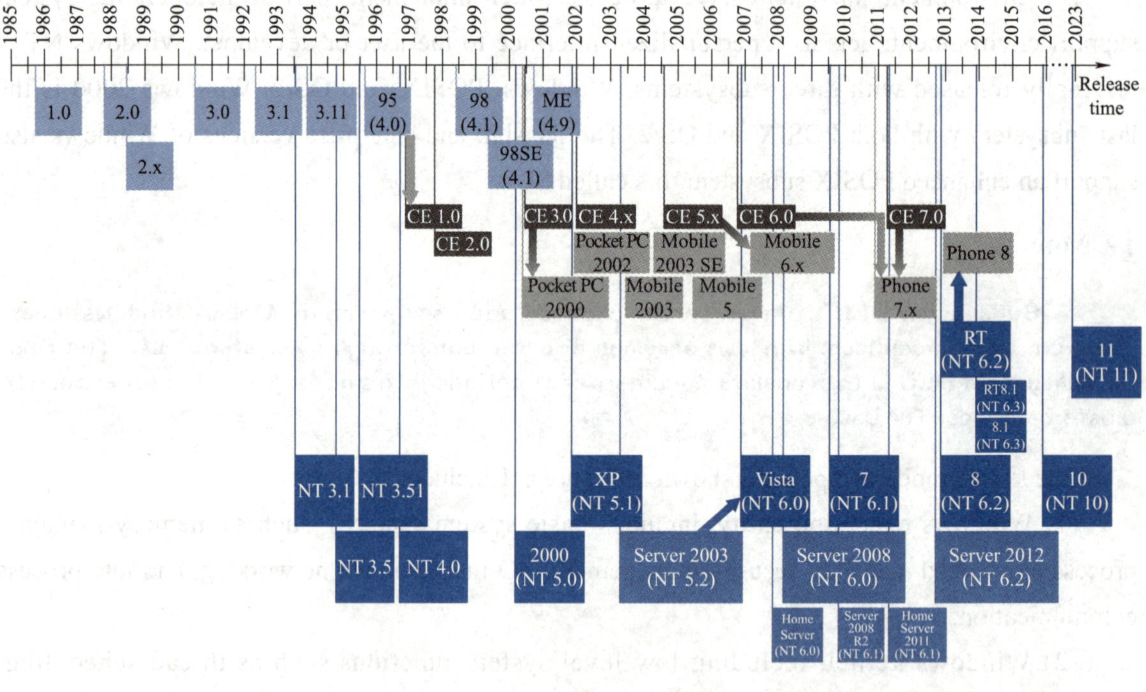

Figure 6-2 Windows family

The differences between server operating systems and individual user operating systems are as follows:

(1) The performance stability of the server is much higher than that of the individual user operating system.

(2) Servers work better for file management or web applications.

(3) In terms of graphics and entertainment, individual user operating systems are better than servers.

(4) Server operating systems are more secure and coordinable than individual user operating systems.

(5) The server version of the operating system includes server-side features and management tools, as well as more stringent requirements for security and stability, and is relatively slow. The server version also support special hardware at the same time.

(6) The system for ordinary users needn't have professional functions. Instead, it will include software and functions for managing media for individual users.

Subtask 6.2.5 Master Windows features

Windows is the operating system with the largest number of users and the strongest compatibility in the world. Its features are as follows:

1. Intuitive, efficient object-oriented graphical user interface, easy to learn and use

In a sense, both the Windows user interface and the development environment are object-oriented. The user works in the "select object-operate object" mode. For example, to open a document, first select the document with the mouse or keyboard, and then select the "Open" command from the right-click menu to open the document. This mode of operation simulates real-world behavior and is easy to understand, learn and use.

2. The user interface is unified, friendly and beautiful

Most Windows applications comply with the common user access (CUA) standard proposed by IBM. All programs have the same or similar basic appearance, including Windows, menus, and toolbars. As long as users master one, it is not difficult to learn other software, thus reducing the cost of learning.

3. Rich device-independent graphical manipulation

Windows graphics device interface provides rich graphical manipulation functions to draw geometric shapes such as lines, circles, boxes, and so on, and supports a variety of output devices. The device can display the same geometry on a printer or monitor.

4. Multi-user, multi-task

Windows is a multitasking operating environment that allows users to run multiple applications at the same time, or to do several things simultaneously within a program. Each program occupies a rectangular area of the screen called a window, which can be overlapped. Users can move these windows around or switch between different applications, and exchanging data and communication between programs can be done manually and automatically. Although multiple applications can be running on a computer at any one time, only one is active and its title bar is highlighted. An active program is one that is currently capable of receiving user keyboard input.

Task 6.3 Master Linux operating system

Subtask 6.3.1 Understand the history of Linux

Before Linux, the computer operating system market is dominated by a large number of systems: Unix and Microsoft Windows. Unix is the mainstream operating system of large and medium minicomputers, workstations and high-grade microcomputers, with a large number of enterprise users, and has become the operating system standard in fact. Windows dominates the microcomputer operating system market for ease of use.

However, both operating systems are commercial software, and the Unix operating system in particular, is too expensive to gain popularity among ordinary users. As a result, several free operating systems have emerged with most of the features of the Unix operating system.

Dutch computer scientist Andrew S. Tanenbaum developed Minix, a Unix-like operating system, primarily for teaching experiments. Linus Torvalds, a Finnish university student, wrote an operating system core software based on Minix and released Linux version 0.0.2 in October 1991. Subsequently, many Unix enthusiasts and hackers in the world worked together to perfect the rest.

The logo (mascot) of Linux is a penguin named Tux, as shown in Figure 6-3. When Linux core 1.0 was released in 1994, Linus Torvalds is asked to think of a mascot. He remembered being bitten by a penguin in an Australian zoo, so he chose a penguin as his mascot! It is easier to accept that penguins represent the South Pole, and that the South Pole is a land mass shared by the whole world and not owned by any country. In other words, Linux does not belong to any commercial company, but is a technological achievement that can be shared by all human beings.

Figure 6-3 The logo of Linux

Subtask 6.3.2 Analysis of Linux system architecture

As shown in Figure 6-4, a Linux system generally consists of four parts: kernel, shell, file system, and applications. Kernels, shells, file systems, and applications are used to manage files and use the system.

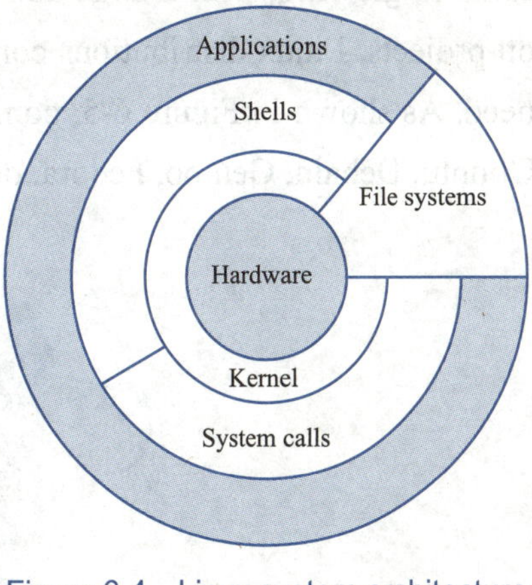

Figure 6-4 Linux system architecture

1. Linux kernel

The kernel is the core of the operating system, with many of the most basic functions, it is responsible for the management of system processes, memory, device drivers, files and network systems, determines the performance and stability of the system.

The Linux kernel consists of the following parts: memory management, process management, device driver, file system, and network management.

2. Linux shell

Shell is the user interface of the system, which provides an interface for users to interact with the kernel. It receives user-typed commands and sends them into the kernel to execute. It is a command interpreter. In addition, shell programming languages have many of the features of ordinary programming languages, and the shell programs written in these programming language have the same effect as other applications.

3. Linux file system

A file system is a method of organizing files on storage devices such as disks. Linux supports many popular file systems, such as EXT2, EXT3, FAT, FAT32, VFAT.

4. Linux applications

Standard Linux systems typically have a set of programs called applications, which include a text editor, programming language, X-Window, office suite, Internet tools and databases.

Subtask 6.3.3 Understand the basic idea of Linux

The basic idea of Linux has two things: everything is a file, and every piece of software has a defined purpose. The first of these is that everything in the system boils down to a single file, including commands, hardware and software devices, operating systems, processes, etc., which are treated as files with their own characteristics or types for the operating system kernel. Linux is based on Unix in large part because the basic ideas are very similar.

Subtask 6.3.4 Understand common Linux operating systems

At present, many commercial companies can obtain the core source code of Linux for free, and form their own products after packaging and improving, namely Linux distribution kits. The Linux camp is growing, with a large audience for each distribution and developers volunteering to work on projects. Linux distributions come in all shapes and sizes, designed to satisfy every conceivable need. As shown in Figure 6-5, common Linux operating systems include RedHat, CentOS, SUSE, Ubuntu, Debain, Gentoo, Fedora, deepin Linux, and EulerOS developed by HUAWEI.

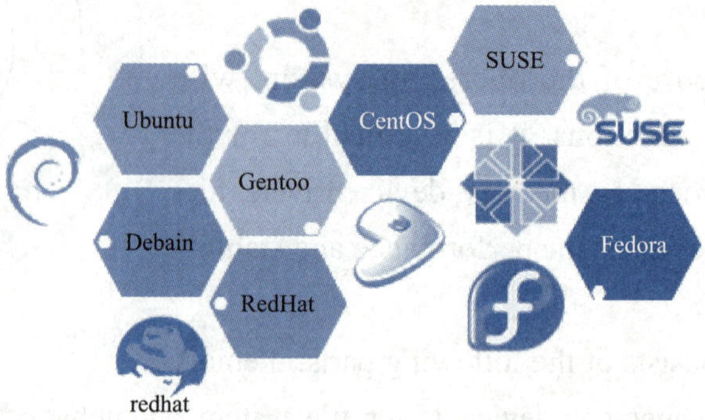

Figure 6-5 Common Linux operating system

Subtask 6.3.5 Master the main features of Linux

The main features of Linux are:

1. Completely free

Linux is a free operating system that is freely available on the web and elsewhere and can be modified at will. This is something no other operating system can do. Because of this, Linux grows and grows with the help of countless programmers from all over the world who can modify and write Linux to follow their own interests and inspirations.

2. Multi-user, multi-task

Linux supports multi-user, and each user has its own special rights over its own file device, ensuring that each user does not interfere with each other. Multitasking is one of the most important features of today's computers. Linux allows multiple programs to run simultaneously and independently.

3. Good interface

Linux has both character interface and graphical interface. On the character interface, users can enter corresponding commands through the keyboard to perform operations. It also provides an X-Window system with a Windows-like graphical interface that users can operate with a mouse. The X-Window environment is similar to Windows, a Linux version of Windows, so to speak.

4. Support for multiple platforms

Linux can run on a variety of hardware platforms, such as x86, 680x0, SPARC, Alpha, etc. Linux is also an embedded operating system that can run on a personal digital assistant, set-top box or game console. The Linux 2.4 kernel, released in January 2001, fully supports Intel's 64-bit chip architecture. Linux also supports multiprocessor technology. Multiple processors work at the same time, so that the system performance is greatly improved.

Subtask 6.3.6 Understand the application fields of Linux

Linux is an operating system software like Windows operating system software. But different from Windows, Linux is a set of open source code programs, and can freely spread the Unix-like operating system software, with the update of information technology changes, Linux application area have tended to be widespread.

1. In the military field

One of the core requirements of military computer operating system must be security. Using Linux operating system is undoubtedly reliable in the information security of military users.

2. Corporate business server

Today's IT server field is Linux, Unix, Windows three parts of the world, Linux system can be described as a rising star, especially in recent years, server-side Linux operating system is constantly expanding market share, rapid growth every year, and poses a serious threat to Windows and Unix server market position.

Linux is widely used as an enterprise server. The Linux system can be used as WWW server, database server, load balancing server, mail server, DNS server, proxy server (transparent gateway), router, etc., which not only reduces the operating cost of enterprises, but also obtains the high stability and reliability brought by the Linux system.

With the wide application of Linux in the field of server, the system has penetrated into various industries such as telecommunications, finance, government, education, banking, oil and other industries from the development of recent years. At the same time, various hardware manufacturers have also supported the Linux operating system. All this suggests that Linux has a bright future in the server market. At the same time, large and super large Internet enterprises (Baidu, Sina, Taobao, etc.) are using Linux system as their server-side program running platform. Almost all the top ten websites in the world and China use Linux system. Linux has gradually penetrated into enterprises in various fields.

3. Network equipment, game consoles

Because Linux system is open source, powerful, reliable, stable, flexible, and has great scalability, plus it widely supports a large number of microprocessor architecture, hardware devices, graphics support and communication protocols. Therefore, in the field of embedded applications, from Internet devices (routers, switches, firewalls, load balancers, etc.) to specialized control systems (vending machines, mobile phones, PDA, various household appliances, etc.), Linux operating system has a wide application market. Especially after several years of development, it has successfully entered the mainstream embedded development platform. In the smartphone field, for example, Android Linux has firmly established itself as a smartphone development platform.

Subtask 6.3.7 Comparison between Windows and Linux

Windows and Linux are both working hard to drive computer operating systems, but despite their different features and application scenarios. Users still face the awkward situation of choosing an operating system in a practical environment, which requires full consideration of specific requirements. Table 6-1 compares the two services in terms of fees, software support, security, usage habits, and customization.

Table 6-1 Comparison between Windows and Linux

Operating system	Free or charged	Software support	Security	Usage habits	Customization
Linux	Free or for a small fee	The number of software that can run directly under Linux is small	Less virus damage	Getting started is difficult and requires some learning and guidance	Open source code
Windows	Fees	Windows can be compatible with most of the software, play most games	System patches are frequently updated and viruses are freq-uently attacked	Easy for users to get started, simple to get started	System cust-omization is poor

Project 7 Common server types and threats

Broadly speaking, a server is a computer system in a network that can provide services to other machines.

In a narrow sense, a server refers to some high-performance computers that provide some business services to external computers through a network. Therefore, the server has higher requirements than ordinary PCs in terms of stability, security, performance, etc.

Task 7.1 Understanding servers

Subtask 7.1.1 Understand server definitions

Server is a type of computer, it is a high-performance computer for the client to provide services (including query, storage, computing, etc.) in the network environment.

The server mainly provides Web applications, file download, database storage, printing and other services for clients. The role of the server is very extensive, online games, websites, most of the software are needed to save to the server. Some enterprises deploy their own servers, they usually work on the most important information stored in the server's hard disk.

Simply speaking, all servers are similar to our daily use computers, just more powerful on the stability, security and data processing performance. In fact, our home computers can also be used as a server, you just need to install the server system. But servers have higher requirements for hardware stability and quality, ordinary computers are generally unable to switch on for a long time. We must know that the important data is generally stored on the server, so the ordinary computer is not suitable for the server.

Subtask 7.1.2 Master the characteristics of the server

It should have "availability", "exploitability", "scalability" and "manageability" as a server.

Firstly, it must be reliable, that is "usable". This is because the server serves clients across the network, not natively logged users. As long as there are users on the network, the server cannot be interrupted. In some special scenarios, the server cannot be interrupted even if no user is using it, because it must continue to serve the user. Some large enterprise servers are required to provide 7×24 hours of service, such as web servers and Web servers for public use.

Secondly, the server needs to serve multi-user, so it needs high connectivity and computing performance, which is the "availability" of the server. We usually feel slow in the use of PC, as if the server performance same as the PC we use, can the server bear so many users access at the same time? The answer is clearly no. So compared to the PC, the performance of the server is definitely much better than the PC we use. In order to achieve high-speed performance, the processing power will be adopted by installing symmetric multi-processors and inserting a large amount of high-speed memory.

Thirdly, with the continuous development of services and the increasing number of users, the server also needs to ensure a certain "scalability". To maintain high scalability, the server must have expandable space and redundant components (such as redundant array of independent disks, PCI-E, memory slots, etc.).

Finally, to ensure high service reliability, the server needs to be equipped with technologies that common PC do not have, such as dual-host backup, system backup, online diagnosis, and fault warning. Ensuring that a server failure can be fixed without shutting down the device is what we call "manageability".

Subtask 7.1.3 Understand server classification

There are many types of servers, with many different classifications.

1. Divided by application level

(1) Entry level server: For a small department's office requirements, the main function of the server is to complete the file and print. Small departments generally use entry level servers.

(2) Workgroup server: In general, if the application is not complex, such as there is no large database to manage, the workgroup server will be used.

(3) Department level server: It is suitable for web sites and data centers of medium-sized enterprises with high availability, reliability, scalability, and manageability.

(4) Enterprise level server: Enterprise level server is mainly applied to large enterprises and important industries (such as finance, transportation, and communication) that need to process a large amount of data and have high requirements on processing speed and reliability.

2. Divided by architecture

(1) x86 server: A CISC server, also known as a PC server, that uses Intel or other processors compatible with the x86 instruction set.

(2) Non-x86 server: It includes mainframes, minicomputers, and Unix servers that use RISC or EPIC processors.

3. Divided by shape

(1) Tower server: Tower server is an early form of server. As shown in Figure 7-1, it appearance and structure are similar to a vertical PC. The motherboard of the tower server is more scalable and has more slots, so the size is bigger than ordinary motherboard and the tower server case is bigger than standard case, usually sets aside enough internal space for future expansion of the hard disk and power supply. In terms of use object and use level, the current common entry level server and workgroup server basically use tower server structure type, also will use some departmental applications, but since there is only one host, even upgrade expansion is limited, so in some enterprises with higher application requirements, single server will not be able to meet the requirements, need more machines to work together. The tower server size is too large, too independent, collaborative work in space occupation and system management are not convenient, which is also the limitations of the tower server. However, in general, the functions and performance of this kind of server can basically meet the needs of most enterprise users, and its cost is usually relatively low, so this kind of server still has a very wide range of application support.

Figure 7-1 Tower server

(2) Rack server: Rack server looks more like a networking device than vertical PC. As shown

in Figure 7-2, rack server chassis are small. Multiple chassis can be installed in a standard 19-inch cabinet to increase the server density. Common chassis sizes include 1U (U is the measurement unit of rack server height, 1U=1.75 inch=1.75×25.4 mm=44.45 mm), 2U, and 4U. Rack servers are mostly functional servers, which are the most sold servers at present. Huawei rack servers include RH1288H, RH2288H, RH5288, RH2488/2488H, and RH5885H.

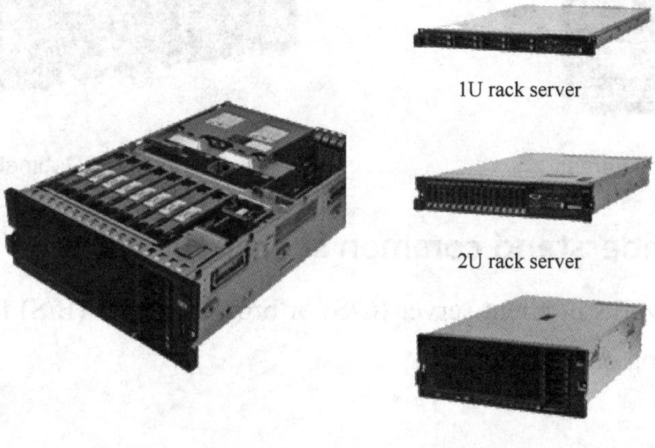

Figure 7-2 Rack server

(3) Blade server: A blade server is a server unit that can be inserted into a rack chassis of standard height. It is a low-cost server platform that implements high availability and high density (HAHD) and is specially designed for special application industries and high-density computing environments. A blade server is like a blade, each blade is a system mainboard, as shown in Figure 7-3. It has the features of ultra-high density, energy saving, centralized management and rapid deployment.

(4) Cabinet server: Some high-end enterprise servers have a complex internal structure and many devices. Some servers have many different device units or several servers are placed in a cabinet. Such servers are cabinet servers, as shown in Figure 7-4. For banks, securities, posts and telecommunications and other important enterprises, should adopt a complete fault self-healing ability of the system. The key components of the system should adopt redundancy measures. For the servers used in key business, dual-system hot-standby high-availability systems or high-performance computers can also be adopted. This system can better guarantee the availability.

4. Divided by use

(1) Generic server: A server that is not specifically designed for a service and can provide a variety of service functions.

(2) Function server: A server specially designed for one or several functions can be plugged and played without specialized hardware and software configurations.

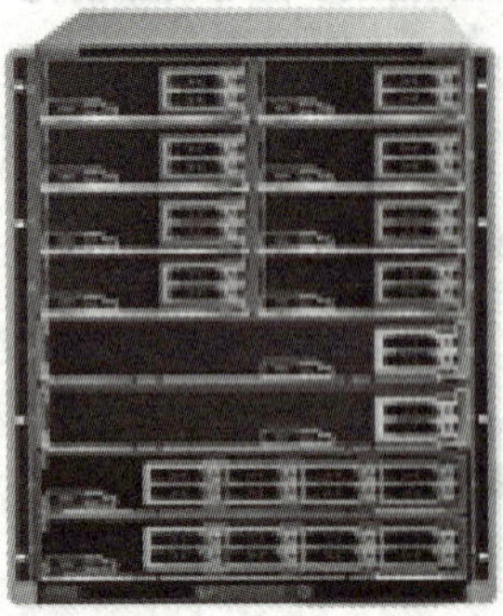

Figure 7-3 Blade server

Figure 7-4 Cabinet server

Task 7.2 Understand common server software

Server software works in client-server (C/S) or browser-server (B/S) fashion. There are many forms of server, common ones include:

(1) File server.

(2) Database server.

(3) Mail server.

(4) Web server.

(5) FTP server.

(6) DNS server.

(7) NTP server.

(8) Proxy server.

Subtask 7.2.1 Analysis of file working process

File server is a computer with special functions. Its main purpose is to provide file sharing services to clients. A file server can be a general-purpose server that can run other applications, or a functional server that provides file services.

As shown in Figure 7-5, in C/S mode, a file server is a computer responsible for central storage and data file management, so that other computers on the same network can access these files. File server allows users to share information over the network without physically moving files through floppy disks or some other external storage device. Any computer can be set up as a file server. In its simplest form, a file server can be an ordinary personal computer that handles file requests and sends them over a network. On more complex networks, a file server can also be a dedicated network attached storage (NAS) device that can also operate as a remote hard drive for other computers and allow people in the network to store files on the server as if they were on their own hard drives.

Figure 7-5 File server

Subtask 7.2.2 Analysis of working process of database

One or more computers running on the LAN and database management system software form a database server to provide services for customer applications, such as query, update, transaction management, indexing, caching, query optimization, security, and multi-user access control, as shown in Figure 7-6.

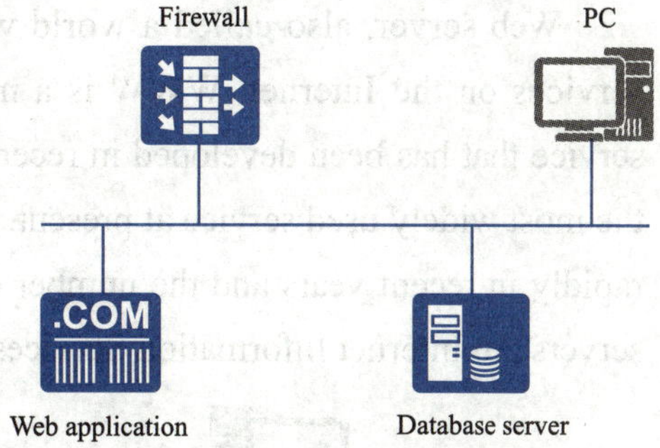

Figure 7-6 Database server

Database server is built on the basis of database system, has the characteristics of database system, and has its unique characteristics. The main functions are as follows:

(1) Database management functions, including system configuration and management, data access and update management, data integrity management and data security management.

(2) Database query and manipulation function, this function includes data retrieval and modification.

(3) Database maintenance functions, including data import/export management, database structure maintenance, data recovery and performance monitoring.

(4) The database runs in parallel. Because more than one user accesses the database at the same time, the database server must support the parallel running mechanism to handle the simultaneous occurrence of multiple events.

Subtask 7.2.3 Analysis of mail working process

Mail server is a device that manages e-mail sending and receiving. It is more secure and efficient than free email on the Internet, so it has been a must-have device for businesses. Mail server adopts client server mode, mail system consists of three components: user agent, mail server and mail transfer protocol.

(1) User agent: An application that handles the action of sending and receiving mail.

(2) Mail server: Used to receive mail from the user agent and send it to the recipient user agent.

(3) Mail transfer protocol: protocol used for mail transfer.

The working process is as shown in Figure 7-7.

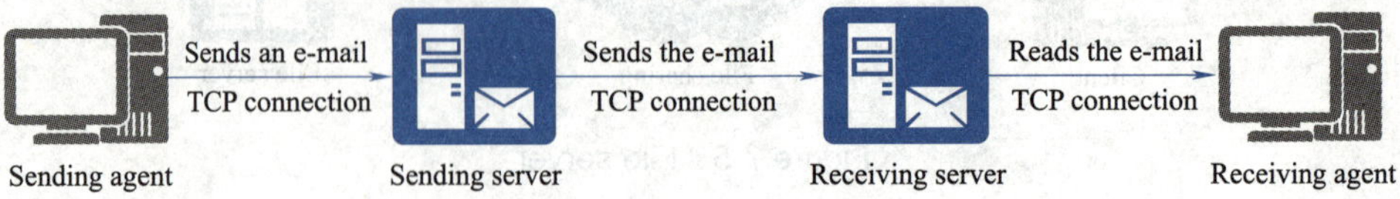

Figure 7-7 Mail server

Subtask 7.2.4 Analysis of working process of web page

Web server, also called a world wide web (WWW) server, provides information browsing services on the Internet. WWW is a multimedia information query tool on the Internet. It is a service that has been developed in recent years on the Internet. It is also the fastest developing and the most widely used service at present. It is because of WWW tools that the Internet has developed rapidly in recent years and the number of users has grown rapidly. The most commonly used Web servers are Internet Information Services (IIS) from Apache and Microsoft, as shown in Figure 7-8.

Figure 7-8 Web server

Subtask 7.2.5 Working process analysis of FTP server

FTP server is a computer that provides file transfer and access services on the network. FTP provides the following functions:

(1) Document transfer and management.

(2) Different levels of user identity: user, guest, anonymous.

(3) Command record and login file record.

(4) Directories that restrict user activity : (change root, chroot for short).

As shown in Figure 7-9, FTP also works in client/server mode. The client and server use TCP connections. The FTP server mainly uses ports 21 and 20. Port 21 is used to send and receive FTP control information and keep the FTP session open. Port 20 is used to send and receive FTP data. The FTP server provides the following functions:

(1) Upload: Sends files from your computer to the FTP server.

(2) Download: Download files from the FTP server to your computer.

Figure 7-9 FTP server

Subtask 7.2.6 Analysis of working process of domain name server

As everybody knows, we usually type in the url directly when we surf the Internet, this is actually what we call a domain name, but our computer actually can only use IP address to identify each other, and it is difficult to remember so many IP addresses we have its corresponding Web service. So there's a way that domain names correspond to IP addresses, because domain names make it easier to remember the corresponding Web service.

The DNS consists of a domain name resolver and a domain name server (DNS). The DNS is a computer that stores the domain names and corresponding IP addresses of all hosts on the network and translates domain names into IP addresses, as shown in Figure 7-10.

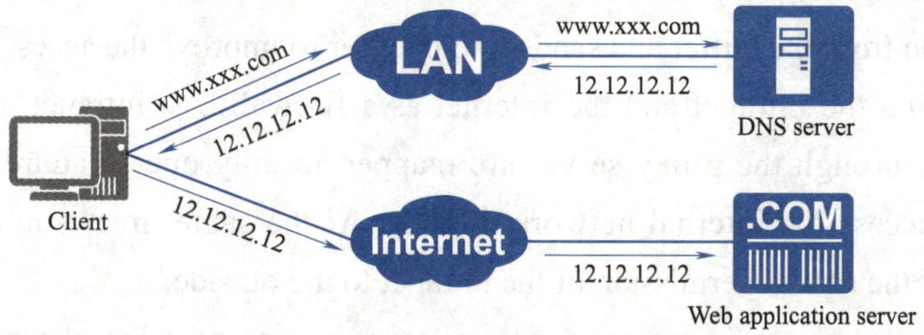

Figure 7-10 Domain name server

A domain name must correspond to a unique IP address. An IP address can correspond to multiple domain names, but it is not necessary for IP address to have a corresponding domain name.

Subtask 7.2.7 Analysis of working process of time synchronization server

In the world of computers, time is very important. For example, scientific research activities such as rocket launch, the requirements for the consistency and accuracy of time are very high. Is it according to the time of A computer or B computer? Network time protocol (NTP) is a protocol used to synchronize the time of each computer on a Network. It is used to synchronize the computer's clock to the universal coordinated time (UTC) with an accuracy of 0.1ms for local area networks and 1~50 ms for most places on the Internet.

It allows the computer to synchronize time with its server or clock source (such as quartz clock, GPS, etc.), it can provide high precision time correction, and it can use cryptographic confirmation to prevent protocol attacks by viruses.

Subtask 7.2.8 Analysis of proxy working process

Proxy server, its function is to proxy network users to obtain network information. Figuratively speaking: it is the transfer station of network information. A proxy server acts like a large cache, which dramatically improves browsing speed and efficiency. More importantly, the proxy server is an important security function provided by the Internet link-level gateway. The main functions are as follows:

(1) Set up the user authentication and billing functions, billing according to users, and unregistered users are not allowed to access the Internet through a proxy server. And collect statistics on users' access time, access location, and information traffic.

(2) Manage users at different levels and set access permissions for different users, filters out external or internal Internet addresses and sets different access permissions.

(3) Increase buffer (Cache), improve access speed, create buffer to frequently visited address, greatly improve the efficiency of popular site access. Generally, the proxy server sets a large disk buffer (up to several GB or larger), it is saved external information in the buffer when external information passes. When other users access the same information, the proxy server directly takes out the information from the buffer and sends it to the user to improve the access speed.

(4) Connecting the intranet and the Internet as a firewall: All intranet users accessing the external network through the proxy server are mapped to only one IP address, so the external network cannot access the internal network directly. At the same time, you can set IP address filtering to restrict the access permission of the intranet to the outside.

(5) Save IP overhead: the proxy server allows the use of a large number of pseudo IP addresses, saving network resources, that is, the proxy server can reduce the need for IP addresses. For the use of LAN access to the Internet, if apply an IP address for each user of the local area network (LAN), the cost can be imagined. However, if a proxy server is used, only a valid IP address is required on the proxy server. Other LAN users can use a private IP address, such as 10.X.X.X. In this way, a large number of IP addresses are saved and network maintenance costs are reduced.

Task 7.3 Understand server security threats

There are various security threats in the process of using the server. If the server is under attack, it may not function properly. So what security threats can affect the normal operation of the server? As shown in Figure 7-11, server security threats include malicious programs, brute force cracking by hackers, SQL injection attacks, DDoS attacks, and system vulnerabilities.

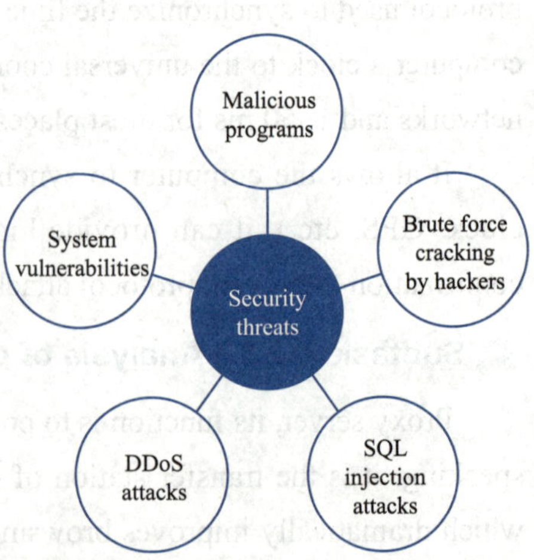

Figure 7-11 Server security threats

Subtask 7.3.1 Malicious program attack

A malicious program is generally a program written with the intention of attacking. These threats can be divided into two categories: require a host program and those that are independent of each other. Threats that require a host program cannot be independent of an actual application; Separate threats are self-contained programs that can be scheduled and run by the operating system.

Malicious programs include Trojan horses, worm, and virus.

(1) Trojan horse: Trojan horse is a useful or seemingly useful program or command process that contains a hidden piece of code that performs a harmful function when activated.

(2) Worm: A worm program is an infectious virus program that spreads from one system to another over the network.

(3) Virus: A virus is an offensive program that infects a computer system by embedding copies of itself in other files.

Subtask 7.3.2 Hacker brute force cracking

When they are cracking passwords, hackers are limited by their hacking abilities. After trying everything but finding nothing, finally have to use violence cracking this method.

Brute force cracking is the exhaustive attack method, which determines the range of answers based on the conditions of the problem and verifies the answers one by one until an answer meets the conditions of the problem, as shown in Figure 7-12.

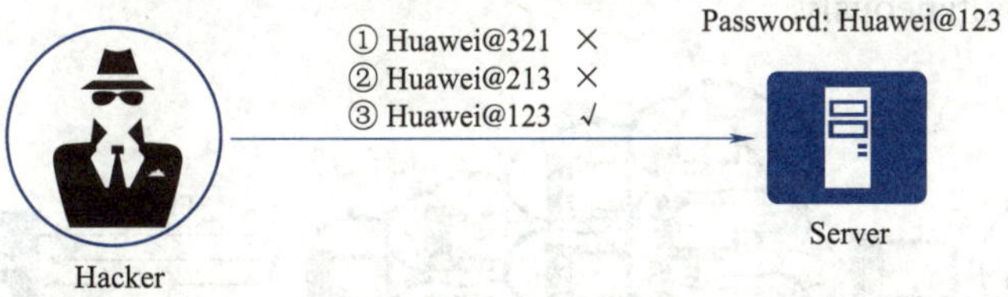

Figure 7-12 Brute force cracking

Subtask 7.3.3 SQL injection attack

SQL injection attack is one of the common methods used by hackers to attack databases. With the development of B/S pattern application develop, more and more programmers use this pattern to write applications. However, due to the level and experience of programmers are also uneven, quite a large number of programmers do not judge the validity of user input data when writing code, so that the application program has security risks.

SQL injection attacks, in which an attacker inserts SQL commands into the input field of a web form or the query string of a page request, tricking the server into executing malicious SQL commands. Forms in which user input is directly used to construct (or influence) dynamic SQL commands or as input parameters to stored procedures are particularly vulnerable to SQL injection attacks. There is also the risk of SQL injection for non-Web CS framework systems.

SQL injection attacks are one of the database security attacks, which can be effectively defended by database security protection technologies, including database miss sweep, database encryption, database firewall, data desensitization, and database security audit system. Database security risks caused by SQL injection attacks include brushing, dragging and bumping databases.

Subtask 7.3.4 DDoS attack

DoS is short for denial of service. An attack that causes DoS is called a DoS attack. It is intended to disable a computer or network from providing normal services. Most DoS attacks still require a considerable amount of bandwidth, and it is difficult for individual hackers to consume high bandwidth resources. To overcome this shortcoming, DoS attackers developed distributed attacks.

Distributed denial of service(DDoS) : A special form of DoS based denial of service attack that targets large sites such as commercial companies, search engines, and government departments. DDoS attacks attack a specific machine by using multiple controlled machines. They are unpredictable and destructive, as shown in Figure 7-13.

Trojan has become a tool for hackers to control puppets, and more and more computers become chickens, which are used by hackers and become their attack tools. A distributed denial of service (DDoS) attack is a simple tool used by hackers to launch a large number of attack requests to the same target simultaneously.

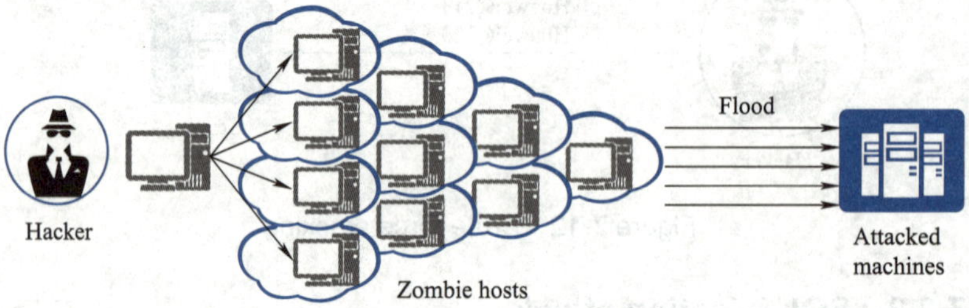

Figure 7-13 Distributed denial of service

Subtask 7.3.5 Master security threat defense methods

There are many security threats to the server. Different defense methods are required for different security threats, as listed in Table 7-1.

Table 7-1 Server security threat defense

Security threats	Defense method
Malicious programs	(1) Do not install suspicious software. (2) Don't open strangers' links. (3) Install antivirus software. (4) Don't visit informal websites

continued

Security threats	Defense method
Brute force cracking by hackers	(1) Passwords are complex enough. (2) Change the default port number. (3) Password login is not allowed. Only authenticated secret key login is allowed
SQL injection attacks	(1) Use regular expressions to filter incoming parameters. (2) Use PreparedStatement instead of Statement. (3) This function is called in the JSP to check for package invalid characters
DDoS	(1) Source authentication. (2) The fingerprint study. (3) Redirecting domain names

Task 7.4 Understand vulnerabilities and patches

Subtask 7.4.1 Understand the definition of vulnerabilities

Vulnerabilities are also called fragility, which refers to the defects and deficiencies of computer systems in specific of hardware, software, and protocols or system security policies.

Vulnerability refers to the weakness or defect of a system, the sensitivity of the system to a specific threat attack or dangerous event, or the possibility of the threat effect of the attack. Vulnerabilities may come from defects in the design of the application software or operating system, or errors in the coding process, or from design defects in the interactive processing of the business or unreasonable logical process.

These defects, errors or irregularities may be intentionally or unintentionally exploited to adversely affect an organization's assets or operations, such as information systems being attacked or controlled, important data being stolen, user data being tampered with, or systems being used as a springboard to invade other host systems. From the current vulnerabilities discovered, the vulnerabilities in the application software are far more than the vulnerabilities in the operating system, especially the vulnerabilities in the web application system is the vast majority of information system vulnerabilities.

In 1990, in *The Data and Computer Security Dictionary of Standard, concepts, and Terms*, Dennis Longley and Michael Shain defined vulnerabilities as follows: In computer security, vulnerabilities are defects in automated system security processes, management controls, and internal controls that can be exploited by threats to gain unauthorized access to information or disrupt the processing of critical data. Vulnerabilities have the following characteristics:

(1) Vulnerabilities are unknown beforehand and discovered afterwards;

(2) Vulnerabilities are security risks that can make computers subject to virus and hacker attacks;

(3) If the vulnerability is exploited, its consequences are unpredictable;

(4) Vulnerabilities can generally be exploited remotely;

(5) Bugs are usually fixable.

Subtask 7.4.2 Understand the hazards of loopholes

The existence of vulnerability is one of the necessary conditions for successful network attack. The key to successful invasion is to discover and exploit the vulnerability of the target network system as early as possible.

Vulnerabilities pose security threats to the network system as follows: permission promotion for common users, access to local administrators, access to remote administrators, local denial of service, remote denial of service, server information leakage, remote access to unauthorized files, access to restricted files, and spoofing.

(1) The purpose of permission bypass and permission promotion is to obtain the desired data operation capability, such as the promotion of ordinary user rights and the acquisition of administrator rights.

(2) Denial of service attacks are used to obtain control rights for certain services in the system to make the services stopped.

(3) Data leakage is mainly caused by hackers being able to access otherwise inaccessible data, such as reading restricted files and server information leakage.

(4) The execution of unauthorized instructions is basically to have the program execute the input as code to gain access to remote systems or higher access to local systems, such as SQL injection and buffer overflows.

Subtask 7.4.3 Understand the classification of vulnerabilities

According to the way to be exploited by an attacker, vulnerabilities are divided into local attack vulnerability and remote attack vulnerability. The attackers of the local attack vulnerability are legitimate local users or illegal users who obtain local permissions through other means. The attacker of the remote attack vulnerability uses the network to attack the remote host connected to the network.

According to the location where the target vulnerability refers to, it can be divided into operating system, network protocol stack, non-server program, server program, hardware, communication protocol, password recovery and other types of vulnerability.

Various system security threat vulnerabilities can be classified according to their potential threats to the system and the possibility of being exploited:

(1) High level vulnerability: Vulnerability that enables access to administrator rights.

(2) Medium level vulnerability: Vulnerability that can obtain common user permissions, read

restricted files, and deny service.

(3) Low level vulnerability: Vulnerability that can read unrestricted files and leak server information.

Of course, the classification of vulnerabilities is not only these, for example, they can also be classified according to the situation that vulnerabilities are mastered, and can be divided into known vulnerabilities, unknown vulnerabilities, 0day and other types. According to the classification of user groups, vulnerabilities are also divided into Windows vulnerability, Linux vulnerability, IE vulnerability, Oracle vulnerability.

Subtask 7.4.4 Analyze the causes of system vulnerabilities

Vulnerabilities are "inevitable" due to the high complexity of the system.

1. Defects in the design of software or protocols

Disadvantages of TCP/IP: TCP/IP is widely used today, but it was designed long before today when invaders are rampant. Therefore, there are many deficiencies that make security vulnerabilities inevitable, such as smurf attacks, ICMPUnreachable packet disconnection, IP address spoofing, and SYNflood. However, the biggest problem is that IP protocols are very "gullible", meaning intruders can forge and modify IP packets at will without being detected. Now Ipsec protocol has been developed to overcome this deficiency, but it has not been widely used.

2. There are bugs in the software writing

Whether it's server programs, client software, or operating systems, anything written in code will exist bugs to varying degrees. Bugs fall into the following categories:

(1) Buffer overflow: It means that an intruder enters a string of more than a specified length in the program's relevant input items. The excess is usually the attack code that the intruder wants to execute, and programmers don't check the input length, the result is that the extra attack code takes up memory behind the input buffer and executes. Don't think setting aside 200 characters for login user name is enough and no longer do length check. There is a saying that prevent the villain not gentleman, intruders will try their best to attack.

(2) The problem of unexpected conjunctive use: A program often consists of layers of code that function differently, even at the lowest operating system level. Intruders often take advantage of this feature to enter different content for different layers in order to steal information.

(3) Haven't check the input content in anticipation: Some programmers are afraid of trouble, they do not perform the expected match check on the input, making the job of delivering "bombs" easy for intruders.

(4) Raceconditions: There are more and more multi-task and multi-threaded programs. While improving operation efficiency, we should also pay attention to the problem of Raceconditions.

For example, both program A and program B operate on a file in "read/modify/write" order. When program A finishes reading/modifying, program B starts to complete all the "read/modify/write" tasks immediately. At this point, A continues to write, and as A result, B's operation has no performance! Intruders may use this vulnerability to rewrite some important files to break into the system, so programmers should pay attention to the sequence of file operations and locking problems.

3. Improper system configuration

(1) Fault for default configuration: Many systems install a default security configuration, often referred to easy to use. But unfortunately, easy to use also means easy to break in. So, be sure to selectively cancel the default configuration.

(2) Lazy administrator: One way to act lazy is to leave an administrator's password empty after a system is installed, and not change it later. You have to know, the first thing the intruder will do is to search the network for such a machine with a empty administrator password.

(3) Temporary port: Sometimes administrators open a temporary port on the machine to test something, but forget to disable it afterwards, leaving intruders with vulnerabilities to access. The usual resolution strategy is to disable a port unless it is required! Typically, security audit packets can be used to discover such ports and notify administrators.

(4) Trust relationship: Systems between networks often establish trust relationship to facilitate resource sharing, but this also brings the possibility of indirect attack, for example, as long as one machine in the trust group is breached, it is possible to further attack other machines. Therefore, trust relationships should be strictly reviewed to ensure a true security alliance.

4. Weak security awareness

When dealing with some emergency accidents, it is found that really difficult intrusion behavior is very few, the reason leading to frequent accidents is that the user password setting is too simple or the operation is wrong. For example, a listed company does a website security test, their system is very secure, and no problems were found after testing for a long time. Later I found out that the company has a QQ group for internal communication, click to apply to join, and soon I joined the group. After joining the group, I found a folder in the group share. The name of the folder is called the company's various system passwords. It's as simple as that, I easily get the company's system password. This is a very common problem. Internet users have weak security awareness and poor control over security.

Subtask 7.4.5 Master vulnerability scanning technology

Vulnerability scanning is a technology based on the network to remotely detect the security vulnerabilities of the target network or the local host. Vulnerability scanning is used for simulated attack experiments and security audits. Vulnerability scanning is an active preventive measure, which can effectively prevent hacker attacks. Of course, hackers can also use vulnerability scanning to discover vulnerabilities and launch attacks. For example, Ping scan determines the address of the target host, and port scan determines the open port of the target host. Then detect the operating system based on the results of the port scan, and vulnerabilities are scanned based on the obtained information.

(1) Ping scan: The Ping scan is to check whether the TCP/IP network of the target host is connected, that is whether the scanned IP address is in use.

(2) Port scan: Port scan is used to detect open ports on the host. Generally speaking, port scanning is to scan a certain port value range or port for a specified IP address. Port scan is classified into the following types:

① Full connection scan: The scanning host establishes a complete connection with a specified port on the target host through TCP/IP three-way handshake. If the port is listening, the connection succeeds; otherwise, the port is unavailable.

② SYN scan: The scanner sends a SYN packet to the target host. If the response is an RST packet, the port is closed. If the reply packet contains SYN and ACK, the port is listening. In this case, the scanner sends another RST packet to the target host to stop establishing the connection.

③ Covert scan: the scanner sends a FIN packet to the target host port. When the FIN packet reaches a closed port, the packet is discarded and an RST packet is returned. If the port is open, the FIN packet is simply discarded.

(3) Operating system detection: Operating system detection is used to detect information about the target host's operating system and computer programs that provide services.

(4) Vulnerability scanning: Vulnerability scanning is used to detect whether the target host system has vulnerabilities. It is generally used to scan the target host for specific vulnerabilities. Vulnerability scanning is divided into the following two types:

① Passive scan is host-based detection that checks the system for inappropriate settings, weak passwords, and other objects that conflict with security rules.

② Active scan is a network-based detection that attacks the system by executing some script files and recording the system response to discover vulnerabilities.

Subtask 7.4.6 Understand the definition of patch

Patches are small pieces of cloth used to cover up holes in clothes and bedding. It also refers to small programs released to solve problems (usually discovered by hackers or virus designers) exposed during use of large software systems (such as Microsoft's operating system). It's like putting a patch on a tattered shirt. People write programs can not be perfect, so the software will inevitably have bugs, the patch is to fix these bugs, that is, the original released software has defects. After the defects are found, another small program is compiled to make it perfect. This small program commonly known as patch. Patches are made by the original author of the software and can be downloaded from the website.

Patches also refer to small programs that are released to solve problems (General problems are discovered by hackers or virus designers, or are generated during the user's use. Patches refer to a work performed by developers to correct errors after the error information is sent to the server) that are exposed during use of large software systems (such as Microsoft operating systems).

Generally, in the development process of a software, many factors are not considered at the beginning, but as time goes by, the problems existing in the software will be gradually discovered. In order to fix the problems in the software itself, the software developer will release the corresponding patches. In other words, we originally bought a piece of clothing with problems, but we didn't pick it up carefully enough, so we didn't find the holes in it. But occasionally one day, there is a hole in the clothes, at this time, the manufacturer will repair the clothes with quality problems in order to maintain the legitimate rights and interests of consumers and the reputation of the company, which is called patch. That is, to fix the hole (defect) in itself. It's like a patient with a cold. An experienced physician can quickly tell if it's a wind-cold or a wind-heat cold and treat it appropriately. Inexperienced physicians may prescribe a variety of cold medicines, but some of them may not only fail to work and may have side effects.

Subtask 7.4.7 Analysis of typical vulnerability attack case

In March 2016, the open source encryption tool OpenSSL used by about two-thirds of web servers in the world has revealed new security vulnerabilitie named "Decrypting RSA with Obsolete and Weakened eNcryption". This vulnerability allows "hackers" to attack websites and read encrypted information such as passwords, credit card account numbers, trade secrets and financial data, which poses a huge security test for global websites.

Subtask 7.4.8 Master vulnerability defense methods

In the face of these vulnerabilities, as a user, you must scan regularly, update the software to the latest version in time, update system patches in time, check and kill viruses and Trojan horses in time, carefully check the software configuration, turn off unsafe options, and pay attention to the threat intelligence broadcast of the security company, close specific ports or services. Improving security awareness is an effective way to prevent such vulnerabilities from harming your system.

Project 8 Host firewall and antivirus software

Wall, its initial role is to protect, and its main role is to guard. Since ancient times, walls can give people a sense of security. A firewall, as its name suggests, blocks fire. This term comes from the field of architecture, and its function is to isolate a fire and prevent it from spreading from one area to another.

Introduced into the communications field, firewalls are mainly used to protect one network from attacks and intrusions from another network.

Anti-virus software is a program tool that can remove all known programs such as virus and Trojan horse that are harmful to the computer.

Task 8.1 Know firewall

Subtask 8.1.1 Understand firewall definition

Firewall technology is a concrete embodiment of security technology. A firewall was originally a wall built between houses to prevent the spread of a fire.

The so-called “firewall” refers to a method to separate the internal network from the public access network (such as the Internet), it is actually a kind of isolation technology. Firewall is a scale of access control implemented in the communication between two networks. It can allow the people and data you “agree” to enter your network, while keeping the people and data you “disagree” out of the door, to prevent the hackers in the network from accessing your network to the maximum extent. In other words, without the firewall, intranet users cannot access the Internet, and Internet users cannot communicate with intranet users, as shown in Figure 8-1.

Figure 8-1 Firewall

Firewall is a collection of hardware, software and control policies. Control policies can be divided into two kinds in the form of expression: permitted unless expressly prohibited; prohibit unless expressly permitted.

Subtask 8.1.2 Master firewall classification

Firewall can be classified according to form and protected object, according to the form it can be divided into:

(1) Hardware firewall: The system is an embedded system, which is usually open source. Hardware firewall is a combination of hardware and software to isolate internal and external networks.

(2) Software firewall: Generally parasitic on the operating system platform. Software firewall realizes the purpose of isolating internal and external networks through pure software.

Protected objects can be classified into:

(1) Standalone firewall: Provides services for the current host.

(2) Network firewall: The service scope of the firewall is the network on the side of the firewall.

Task 8.2 Windows firewall settings

Firewall is a device that helps ensure information security, it will allow or restrict the transmitted data to pass through according to specific rules. A firewall can be a proprietary hardware or a set of software built on general hardware. Windows firewall, as its name implies, is a software firewall in the Windows operating system.

The importance of firewall for every computer user is self-evident, especially in the current network threat flooding environment, through professional and reliable tools to help protect their computer information security is very important. There are many brands of anti-virus software on the market but not every one provides a firewall function for users, so a lot of users are installed anti-virus software and look for a professional firewall, this approach rejects what is near at hand and seeks what is far away, because the Windows operating system has its own firewall.

Since Windows XP, Microsoft has built firewalls into the system. Of course, the early built-in firewall functions are relatively weak. Windows XP only provides simple and basic functions, and can only protect inbound traffic and block any non-native machines. By default, the firewall is turned off, and the firewall of the Windows XP SP2 system is turned on by default. The system administrator can enable the firewall software through Group Policy. The firewall of Windows Vista is built on the new Windows filtering platform, and adds the ability to filter outbound traffic through the advanced security MMC snap-in. In Windows 7, Microsoft has further adjusted the function of the firewall to make the firewall more user-friendly, especially in mobile computers, which can support multiple firewall policies.

The path to open the Windows firewall page is choose "Control Panel" → "System and Security" → "Windows Firewall", as shown in Figure 8-2. You can select public network or private network when accessing a network. "Public" (That is, the discovery of the local computer is restricted and the discovery of other computers is restricted, which greatly improves security, illegal remote users cannot discover the IP address and MAC address of the computer on the network, that is, the local computer is hidden), and "private" is to to make other computers find the local computer. This is available in the "Control Panel" → "Network and Sharing Center" → "Network" → "Custom" (on the right, click on it) and then can be set. In fact, Windows 7 already supports independent configurations of different network types without affecting each other. In short, the public security level is definitely higher than the private. On this page, you can also set the following parameters:

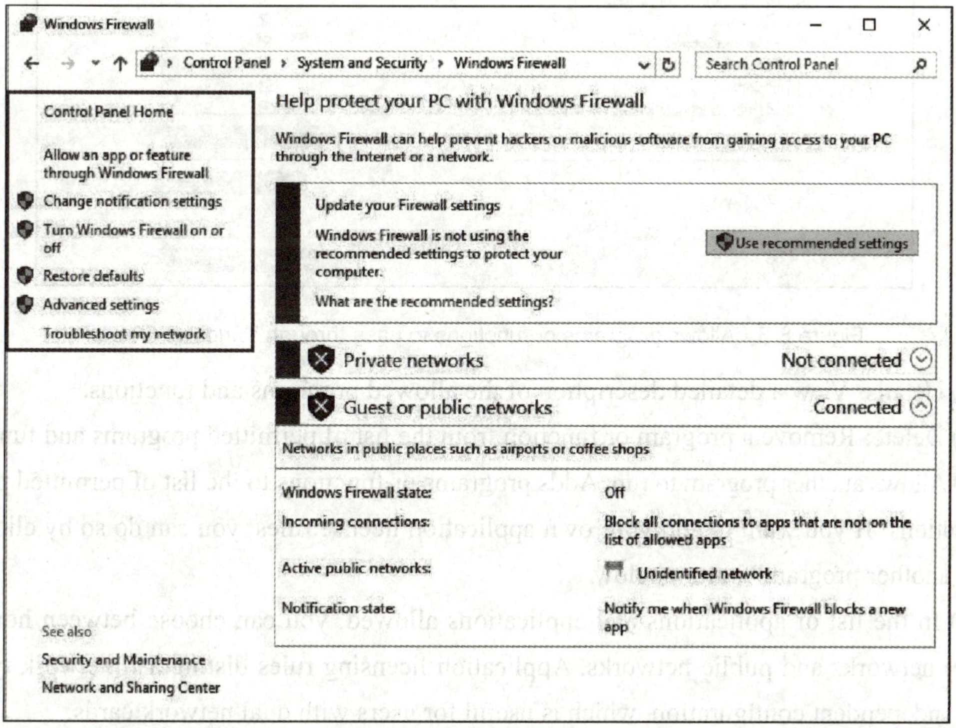

Figure 8-2 Windows firewall

1. Allows programs or functions to pass through Windows firewall

Allows programs or functions to set data traffic rules through the Windows firewall, as shown in Figure 8-3.

(1) Change settings: Add, change, or remove allowed programs and ports. There is no port configuration in the general configuration, so there is no need to manually specify TCP and UDP,

because for many users who do not know what these two things are, these configurations have been moved to the advanced configuration, for ordinary users are generally used to increase the application license rules.

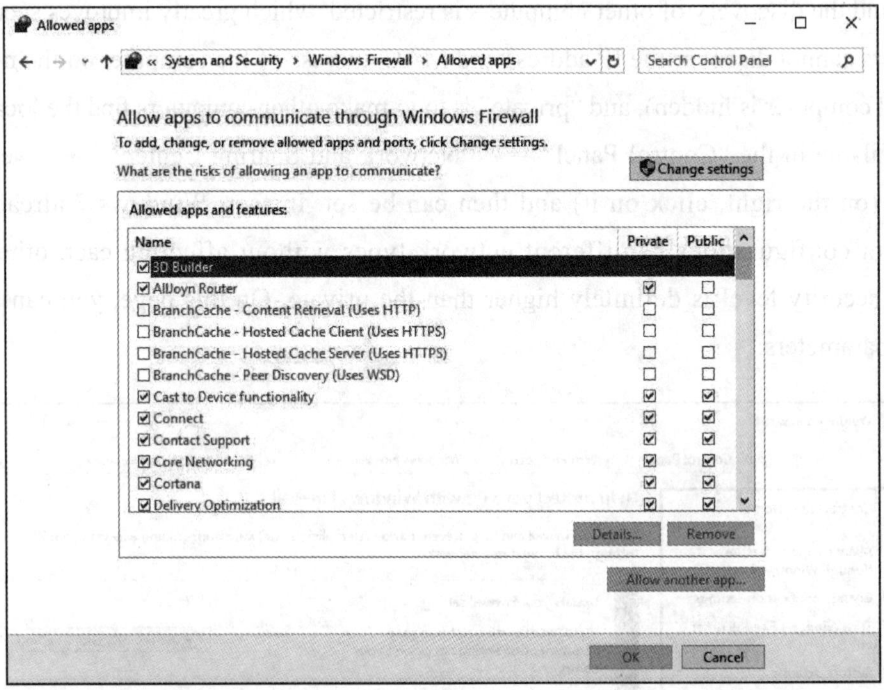

Figure 8-3 Allows programs or functions to pass through Windows firewall

(2) Details: View a detailed description of the allowed programs and functions.

(3) Delete: Remove a program or function from the list of permitted programs and functions.

(4) Allows another program to run: Adds programs or functions to the list of permitted programs and functions. If you want to add your own application license rules, you can do so by clicking the "Allow another program" button below.

(5) In the list of applications and applications allowed, you can choose between home/work (private) networks and public networks. Application licensing rules distinguish network types and support independent configuration, which is useful for users with dual network cards.

2. Change notification settings

If the firewall is enabled, select whether to notify new applications blocked by the firewall, as shown in Figure 8-4.

3. Enable or disable Windows firewall: firewall switch interface

Turn on the firewall for security protection or turn off the firewall to allow all programs to pass through the Windows firewall. Enabling change notification settings is the same as enabling or disabling Windows Firewall, as shown in Figure 8-4.

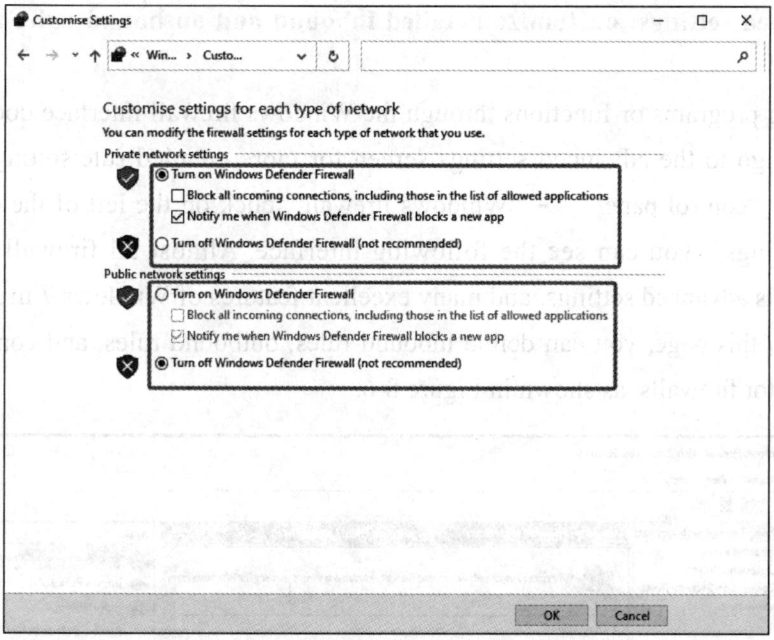

Figure 8-4 Windows firewall change notification settings

4. Restore default settings: initialize Windows firewall

If the firewall rules are not set up, it may not only fail to block malicious attacks on the network, but also cause users to be unable to access the Internet normally. If you fail to access the Internet due to misoperations, do not worry. The Windows firewall has the default restoration function. You can click the mouse to restore the firewall to its original state, as shown in Figure 8-5.

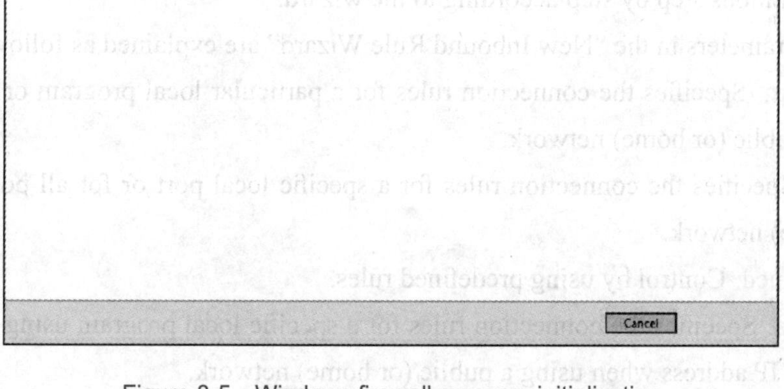

Figure 8-5 Windows firewall recovery initialization

5. Advanced settings: customize detailed inbound and outbound rules and connection security rules

If allowing programs or functions through the Windows firewall interface does not meet your needs, you can go to the advanced settings screen for more detailed rule settings. Click in turn "computer" → "control panel" → "Windows firewall", click on the left of the control interface "advanced settings", you can see the following interface. Almost all firewall settings can be completed in this advanced settings, and many excellent features of Windows 7 firewall can also be shown here. On this page, you can define inbound rules, outbound rules, and connection security rules, and monitor firewalls, as shown in Figure 8-6.

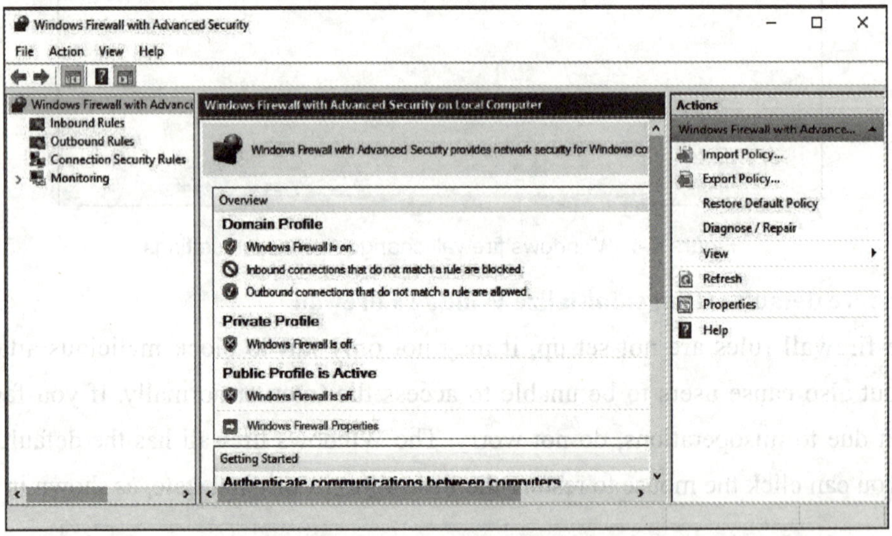

Figure 8-6 Windows firewall advanced settings

The following uses creating an inbound rule as an example to describe how to configure Windows firewall rules, as shown in Figure 8-7. Click "Inbound Rules" on the right of the page, find "New Rule" and click. The "New Inbound Rule Wizard" is displayed. You can perform subsequent operations step by step according to the wizard.

Several parameters in the "New Inbound Rule Wizard" are explained as follows:

(1) Program: Specifies the connection rules for a particular local program or for all programs when using a public (or home) network.

(2) Port: Specifies the connection rules for a specific local port or for all ports when using a public (or home) network.

(3) Predefined: Control by using predefined rules.

(4) Custom: Specifies the connection rules for a specific local program using a specific source port and source IP address when using a public (or home) network.

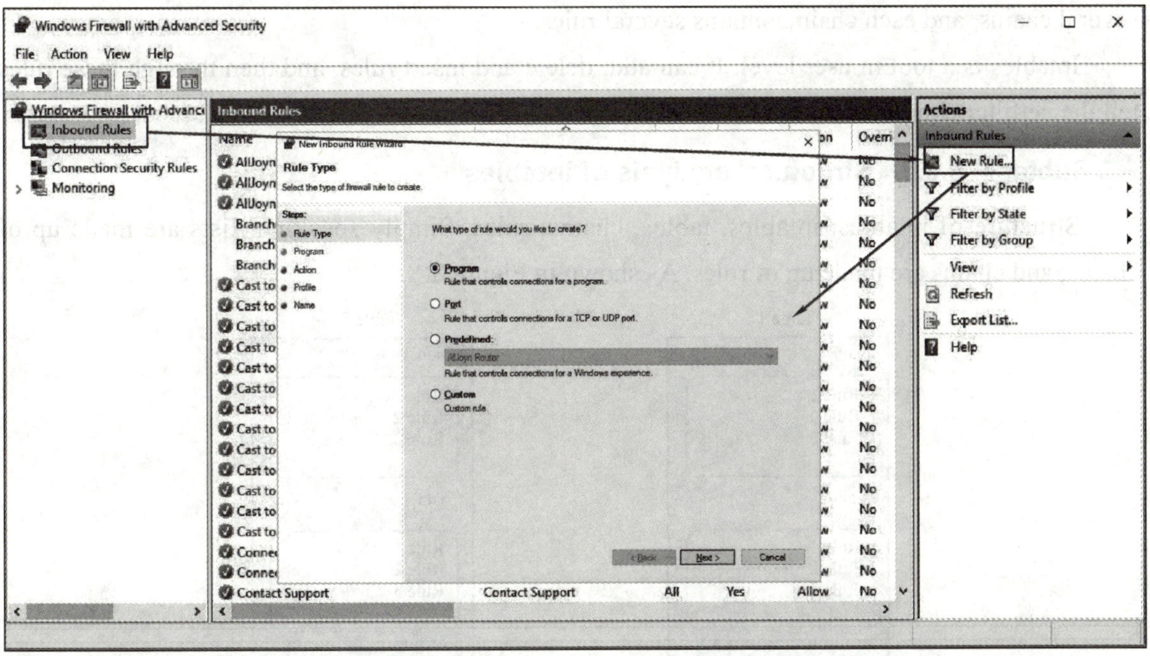

Figure 8-7 Windows firewall rule settings

6. Troubleshoot the network

Troubleshooting the network can detect network problems and help users repair network errors when the system or network fails.

Task 8.3 Linux firewall settings

Iptables is a built-in firewall in the Linux kernel. It allows the administrator to filter packets and implement NAT by setting tables, chains, and related rules. Generally speaking, iptables firewall is built into CentOS 6 and other Linux versions, and the iptables service is enabled by default. Iptables is used for IPv4, but ip6tables is required for IPv6.

Subtask 8.3.1 Meet iptables

Iptables is a free packet filter firewall that has evolved with the development of the kernel through four basic phases:

(1) 1.1 kernel, ipfirewall is used to operate kernel packet filtering rules.

(2) 2.0 kernel, using ipfwadm to operate kernel packet filtering rules.

(3) 2.2 kernel, ipchains is used to operate kernel packet filtering rules.

(4) 2.4 kernel, using iptables to operate kernel packet filtering rules.

Iptables is just an interface between the firewall and users. The real firewall is netfilter running in the Linux kernel. A Linux firewall consists of netfilter and iptables.

Netfilter is a framework in the Linux kernel that provides a series of tables, each consisting of several chains, and each chain contains several rules.

Iptables is a tool in user level. It can add, delete and insert rules, and then through these rules tell the netfilter component how to deal with.

Subtask 8.3.2 Structural analysis of iptables

Structure of iptables: iptables, tables, chains, rules. Simply speaking, lists are made up of chains, and chains are made up of rules. As shown in Figure 8-8.

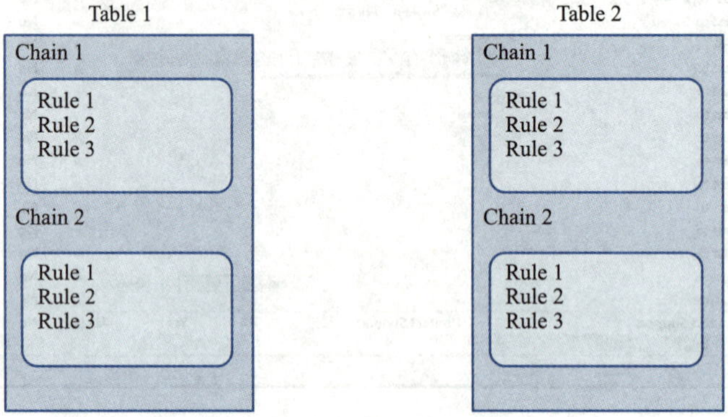

Figure 8-8 Iptables structure

Subtask 8.3.3 Understand iptables rules

A rule is generally defined as "if a packet meets such a condition, the packet is processed as such". Rules in iptables specify the five-tuple information such as source address, destination address, source port, destination port, and protocol. Iptables processes the packet according to the method defined by the rule when a packet matches a rule, such as allowing it through or discarding it. As shown in Figure 8-9, traffic whose source IP address is 192.168.1.1 and destination port is 80 (HTTP) is allowed to pass.

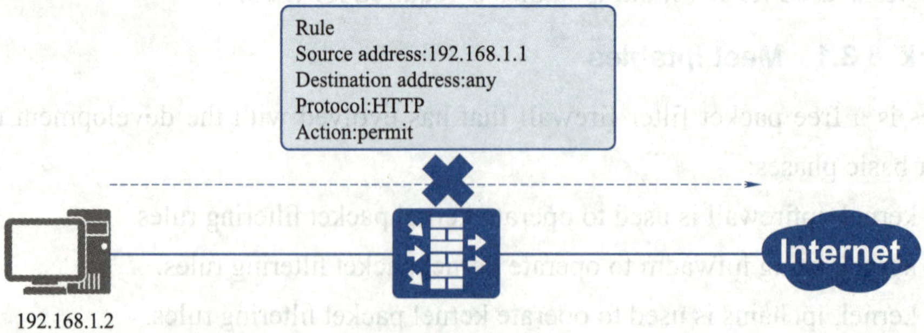

Figure 8-9 Iptables rule

Subtask 8.3.4 Understand iptables chain

A chain is the path through which data packets are propagated, and each chain contains one or more rules. When a data packet reaches a chain, iptables will use the first rule in the chain to match the packets, and check whether the data packet meets the conditions defined by this rule. If the condition is meet, the packet will be processed according to the action defined by the rule; otherwise, iptables will continue to use the next rule to match. If the packet does not meet all the rules in the chain, the default policy of the chain will be used. This is the same idea as the ACL defined in routing switching, as shown in Figure 8-10.

```
Chain
Rule 1 Source address:192.168.1.1 Destination address:any Protocol:HTTP Action:permit
Rule 2 Source address:192.168.1.2 Destination address:any Protocol:DNS Action:permit
Rule 3 Source address:192.168.1.3 Destination address:any Protocol:ICMP Action:deny
Rule 4 Source address:192.168.1.4 Destination address:any Protocol:ARP Action:deny
Rule 5 Source address:192.168.1.5 Destination address:any Protocol:IGMP Action:deny
```

Figure 8-10 Iptables chain

Netfilter stipulates five rule chains, as long as any packet passes through the local computer, it will pass through one of the five chains. The five rule chains in iptables are:

(1) PREROUTING (before routing).

(2) INPUT (packet flow entry).

(3) FORWARD (forwarding checkpoint).

(4) OUTPUT (packet export).

(5) POSTROUTING(after routing).

Subtask 8.3.5 Understand iptables table

Iptables table provides a specific function, and contains four tables:

(1) Filter table: Allowed or disallowed policies in the data packet.

(2) NAT table: The function of address translation.

(3) Mangle table: Modify message data.

(4) Raw table: Determine whether the data packet is processed by the state tracking mechanism.

Processing priority of the table: Raw >Mangle >NAT >Filter.

Generally speaking, filters can only be done on 3 chains: INPUT, FORWARD, and OUTPUT. For NAT, generally it can only be done on 3 chains: PREROUTING, OUTPUT, and POSTROUTING. And Mangle can be done with all 5 chains: PREROUTING, INPUT, FORWARD, OUTPUT, POSTROUTING.

Subtask 8.3.6 Analysis of the process of iptables transmitting data packets

When a data packet enters the network interface controller, it first matches the PREROUTING chain, and the system will determine the next processing process based on the destination address of the data packet. There may be the following three situations, as shown in Figure 8-11.

(1) If the destination address of the data packet is the local machine, the system will send the data packet to the INPUT chain, matching the rules in the INPUT chain, and if it passes the rule check, the data packet will be sent to the corresponding local process for processing. If the packet does not pass the check, then it will be discarded.

(2) If the destination address of the data packet is not the local machine, the packet will be forwarded, and the system will send the data packet to the FORWARD chain to match the rules in the FORWARD chain, if it passes the rule check, the data packet will be sent to the corresponding local process for processing. If the packet does not pass the check, then it will be discarded.

(3) If the data packet is generated locally, the system sends the data packet to the OUTPUT chain, matching the rules in the OUTPUT chain, if it passes the rule check, the data packet will be sent to the corresponding local process for processing; if it does not pass the check, the data packet will be discarded.

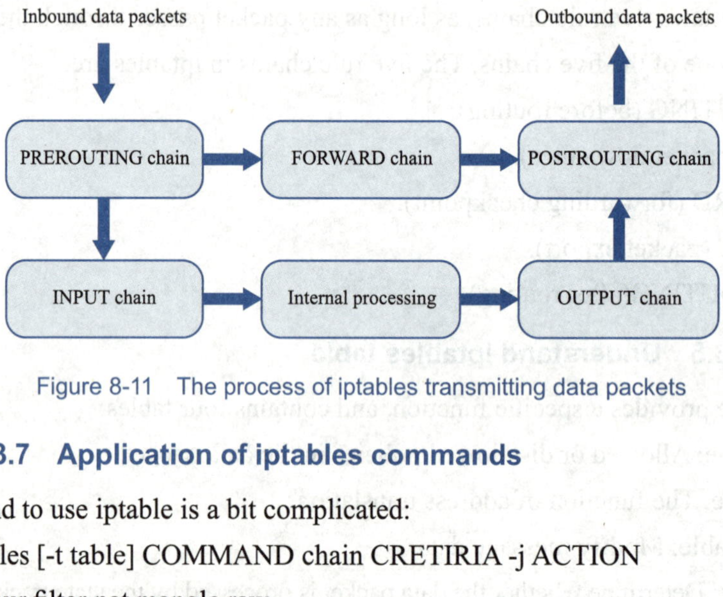

Figure 8-11 The process of iptables transmitting data packets

Subtask 8.3.7 Application of iptables commands

The command to use iptable is a bit complicated:

Format:iptables [-t table] COMMAND chain CRETIRIA -j ACTION

(1) -t table:four filter nat mangle raw.

(2) COMMAND: Define how to manage the rules.

(3) chain: Specify the chain on which your next rule is to be operated. When defining the strategy, you can omit it.

(4) CRETIRIA: Specify matching criteria.

(5) -j ACTION: Specify how to proceed.

For example: Do not allow 172.16.0.0/16 be accessed.

```
iptables -t filter -A INPUT -s 172.16.0.0/16 -p udp --dport 53 -j DROP
```

Task 8.4 Use of host antivirus software

Subtask 8.4.1 Understand the definition of antivirus software

Antivirus software, also called virus protection software or virus firewall, is a type of software used to eliminate computer threats such as common computer viruses, malware, and Trojan viruses.

Antivirus software is usually integrated with monitoring recognition, virus scanning and removal, automatic upgrade, active defense and other functions, and some antivirus software also comes with data recovery, to prevent hacking, network traffic control and other functions, it is an important part of the computer defense system.

Subtask 8.4.2 Understand the basic functions of antivirus software

The basic functions of antivirus software include preventing virus, searching for virus, and clearing virus.

(1) Preventing virus: The prevention of computer viruses.

(2) Searching for virus: Scan the program or file run by the computer for virus, and accurately report the name of the virus compared to the virus library.

(3) Clearing virus: Modifies the infected objects according to different types of viruses and restores the infected objects according to their infection characteristics.

Subtask 8.4.3 Understand the composition of antivirus software

Antivirus software is generally composed of scanner, virus database and virtual machine, and the main program integrates them, as shown in Figure 8-12.

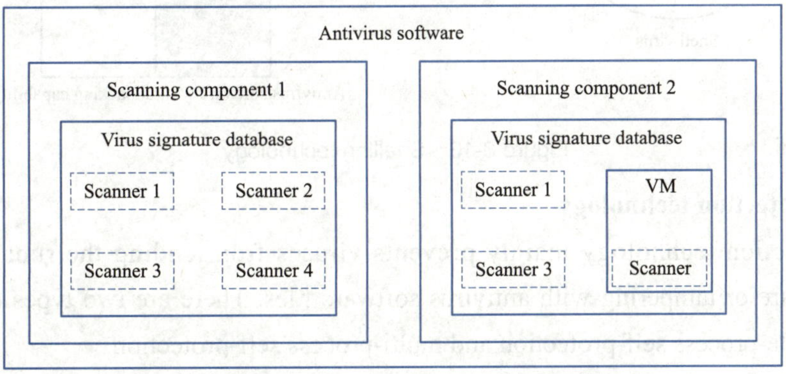

Figure 8-12 Composition of anti-virus software

(1) Scanner is the main body of antivirus software, which is mainly used to scan for viruses. The effect of an antivirus software directly depends on the advanced degree of scanner compilation technology and algorithm. Therefore, most antivirus software have more than one scanner.

(2) The virus database is used to store virus signatures, which are mainly divided into memory

signatures and file signatures. File feature codes generally exist in some unexecuted files. Memory feature codes generally exist in applications that are already running.

(3) The virtual machine can make the virus execute in a virtual environment built by antivirus software.

Subtask 8.4.4 Master the key technologies of antivirus software

1. Unpacking technology

Unpacking technology is a commonly used technology in antivirus software, which can analyze compressed files, extra-flowered files, packed files, and sub-packaged files.

If the hacker uses another packing tool to pack, then for the anti-virus software, this kind of virus becomes another kind of virus, and a new feature record must be added to detect and kill.If the antivirus software has strong unpacking ability, you can unpack the virus files first, and then check and kill them. In this way, only one record can be used to kill these viruses, which not only reduces the antivirus software's occupation of system resources, but also greatly improves its ability to check and kill viruses, as shown in Figure 8-13.

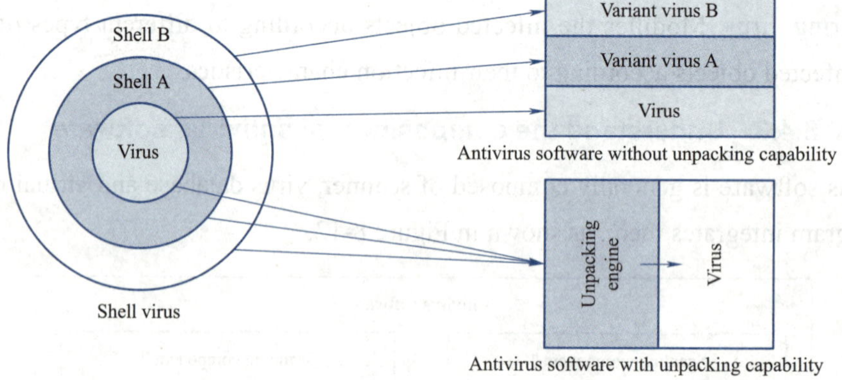

Figure 8-13 Shelling technology

2. Self-protection technology

Self-protection technology mainly prevents viruses from ending the running process of antivirus software or tampering with antivirus software files. There are two types of process self-protection: single-process self-protection and multi-process self-protection.

3. Repair technology

When antivirus software checks and kills viruses, it usually deletes the infected files directly. In this way, some system files may be deleted by mistake, which will eventually cause the system to crash and fail to start. The antivirus software repair technology can repair damaged files. If a virus damages the system files, the antivirus software can repair or download the corresponding files for repair. Antivirus software without this technology often deletes the infected system files and then the computer crashes and cannot be started.

4. Real-time upgrade technology

Viruses and antivirus software are like spears and shields, except that the development of spears is faster, so the virus database of antivirus software generally lags behind computer viruses. Then the real-time update of the virus database is particularly important. Every time you connect to the Internet, the antivirus software will automatically connect to the upgrade server to query for upgrade information, and upgrade if necessary. At present, the more advanced cloud killing technology can access the virus database on the cloud in real time for judgment, and users can defend without frequently updating the virus database.

5. Active defense technology

Active defense technology is the automatic monitoring of various program actions through dynamic simulation of the antivirus expert system, automatic analysis of the logical relationship between program actions, comprehensive application of virus identification rule knowledge and automatic virus identification to achieve the purpose of active defense.

Subtask 8.4.5 Understand common knowledge of antivirus software

When choosing antivirus software, comprehensive consideration can be made according to factors such as antivirus effect, resource occupation, antivirus speed, operation interface and other factors.

For the use of antivirus software, many people have a misunderstanding, thinking that as long as the antivirus software is installed, they can guarantee that their computers will not be infected by viruses, but this idea is wrong. It is necessary for us to understand some common sense of antivirus software:

(1) It is impossible for antivirus software to check and kill all viruses.

(2) Some software can detect viruses but cannot remove them.

(3) It is not necessary to install two or more antivirus software under each operating system of a computer at the same time. In addition, it is recommended to check the list of incompatible programs.

(4) Antivirus software has many ways to kill infected files: Clear; Delete; Forbid access; Isolate; Do not process.

① Clear: Clear the files infected by the worm, and the files will return to normal after the cleanup.

② Delete: Delete virus files. Such files are not infected files, they are poisonous, cannot be cleaned, and can be deleted.

③ Forbid access: Forbid access to virus files. After the virus is found, if the user chooses not to deal with it, the antivirus software may prohibit the virus from being accessed. When the user opens, an error dialog box will pop up with the content “The file is not a valid Win32 file”.

④ Isolate: After the virus is deleted, it is transferred to the isolate area. Users can retrieve deleted files from the isolate area. Files in the isolate area cannot be run.

⑤ Do not process: Do not process the virus. If the user does not know whether it is a virus or not, you can temporarily leave it alone.

Most antivirus software lags behind computer viruses. Therefore, in addition to updating and upgrading the software version and scanning regularly, you should also pay attention to enriching your computer security and network security knowledge, so as not to open unfamiliar files or unsafe web pages at will, not to browse unhealthy sites, and to update your privacy password, supporting the use of security assistants and personal firewalls, etc. Only in this way can we better maintain our computer and network security.

Section 3 Network security foundation

In the last section,we introduced the related knowledge of terminal security. Whether terminal security can solve all information security problems is obviously not enough. Each terminal needs to access the computer network to realize resource sharing. If the network is not secure, the information security cannot be guaranteed.

Network security means that the hardware, software and data in the network system are protected from damage, change and disclosure due to accidental or malicious reasons. The system operates continuously, reliably and normally, and the network service is not interrupted. Firewall is the most commonly used and widely used in network security.

This section will focus on network security and how to improve network security in five projects:

Working principle and configuration of hardware firewall: This project introduces the development history, characteristics, typical networking methods, application scenarios and configuration of Huawei firewall.

Network address translation technology: Describes the technical background and working mechanism of NAT, and the application scenarios and configurations of source NAT, server mapping and destination NAT.

Dual-system hot backup technology of firewall: It describes the application scenario, working mechanism and configuration of Huawei firewall dual-system hot backup.

Firewall user management: Introduce AAA technology, RADIUS, HWTACACS, LDAP and other protocols, and describe the application scenario, working mechanism and configuration of

Huawei firewall user management.

Intrusion prevention: Introduce Huawei's intrusion prevention system, network antivirus technology, configuration, operation and maintenance.

Through these five projects, readers can understand the related technologies of network security, master the working principle and configuration of network security products such as hardware firewall, and have the necessary foundation for network security engineers.

Project 9 Working principle and configuration of hardware firewall

In the process of data communication, because insecure factors in the network will cause information leakage, incomplete information, unavailability and other issues, firewall equipment is required when deploying the network.

This project mainly introduces the development history, characteristics, typical networking methods, application scenarios, and technical indicators of Huawei firewalls.

Task 9.1 Understand hardware firewalls

Subtask 9.1.1 Understand hardware firewall characteristics

Firewall technology is a concrete manifestation of security technology. The firewall originally refers to a wall built between houses to prevent the spread of fire when a fire occurs. What we are discussing here is the hardware firewall, which integrates various security technologies, adopts a dedicated hardware structure, selects high-speed CPU, embedded operating system, and supports various high-speed interfaces (LAN interfaces) to protect private network (computer) security, such equipment is called a hardware firewall. The hardware firewall can run independently of the operating system and computer equipment. It is used to centrally solve network security problems, can be suitable for various occasions, and can provide high-efficiency "filtering". At the same time, it can provide security features including access control, identity verification, data encryption, VPN technology, address translation, etc. Users can configure complex security policies according to the needs of their own network environment, prevent some illegal access and protect their own network security.

The modern firewall system should not be just an "entrance barrier". The firewall should be the access control point of several networks. All data flows in and out of the network protected by the firewall should first pass through the firewall to form a gateway for information entry and exit. The firewall can not only protect the security of the internal network in the Internet, but also protect the security of several hosts in an internal network. In the same network divided by a firewall, all computers are considered "trusted", and the communication between them is not interfered by the firewall. But between the networks divided by the firewall, access must be conducted in accordance with the "policy" stipulated by the firewall.

Subtask 9.1.2 Understand hardware firewall classification

The development of firewalls has gone through three generations, and there are various classification methods. For example, it can be divided into hardware firewalls and software firewalls according to the form; it can be divided into stand-alone firewalls and network firewalls according to the protection objects. But in general, the most mainstream classification method is to classify according to access control methods, which can be divided into packet filtering firewalls, proxy firewalls and stateful inspection firewalls.

1. Packet filtering firewall

Packet filtering refers to checking each data packet at the network layer, and forwarding or discarding the data packet according to the configured security policy. The basic principle of a packet filtering firewall is to implement packet filtering by configuring an access control list (ACL, Access Control List). Mainly based on information such as the source/destination IP address, source/destination port number, IP identification, and message delivery direction in the data packet, as shown in Figure 9-1.

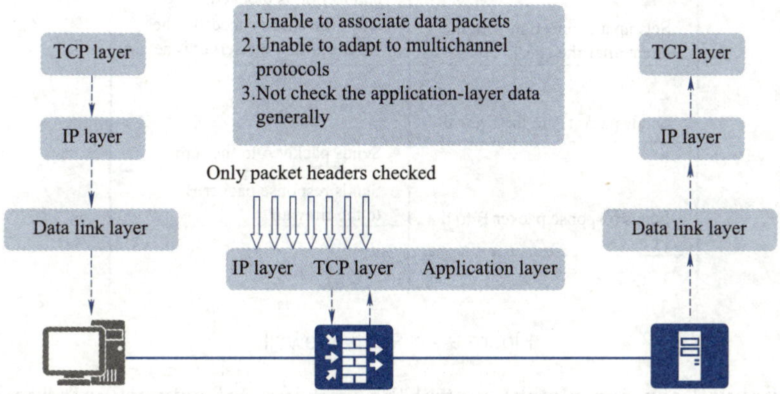

Figure 9-1 Packet filtering firewall

The design of the packet filtering firewall is simple, very easy to implement, and inexpensive. The main disadvantages of packet filtering firewall are as follows:

(1) With the increase of ACL complexity and length, its filtering performance shows an exponential decline;

(2) Static ACL rules are difficult to adapt to dynamic security requirements;

(3) Packet filtering does not check the session state or analyze the data, so it's easy for hackers to get through. For example, an attacker can use a fake address to spoof, and by setting his host IP address to a legitimate host IP address, he can easily pass the message filter.

(4) Not applicable to multichannel protocols, such as FTP protocol. FTP generates dynamic data channel ports on the basis of control channel negotiation, and then data interaction is mainly carried out on the data channel. The packet filtering firewall cannot match the generated dynamic data channel port.

2. Proxy firewall

The proxy service acts on the application layer of the network, and its essence is to take over the business directly between the internal network and external network users by the proxy. As shown in Figure 9-2, the proxy checks the request from the user. After the user passes the security policy check, the firewall will establish a connection with the real server on behalf of the external user, forward the request from the external user, and send the response returned by the real server back to the outside User.

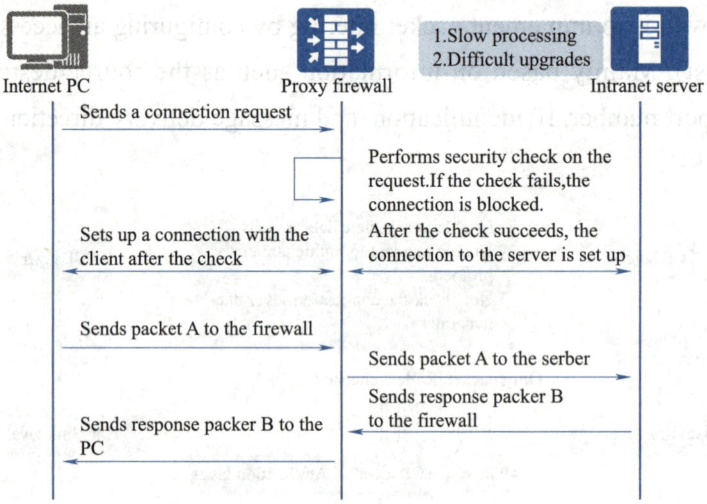

Figure 9-2 Proxy firewall

The proxy firewall can completely control the exchange of network information, control the session process, and has high security. Its shortcomings are mainly manifested in:

(1) Software implementation limits the processing speed and is prone to denial of service attacks;

(2) The application layer agent needs to be developed for each protocol, the development cycle is long, and the upgrade is difficult.

3. Stateful inspection firewall

Status inspection is an extension of packet filtering. Packet filtering based on connection status not only considers each packet as an independent unit, but also considers the historical relevance of the packets before and after the packet. As we know, the establishment of all data streams based on reliable connections (data streams based on the TCP protocol) needs to go through the three processes of "client synchronization request", "server response" and "the client replies again " three process (namely "three-way handshake" process), it shows that each data packet does not exist independently, but has the closed status connection before and after. Based on this status connection, state detection technology is developed.

Figure 9-3 shows the basic principles. Stateful inspection firewall uses various session tables to track activated transmission control protocol (TCP) and user datagram protocol (UDP) pseudo sessions. Access control list (ACL) determines which sessions are established. Packets are forwarded only if they are associated with a session. The UDP pseudo session establishes a virtual connection (UDP is connectionless protocol) for the UDP data stream during UDP packet processing to monitor the status of the UDP connection process. Stateful inspection firewall intercepts data packets at the network layer, then extracts the state information required by the security policy from each application layer and saves it in the session table. The appropriate decision can be made by analyzing the session table and the subsequent connection requests related to the data packet.

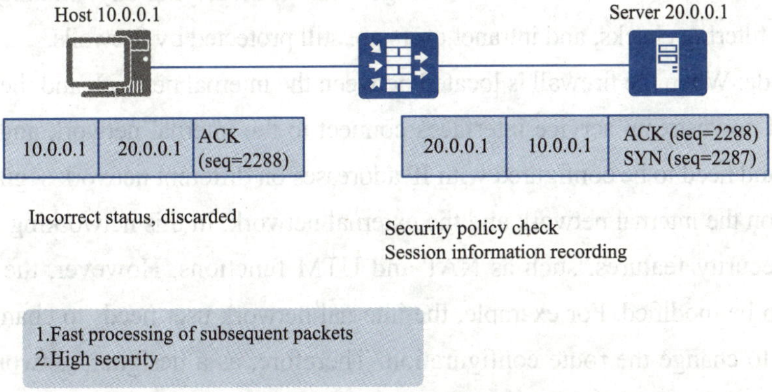

Figure 9-3 Stateful inspection firewall

The stateful inspection firewall has the following advantages:

(1) Excellent performance in subsequent packet processing: The stateful inspection firewall performs ACL checks on data packets, it records the connection state of data flows, the subsequent packets in the data flow do not need to be checked by ACL, but only need to check the connection records of the newly received packets according to the session table. After the check passes, the connection state record will be refreshed to avoid repeated checks on packets with the same connection status. The records in the connection session table can be arranged randomly, which is different from the fixed ACL. Therefore, the stateful inspection firewall can use algorithms such as binary tree or Hash to search quickly, which improves the transmission efficiency of the system.

(2) High security: The connection status list is dynamically managed. After the session is complete, the temporary entry for returning packets created on the firewall is closed, ensuring the real-time security of the internal network. At the same time, the stateful inspection firewall adopts the real-time connection state monitoring technology, which enhances the security of the system by identifying the connection state factors such as reply and response in the session table.

Subtask 9.1.3 Master firewall networking mode

The firewall networking can be divided into transparent mode and routing mode according to whether firewall interfaces work at Layer 2 or Layer 3. As shown in Figure 9-4(a), the upstream and downstream ports of the firewall work at Layer 2, which is called transparent mode. As shown in Figure 9-4(b), the upstream and downstream ports of the firewall work at Layer 3, which is called routing mode.

Transparent mode: The firewall forwards packets but does not address routes, and the two service networks connected to the firewall must reside on the same network segment. In this case, the upstream and downstream interfaces of the firewall work at Layer 2 and have no IP addresses. In this networking mode, the firewall can avoid the trouble caused by changing the topology structure. You only need to connect the firewall to the network like a Bridge without modifying any existing configuration. IP packets also pass filtering checks, and intranet users are still protected by firewalls.

Routing mode: When the firewall is located between the internal network and the external network, the upstream and downstream service interfaces connect to the internal network and external network work at Layer 3 and need to be configured with IP addresses on different network segments. The firewall addresses routes on the internal network and the external network. In this networking mode, the firewall supports more security features, such as NAT and UTM functions. However, the original network topology needs to be modified. For example, the internal network user needs to change the gateway or the router needs to change the route configuration. Therefore, as a designer, network transformation, service interruption and other factors should be considered comprehensively.

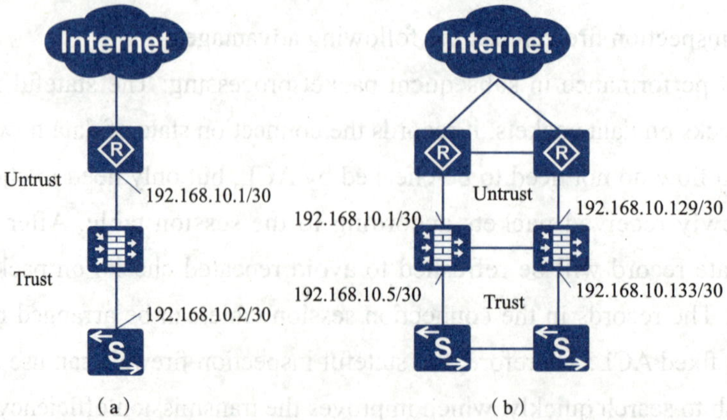

Figure 9-4 Firewall networking mode

Task 9.2 Analysis of firewall forwarding principle

Subtask 9.2.1 Understand packet filtering technology

As a network security protection mechanism, packet filtering is mainly used to control the forwarding of different traffic on the network. The core technology of packet filtering is access control list.

The traditional packet filtering firewall obtains the packet header information before forwarding the required packet, including the source IP address, destination IP address, protocol number of the upper-layer protocol carried by the IP layer, source port number, and destination port number of the packet, and the packets are matched with the predefined filtering rules and forwarded or discarded according to the matching results..

The forwarding mechanism of the packet filtering firewall matches and checks packet filtering rules one by one. Therefore, the forwarding efficiency is low. Currently, the firewall uses the stateful inspection mechanism to filter only the first packet of a connection. If the first packet passes the check of the packet filtering rules and set up a session, subsequent packets are directly forwarded through the session table rather than through the packet filtering mechanism.

Subtask 9.2.2 Master firewall security policy

Security policy controls traffic forwarding and implements integrated content security detection on traffic based on certain rules. The rules in question are essentially packet filtering, matching traffic by defining the five-tuple of data flows.

The basic function of a firewall is to protect a specific network from untrusted networks, but at the same time, it must allow legitimate communication between two networks. The purpose of a security policy is to verify the data flow that passes through the firewall. Legitimate data flows that comply with security policies can pass through the firewall.

Through the firewall security policy, you can control the access rights of the internal network to the external network, and control the access rights between the subnets of different security levels in the internal network, and so on. At the same time, it can also control the access of the device itself, such as restricting which IP addresses can log in to the device through Telnet and Web, and control the mutual access between the network management server, NTP server and the device.

The firewall security policy defines the processing rules of the data flow on the firewall, and the firewall processes the data flow according to the rules. Therefore, the core function of the firewall security policy is to filter the traffic passing through the firewall according to the defined rules, and the keywords to determine how the filtered traffic will proceed, as shown in Figure 9-5.

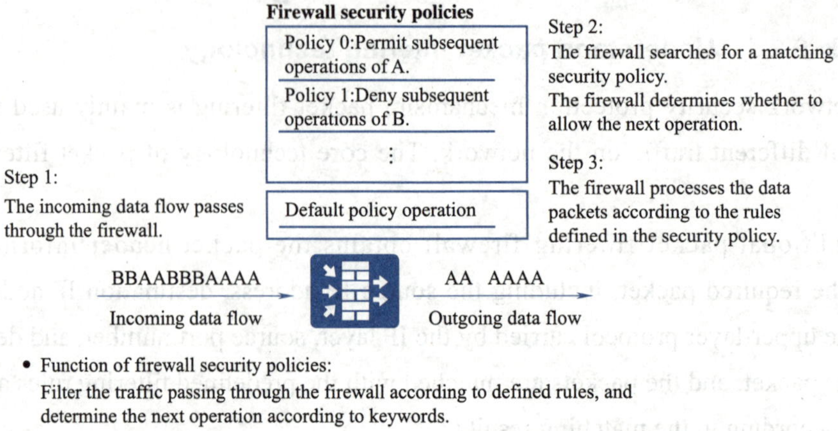

Figure 9-5 Security policy management

In firewall applications, the firewall security strategy is the basic means for network security access to the data flow passing through the firewall, and determines whether the subsequent application data flow is processed. The NGFW will detect the received traffic and detect the attributes of the traffic, including: source security zone, destination security zone, source address/region, destination address/region, user, service (source port, destination port, protocol type), application and time period.

Subtask 9.2.3 Understanding firewall inter-domain forwarding

Early packet filtering firewalls adopted a "packet-by-packet inspection" mechanism, that is, all packets received by the device are checked each time according to packet filtering rules to determine whether to let the packet pass. This mechanism seriously affects the forwarding efficiency of the device, making the packet filtering firewall a forwarding bottleneck in the network.

As a result, more and more firewall products adopt a "stateful inspection" mechanism for packet filtering. As shown in Figure 9-6. The "stateful inspection" mechanism uses the flow as the unit to detect and forward packets, that is, check the packet filtering rules for the first packet of a flow, and record the judgment result as the "state" of the flow. For subsequent packets of the traffic, the firewall directly determine based on this "state" whether to forward (or perform content security detection) or discard. This "state" is what we usually call the session table entry. This mechanism has rapidly improved the detection rate and forwarding efficiency of firewall products, and has become the current mainstream packet filtering mechanism.

The firewall generally checks the five elements in the IP message, also known as the "five-tuple", that is, the source IP address and destination IP address, source port number and destination port number, and protocol type. By judging the five-tuple of the IP data message, an IP data message with the same data flow can be judged. In addition to detecting the five-tuple, next-generation firewalls also detect the user, application, and time period of the message.

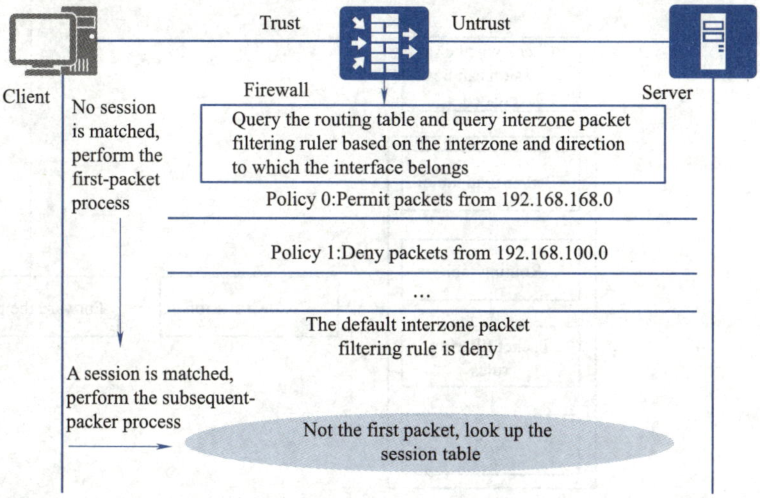

Figure 9-6 Firewall interzone forwarding

Among them, the data message of the TCP protocol generally calculates and checks other fields in addition to the five-tuple in the three way handshake phase. After the three way handshake is successfully established, the device performs matching detection on subsequent packets received by the device through the five-tuple in the session table to determine whether the packet is allowed to pass.

Subtask 9.2.4 Create and query session entries

The process of creating and querying sessions is shown in Figure 9-7. It can be seen that the detection process for a message with an existing session table is much shorter than that for a message without a session table. Generally, after detecting the first packet of a connection and establishing a session, most of the packets of the connection no longer need to be re-detected. This is the improvement of the "stateful inspection mechanism" of the stateful inspection firewall over the "packet-by-packet inspection mechanism" of the packet filtering firewall. This improvement enables the stateful inspection firewall to rapidly improve its detection and forwarding efficiency.

Subtask 9.2.5 Understand the state detection mechanism

The firewall uses the stateful inspection function to check the legitimacy of the link state of the packet, and discards the packet with an illegal link state. The stateful inspection function not only detects ordinary messages, but also detects inner-layer messages (messages after VPN messages are decapsulated).

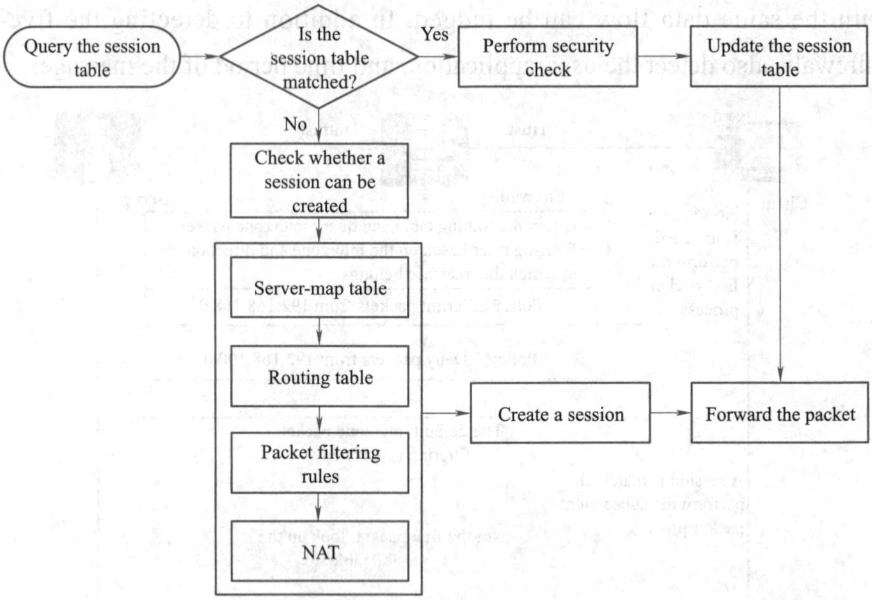

Figure 9-7 Flowchart for creating and querying session table

If the firewall serves as the only egress of the network, all packets must be forwarded through the firewall. In this case, the firewall processes the packets in both directions during one communication. This networking environment is also called the networking environment where the packets have the same inbound and outbound paths. In this case, you can enable the stateful inspection function on the firewall to ensure service security.

However, in a networking environment where the forward and return paths of packets are inconsistent, the firewall may receive only subsequent packets rather than the first packet. In this case, you need to disable the firewall stateful inspection function to ensure normal services. After the stateful inspection function is disabled, the firewall can establish sessions through subsequent packets to ensure normal service running.

When the stateful inspection mechanism is enabled, only the first packet that passes through the device can establish a session entry, and the subsequent packets match the session entry and are forwarded. For example, when the stateful inspection mechanism is enabled for TCP packets, the first packet (SYN packet) establishes a session entry. For packets except SYN packets, if no corresponding session entry exists (the device does not receive any SYN packet or the session entry is aged), the packets are discarded and the session entry is not established.

If the stateful inspection mechanism is disabled, subsequent packets can generate session entries even if the first packet does not pass through the device. For example, when the stateful inspection mechanism is disabled for TCP packets, session entries can be set up for packets of any format that do not have corresponding session tables but pass the checks of various security mechanisms.

Subtask 9.2.6 Understand conversation entries

Session is the basis of the firewall for stateful inspection. Each data flow passing through the firewall creates a session table on the firewall. The session table uses the five-tuple (source and destination IP address, source and destination port, and protocol number) as the Key value to improve the security of interzone forwarding data flows. The next generation firewall adds users on the five-tuple basis and expands the application field to a seven-tuple, as shown in Figure 9-8. The following seven elements are included:

(1) Source IP address.

(2) Source port.

(3) Destination IP address.

(4) Destination port.

(5) Protocol number.

(6) User.

(7) Application.

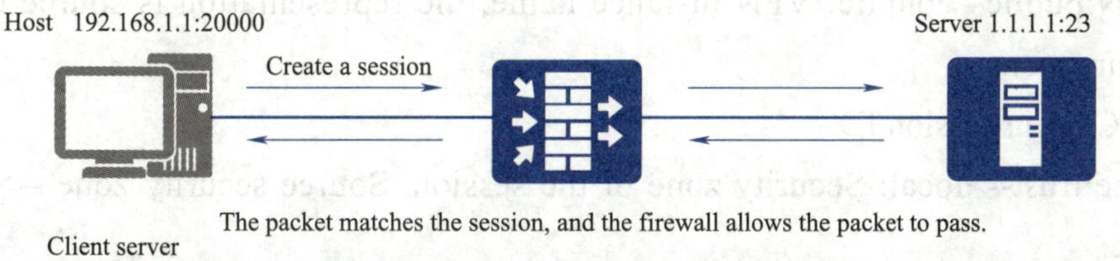

Client server

Source IP address	Source port	Destination IP address	Destination port	Protocol	User	Application
192.168.1.1	20000	1.1.1.1	23	TCP	abc	Telnet

Server client

Source IP address	Source port	Destination IP address	Destination port	Protocol	User	Application
1.1.1.1	23	192.168.1.1	20000	TCP	abc	Telnet

Session:TCP 192.168.1.1:20000 1.1.1.1:23

Figure 9-8 The session table

You can run the “display firewall session table” command on the Huawei firewall to view session information, as shown in Figure 9-9.

```
<sysname> display firewall session table
Current Total Sessions : 2
   telnet VPN:public --> public 192.168.3.1:2855-->192.168.3.2:23
   http VPN:public --> public 192.168.3.8:2559-->192.168.3.200:80
```

Figure 9-9 View the firewall session table

(1) Current Total Sessions: Statistics on the number of current session tables.

(2) telnet/http: The name of the protocol.

(3) VPN:public-->public: VPN instance name, the representation is source direction --> destination direction.

(4) 192.168.3.1:2855-->192.168.3.2:23: Session table information.

You can run the "display firewall session table verbose" command on the Huawei firewall to view detailed firewall session information, as shown in Figure 9-10.

```
<sysname> display firewall session table verbose
Current Total Sessions : 1
   http VPN:public --> public ID: a48f3648905d02c0553591da1
   Zone: trust--> local TTL: 00:20:00 Left: 00:19:56
   Output-interface: InLoopBack0 NextHop: 127.0.0.1 MAC: 00-00-00-
00-00-00
   <--packets:3073 bytes:3251431 -->packets:2881 bytes:705651
   128.18.196.4:1864-->128.18.196.251:80 PolicyName: test
```

Figure 9-10 Display firewall session table verbose

(1) Current Total Sessions: Statistics on the number of current session tables.

(2) http: The name of the protocol.

(3) VPN:public-->public: VPN instance name, the representation is source direction --> destination direction.

(4) ID: Current session ID.

(5) Zone:trust-->local: Security zone of the session. Source security zone --> Destination security zone.

(6) TTL: Total lifetime of the session table.

(7) Left: Remaining lifetime of the session table.

(8) Output-interface: Output-interface.

(9) NextHop: IP address of the next hop.

(10) MAC: Next-hop MAC address.

(11) <--packets:3073 bytes:3251431: Statistics on the number of incoming packets (including fragments) and bytes of the session.

(12) -->packets:2881 bytes:705651: The number of outbound session fragments (including fragments) and bytes.

(13) PolicyName: Name of the matched policy.

Task 9.3 Firewall security policy and its application

Subtask 9.3.1 Understand the matching principles of security policies

As shown in Figure 9-11, security policies are filtered for the first-packet flow, but not for subsequent-packet flows.

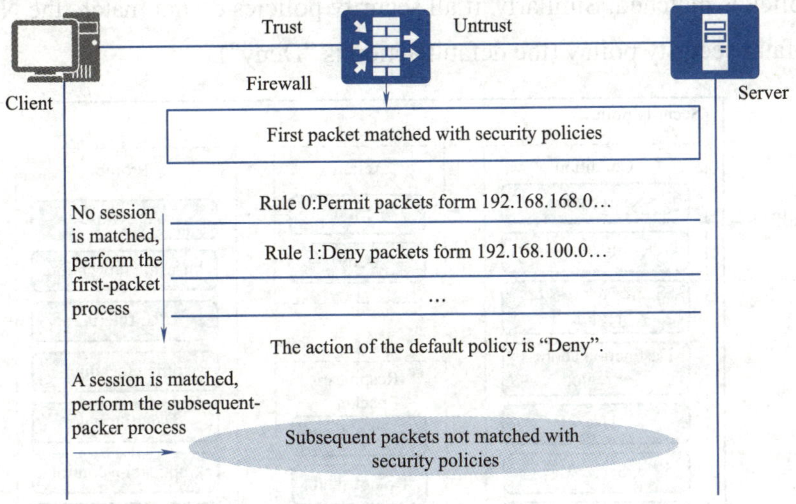

Figure 9-11 Matching rules of security policies

For the same data flow, you only need to configure a security policy for the direction in which the access is initiated. That is the first-packet matches the security policy, and the session table is established after filtering through the security policy. Subsequent-packet directly matches the session table without matching the security policy, improving service processing efficiency. This section uses a client PC accessing a Web server as an example to describe the filtering mechanism of security policies. After the traffic reaches the FW, the FW performs first-packet filtering and matches the security policy. The FW allows the packets to pass and establishes a session. The session contains information about the packets sent from the PC, such as the source and destination IP addresses, source and destination port numbers, and application protocols. The FW searches the session table when the Web server responds to a request, and compares the information in the response request packet with the session information in the session table. If the information in the packet matches the information in the session and complies with the definition of subsequent packets in the protocol specification, the packet is regarded as a follow-up response packet for the PC to access the Web server, and the packet is directly allowed to pass.

Subtask 9.3.2 Analysis of security policy matching process

Figure 9-12 shows the process for processing security policies when traffic passes the NGFW.

The NGFW detects the received traffic and identifies traffic attributes, including source security zone, destination security zone, source address/region, destination address/region, user, service (source port, destination port, and protocol type), application, and time range.

The NGFW matches traffic attributes with security policy conditions. If all conditions match, the traffic successfully matches the security policy. If one of the conditions does not match, the next security policy is matched. Similarly, if all security policies do not match, the NGFW takes the action of the default security policy (the default action is "Deny").

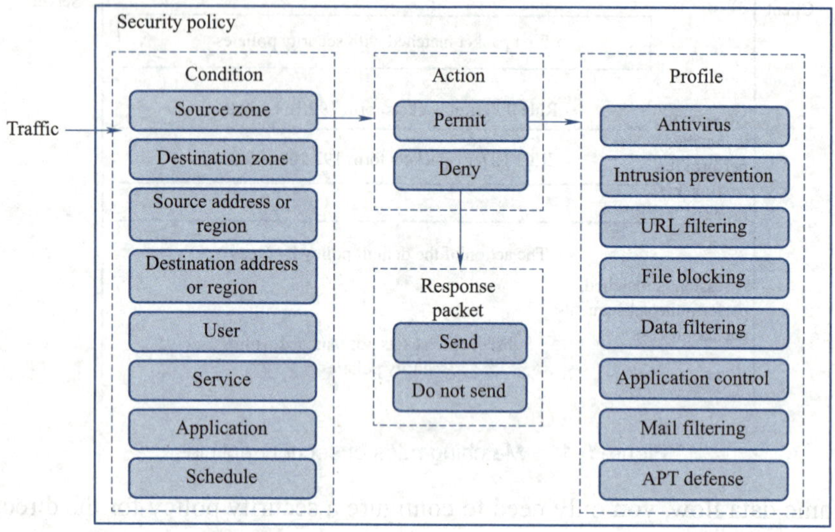

Figure 9-12 Security policy matching process

If the traffic matches a security policy, the NGFW executes the action of the security policy. If the action is "Deny", the NGFW blocks the traffic; If the action is "Permit", the NGFW checks whether the security policy references the security profile. If the security profile is referenced, perform Step four; If no security profile is referenced, the traffic is allowed to pass.

If the action of the security policy is "Permit" and the security profile is referenced, the NGFW implements integration content security inspection on traffic.

Integration inspection detects traffic based on the conditions of the security profile and performs actions in the security profile based on the inspection result. If one of the security profiles blocks the traffic, the NGFW blocks the traffic. If all security profiles allow this traffic to be forwarded, the NGFW allows this traffic to be forwarded.

Subtask 9.3.3 Application of security policy

As shown in Figure 9-13, an enterprise deploys a NGFW to improve network security, forbids the marketing department from accessing IM and Game applications, allows the research and development department to access Web services through http, and deploys security policies in the direction from the trust zone to the untrust zone.

Compared with traditional firewall security policies, the security policies of the next generation firewall have the following advantages:

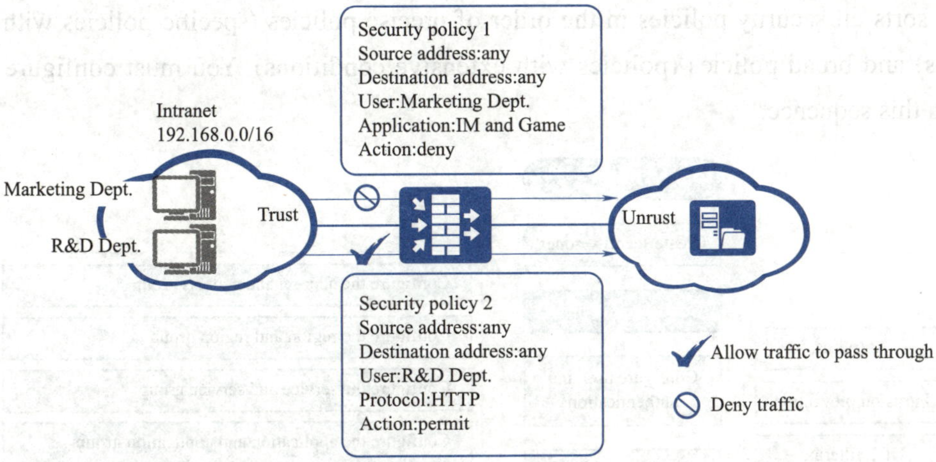

Figure 9-13 Security Policy Application

(1) The next generation firewall can distinguish employees in different departments by "users", and make network management more flexible and visual.

(2) The next generation firewall can effectively distinguish different applications (such as web IM and web games) carried by protocols (such as http), so that the network management is more refined.

(3) The next generation firewall can implement content security detection through security policies, block viruses, hackers and other intrusion, and better protect the internal network.

Subtask 9.3.4 Thinking of configuring security policy

Figure 9-14 shows the roadmap for configuring a security policy:

(1) The administrator should first clarify which security zones need to be divided into, how interfaces are connected, and which security zones are added to.

(2) The administrator selects "source address" or "user" to identify enterprise employees.

(3) The administrator determines the permissions of each user group before determining the permissions of specific users. The information includes the source security zone and address of the user, the destination security zone and address of the user, the services and applications that the user can use, and the period in which the user's network access permission takes effect. If the administrator wants to allow certain network access, set the action of the security policy to "Permit"; If the administrator wants to deny certain network access, set the action of the security policy to "Deny".

(4) The administrator determines which traffic passes the firewall for content security checking and which content security checking is performed.

(5) The administrator lists the parameters of the security policies planned in the preceding

steps and sorts all security policies in the order of precise policies (specific policies with detailed conditions) and broad policies (policies with extensive conditions). You must configure security policies in this sequence.

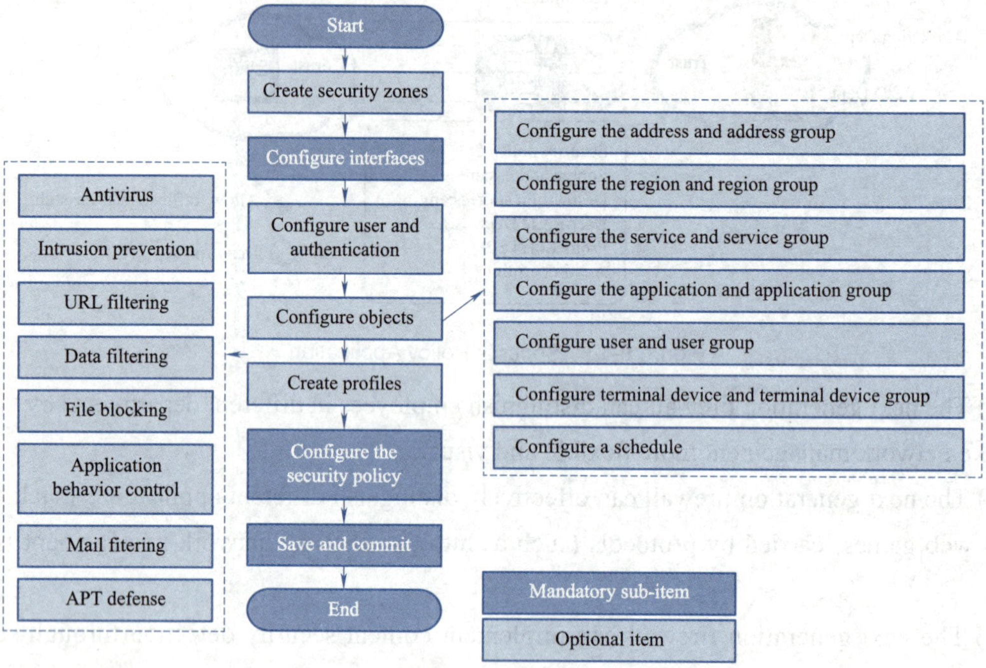

Figure 9-14 Process for configuring security policies

Subtask 9.3.5 Configure security policy in Web mode

Huawei VRP supports CLI mode and Web mode. Web mode is recommended in actual configuration.

1.Security policy introduction

Configuration of security policy is shown in Figure 9-15.

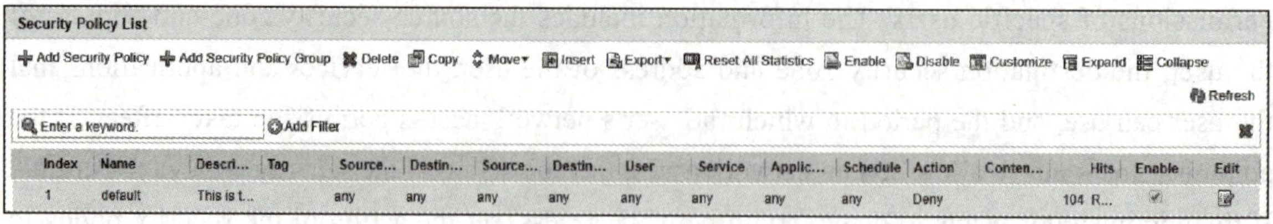

Figure 9-15 Security policy

(1) Matching conditions: Source security zone, Destination security zone, Source address, Destination address, User, Service, Application, and Time range.

(2) Policy action: Permit or Deny.

(3) Content Security Profile (Optional): Antivirus, Intrusion Prevention, URL filtering, File

filtering, Content filtering, Application behavior control, and Mail filtering.

The device can identify traffic attributes and match traffic attributes with security policy conditions. If all conditions match, the traffic successfully matches the security policy. After the traffic matches the security policy, the device takes the action of the security policy.

If the action is "Permit", content security check is performed on the traffic. If the content security check also passes, the traffic is allowed to pass. If the content security check fails, the traffic is denied.

If the action is "Deny", the traffic is prohibited.

2. Security Zone Configuration

By default, the system has created four security zones: Local, Trust, Untrust, and DMZ, as shown in Figure 9-16. If you need to assign more security levels, you can create new security zones and define their security levels. Procedure for creating a security zone:

(1) Choose "Network" → "Security Zone".

(2) Click "Add" to Configure security zone parameters.

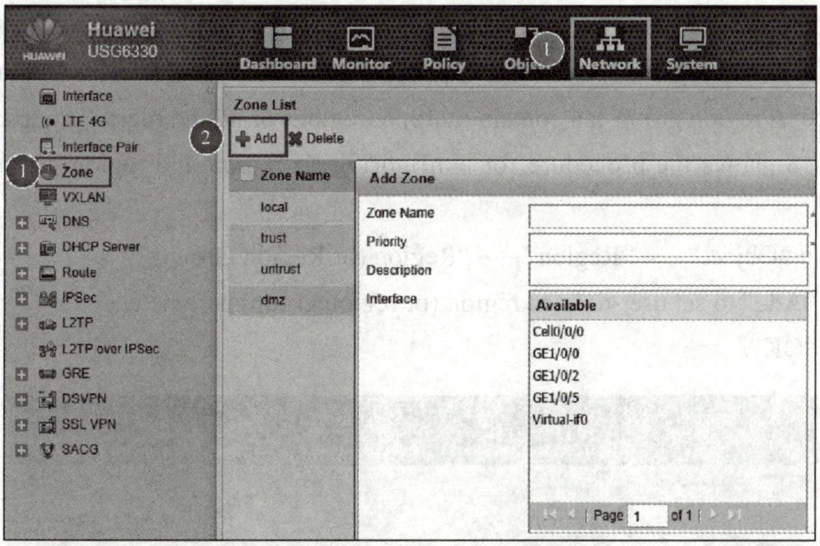

Figure 9-16 The security policy

3. Configure the address and address group

An address is a set of IPv4, IPv6 or MAC addresses, and an address group is a set of addresses. An address contains one or more IPv4/IPv6 addresses or MAC addresses. It is similar to a basic component. After being defined once, it can be referenced by various policies (such as security policies and NAT policies) for multiple times.

Figure 9-17 shows the procedure for configuring an ADDRESS and address group on the Web interface:

(1) Choose "Object" → "Address" → "Address (or Address Group) ".

(2) Click "Add" to set the parameters of the address or address group.

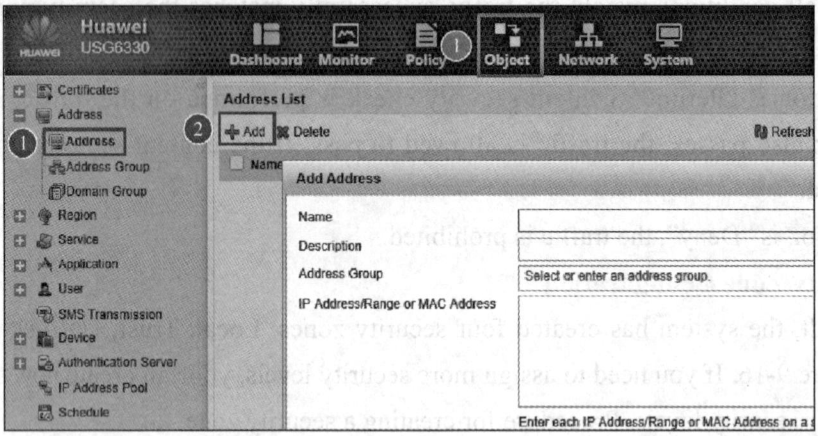

Figure 9-17 Address and address group

4. Configure regions and region groups

Regions are IP address objects in the unit of regions. Each region is a set of public IP addresses in the current region. To facilitate expansion and reuse, you can configure area groups for policy reference. A region group can contain multiple regions or nested region groups.

Figure 9-18 shows the procedure for configuring an region and region group on the Web interface:

(1) Choose "Object" → "Region" → "Region (or Region Group) ".

(2) Click "Add" to set user-defined region (or region group) parameters.

(3) Click "OK".

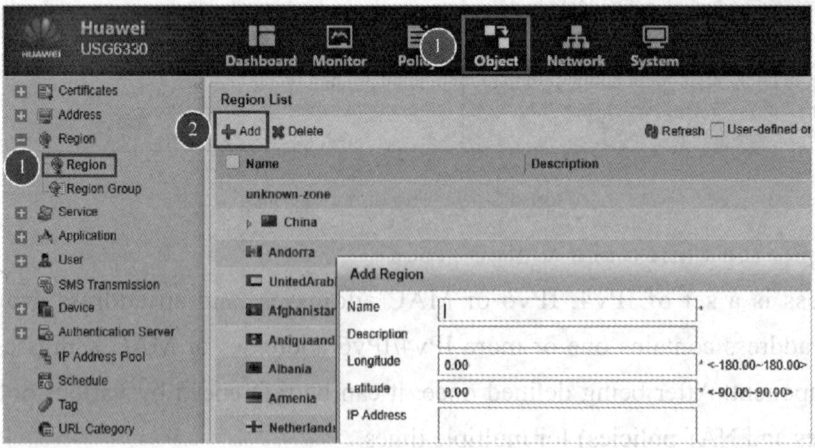

Figure 9-18 Districts and area groups

5. Configure services and service groups

A service is an application protocol type determined by protocol type and port number. A service group is a collection of services and service groups. Huawei NGFW provides predefined services and user-defined services.

Predefined service: Indicates a default service type that can be directly selected. Predefined services are usually well-known protocols, such as HTTP, FTP, and Telnet. Predefined services cannot be deleted.

User-defined service: Defines some application protocol types by specifying the protocol type (such as TCP, UDP, and ICMP) and port number.

Figure 9-19 shows the procedure for configuring services (similar to configuring a service group) on the Web interface.

(1) Choose "Object" → "Service" → "Service Group".

(2) Click "Add" to configure the parameters of the user-defined service.

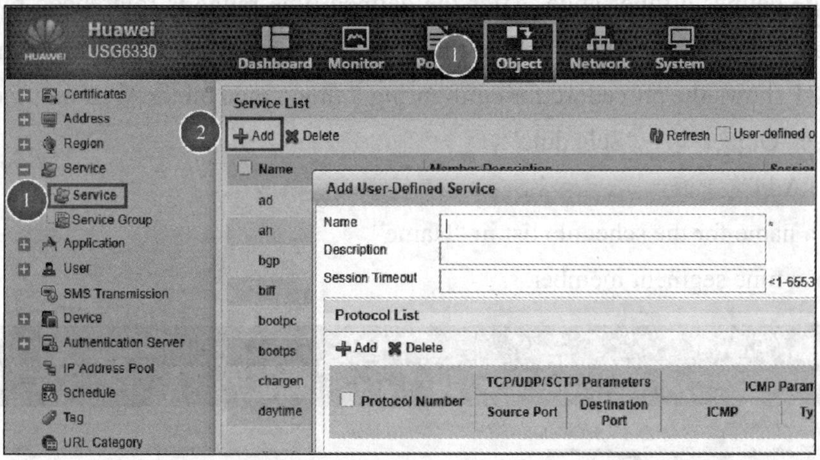

Figure 9-19 Services and service groups

6. Configure applications and application groups

An application is a computer program used to perform a particular task or purpose. An application group is a collection of multiple applications. An application is a computer program used to perform a particular task or purpose. An application group is a collection of multiple applications.

Figure 9-20 shows the procedure for configuring an application on the Web interface (similar to configuring an application group):

(1) Choose "Object" → "Application" → "Application".

(2) Click "Add" to set parameters for the user-defined application.

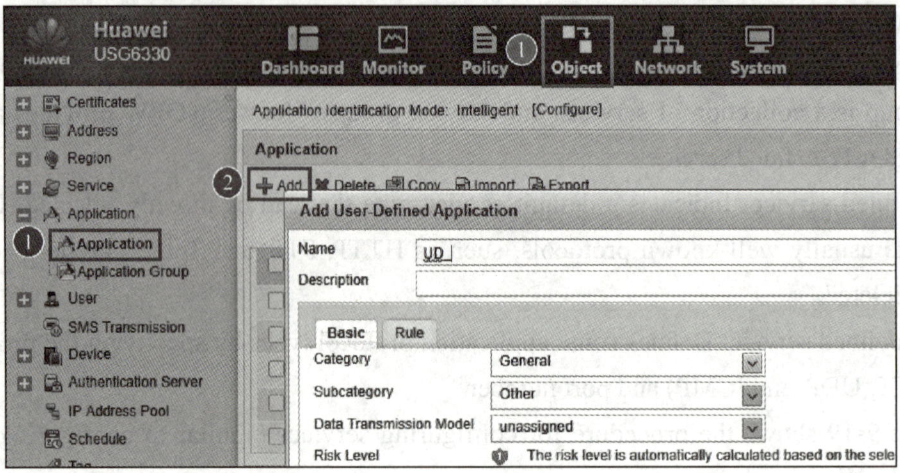

Figure 9-20 Application and application groups

7. Configuring a Time Range

Time range defines a time range. After the defined time range is referenced by a policy, the NGFW can match and control the traffic that flows through the NGFW within a certain time range.

Figure 9-21 shows the procedure for configuring a time range on the Web interface:

(1) Choose "Object" → "Schedule".

(2) Click "Add".

(3) Enter a name for the schedule list in "Name".

(4) Create a time segment member.

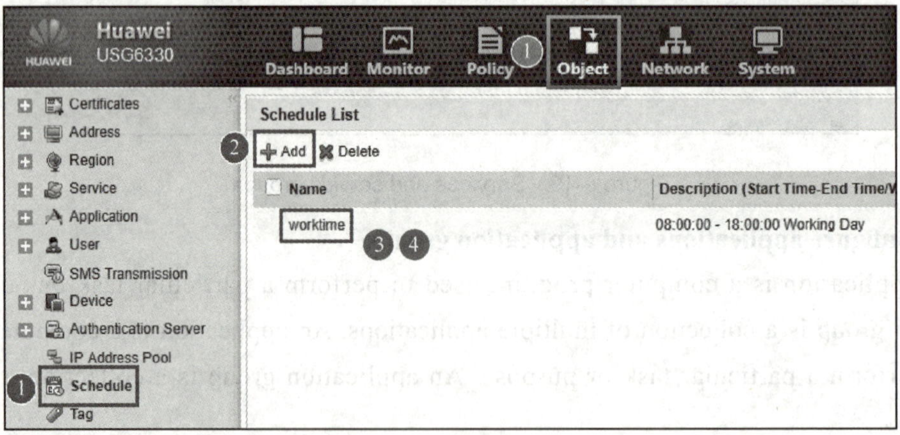

Figure 9-21 Configuring a Time Range

8. Configuring security Policies

Existing objects can be referenced in a security policy.

Figure 9-22 shows the procedure for configuring security policies on the Web interface:

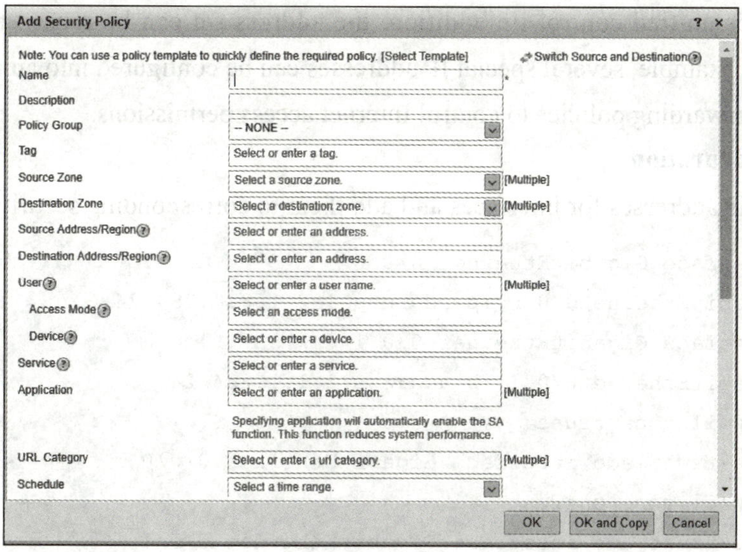

Figure 9-22 Configuring security policies

Task 9.4 Firewall security policy configuration case

As shown in Figure 9-23, employees in network segment 192.168.5.0/24 are allowed to access the Internet. However, PCs 192.168.5.2, 192.168.5.3, and 192.168.5.6 in this network segment have high security requirements and are not allowed to access the Internet.

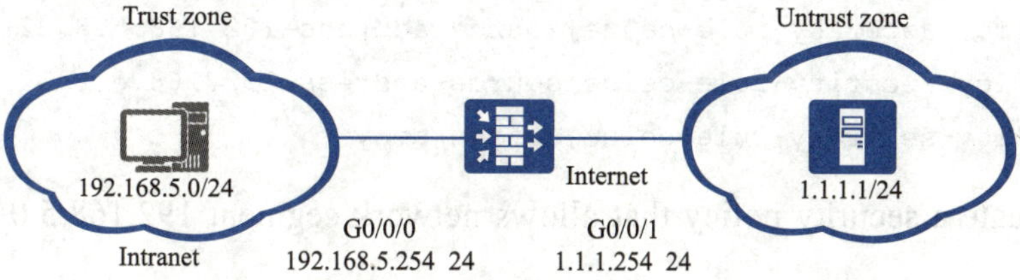

Figure 9-23 Example of configuring security policies

1. Configuration idea

Plan security policies. The requirement is to allow a wide range of IP addresses (192.168.5.0/24) to pass, and then deny several special IP addresses in this range. In this case, you need to configure two forwarding policies. First, configure a forwarding policy to deny special IP addresses, and then configure a forwarding policy to allow the entire network segment to pass. If the configuration is reversed, several special IP addresses will match a wide range of policies and pass the firewall, and will not continue to match the following special policies.

Address set planning. If the access permission is controlled by IP address, specify the IP address as the matching condition in the forwarding policy. You can directly configure continuous address segments in a policy. However, you are advised to configure scattered addresses as an address set for unified control. In addition, the address set can be reused by other policies. Therefore, in this example, several special IP addresses can be configured into an address set.

Configure forwarding policies to control Internet access permissions.

2. CLI configuration

Step 1: Set IP addresses for interfaces and add them to corresponding security zones.

```
[NGFW]interface GigabitEthernet 0/0/0
[NGFW-GigabitEthernet0/0/0]ip address 192.168.5.254 24
[NGFW]interface GigabitEthernet 0/0/1
[NGFW-GigabitEthernet0/0/1]ip address 1.1.1.254 24
[NGFW]firewall zone trust
[NGFW-zone-trust]add interface GigabitEthernet 0/0/0
[NGFW]firewall zone untrust
[NGFW-zone-trust]add interface GigabitEthernet 0/0/1
```

Step 2: Create security policy rules that deny Internet access from specific IP addresses.

```
[NGFW]security-policy
[NGFW-policy-security]rule name celue
[NGFW-policy-security-rule-celue]source-zone trust
[NGFW-policy-security-rule-celue]destination-zone untrust
[NGFW-policy-security-rule-celue]source-address 192.168.5.2 32
[NGFW-policy-security-rule-celue]source-address 192.168.5.3 32
[NGFW-policy-security-rule-celue]source-address 192.168.5.6 32
[NGFW-policy-security-rule-celue]action deny
```

Step 3: Create a security policy that allows network segment 192.168.5.0/24 to access the Internet.

```
[NGFW]security-policy
[NGFW-policy-security]rule name celue2
[NGFW-policy-security-rule-celue2]source-zone trust
[NGFW-policy-security-rule-celue2]destination-zone untrust
[NGFW-policy-security-rule-celue2]source-address 192.168.5.0 24
[NGFW-policy-security-rule-celue2]action permit
```

Step 4: Test whether different IP addresses can access the server through the firewall.

3. Web configuration

Step 1: Set IP addresses for interfaces and add them to corresponding security zones, as shown in Figure 9-24.

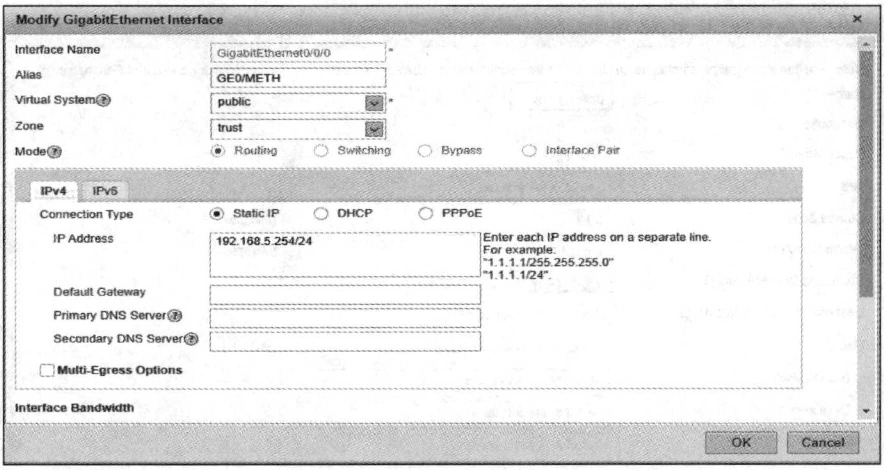

Figure 9-24 Interface configuration example

Select "Network" → "Interface" → "GE0/0/0", set the IP address of the interface, and add the interface to the trust zone.

Select "Network" → "Interface" → "GE0/0/1", set an IP address for the interface, and add the interface to the untrust zone.

Step 2: Configure an address group named ip_deny as shown in Figure 9-25. Add several IP addresses that cannot access the Internet to the address group.

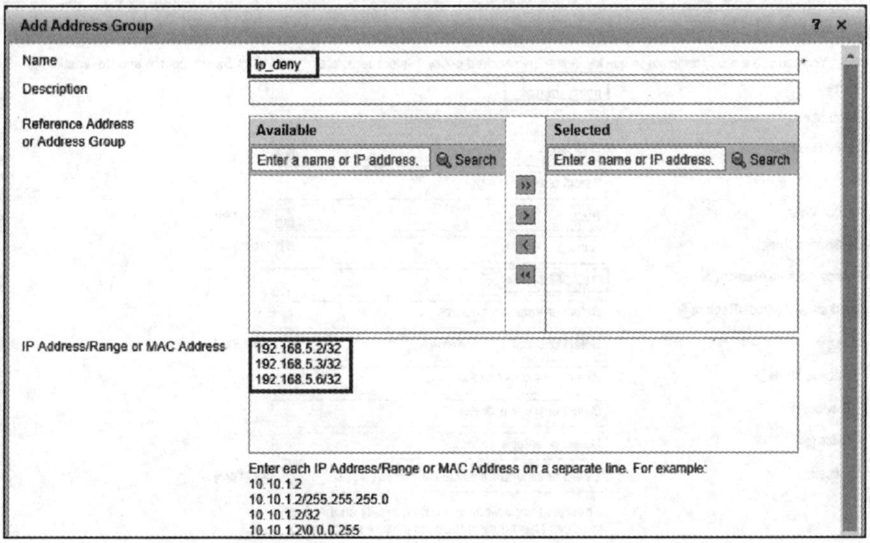

Figure 9-25 Configuring an address group

Step 3: Configure a security policy to deny the Internet access from IP addresses in special address group IP_deny. Select ip_deny and set the action to deny, as shown in Figure 9-26.

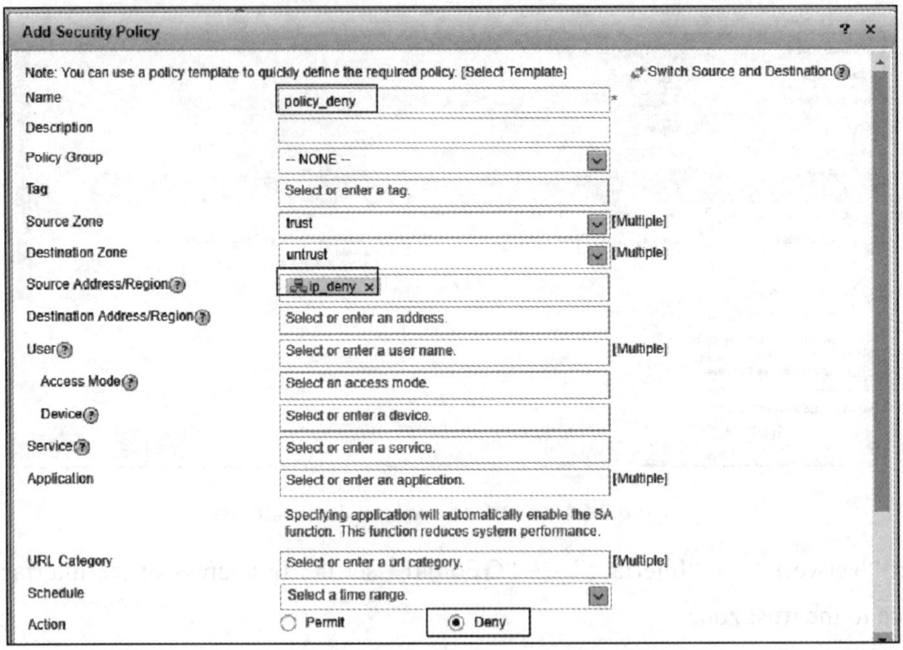

Figure 9-26 Configure a denial security policy

Step 4: Configure a security policy that allows network segment 192.168.5.0/24 to access the Internet, enter the address "192.168.5.0/24" and the action "Permit", as shown in Figure 9-27.

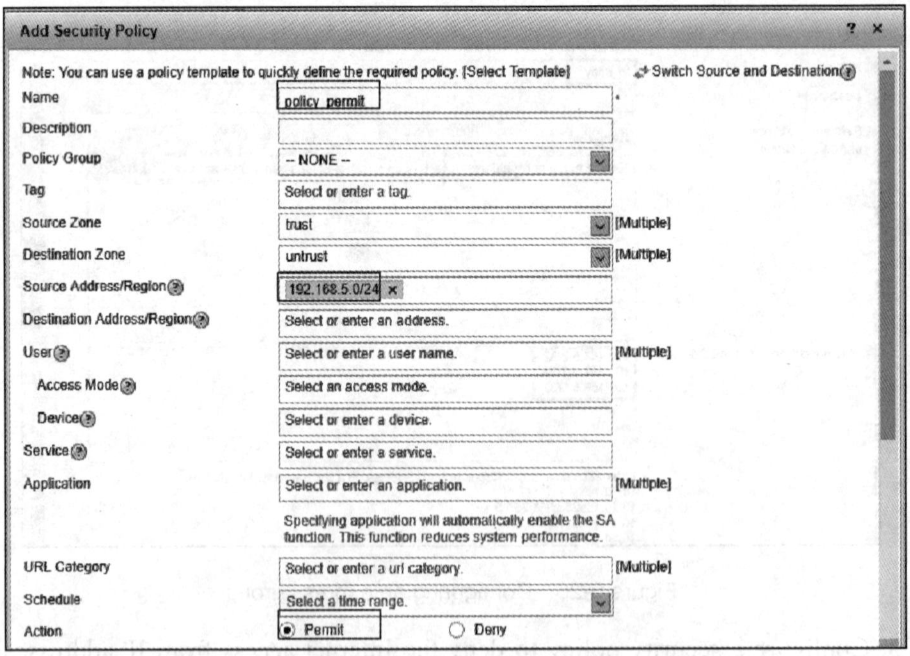

Figure 9-27 Configure permit security policies

Step 5: Test whether different IP addresses can access the server through the firewall.

Task 9.5 ASPF technology

Subtask 9.5.1 Understand multichannel protocols

In computer networks, application layer protocols are divided into single channel protocols and multichannel protocols according to the number of ports occupied during communication. Multichannel is designed to improve protocol performance.

Single channel protocol: A protocol that occupies only one port during communication. For example, WWW only takes up port 80.

Multichannel protocol: A protocol that occupies two or more ports during communication. A multichannel protocol consists of one control channel and several other control or data channels, it means the exchange of control information and the transmission of data are accomplished on different channels. For example, in FTP passive mode, the control information occupies port 21 and a random port.

Most multimedia application protocols (such as H.323 and SIP), FTP, netmeeting, and so on use fixed ports to initiate a control connection, and then dynamically select ports for data transmission. The choice of ports is unpredictable, and some applications may even use multiple ports at the same time. Traditional packet filtering firewalls can match the application transmission of single channel protocols by configuring ACL filtering rules to protect the internal network from attacks. However, the traditional packet filtering firewall can only block some applications that use fixed ports, and cannot match the multichannel protocol applications that use negotiated random ports to transmit data, and it leaves many security risks.

Subtask 9.5.2 Understand ASPF technology

Application specific packet filter (ASPF) is an advanced communication filter that checks application layer protocol information and monitors the state of application layer protocols. For all connections for a particular application protocol, each connection state information is maintained by ASPF and used to dynamically determine whether packets are allowed through the firewall or discarded.

ASPF maintains connection state information in the data structure of the session table and uses this information to maintain session access rules. ASPF holds important state information that cannot be saved by access control list rules. The firewall checks each packet in the data flow to ensure that the packet state and the packet comply with user-defined security rules. The connection state information is used for intelligent permit/deny messages. When a session terminates, the session entry is deleted and the session on the firewall is closed.

ASPF can intelligently detect "TCP three-way handshake information" and "Disconnection handshake information". By detecting the state of the handshake and disconnection, ASPF can ensure that a normal TCP access can be completed. However, it directly rejects incomplete TCP handshake packets.

UDP is a connectionless message, so there is no real UDP "connection". Because ASPF is connection-based, it checks the source IP address, destination IP address, and port of UDP packets. ASPF checks whether the packet is similar to other UDP packets within the specified period to determine whether a connection exists.

In common applications, the IP packet filtering technology based on ACL is generally used. This technology is relatively simple but lacks flexibility. In many miscellaneous applications, ordinary packet filtering cannot protect the network security. For example, it is very difficult to configure firewalls for multichannel protocols such as FTP.

ASPF enables firewalls to support protocols that the presence of multiple data connections on a control connection, the firewall can also be used in the very complex situation of the convenient formulation of various security policies. To implement effective access control for applications that use dynamic ports, ASPF listens on the port used by every connection in every application, and open appropriate channels to allow session data to pass through the firewall, and close the channel at the end of the session.

Subtask 9.5.3 ASPF's support for multichannel protocols

In multichannel protocols, such as FTP, the control channel and data channel are separate. Data channels are dynamically negotiated in control packets. To prevent the negotiated channels from being interrupted by other rules (such as ACL), it need to temporarily enable a channel. Server-map is a data structure designed for this application.

FTP contains a TCP control channel with a predictable port (default: 21) and a TCP data channel with dynamic negotiation. For common packet filtering firewalls, the port number of the data channel cannot be predicted during security policy configuration. Therefore, the entry of the data channel cannot be determined, and accurate security policies cannot be configured. ASPF technology solves this problem by detecting packet information at the application layer above the IP layer and dynamically creating and deleting temporary Server-map entries based on packet content to allow related packets to pass through.

As shown in Figure 9-28, Server-map entries are generated dynamically during dynamic detection of the FTP control channel. When a packet passes the firewall, ASPF compares the packet with the specified access rule. If the rule allows the packet, the packet is checked. Otherwise, the packet is discarded. If the packet is used to open a new control or data connection, ASPF dynamically generates Server-map entries and allows the returned packet to pass through the firewall only if it belongs to an existing valid connection. The state table also needs to be updated when processing returned packets. When a connection is closed or times out, its state table is deleted to ensure that unauthorized packets cannot pass through the firewall. Therefore, ASPF technology can ensure network security in the case of complex applications.

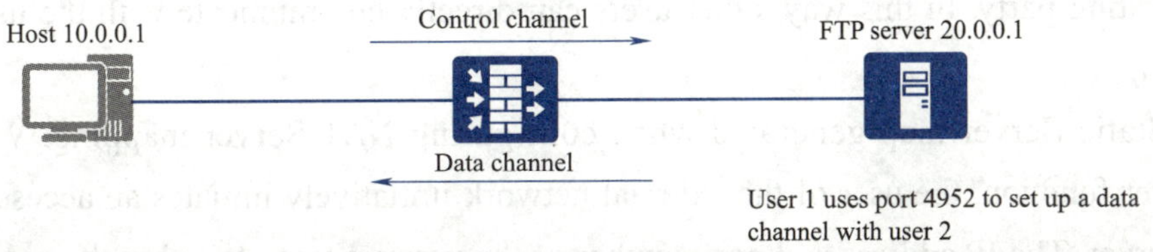


```
Session table: FTP:10.0.0.1:4927 --> 20.0.0.1:21

                   FTP:10.0.0.1:4926 --> 20.0.0.1:4952
```

```
Server-map table
-----------------------------------------------------------
Inside-Address :Port  Global-Address :Port  Pro   AppType  TTL  Left
-----------------------------------------------------------     20.0.0.1      :  4952
---
tcp    FTP   DATA      00:01:00   00:00:47
```

Figure 9-28 ASPF support for multichannel protocols

Server-map is a mapping relationship. When a data connection matches a dynamic Server-map entry, it needn't to search for a packet filtering policy, and ensures normal forwarding for certain applications. In another case, when the data connection matches the Server-map table, the IP address and port in the packet are translated, which is described later.

The Server-map checks only the first packet. After the channel is established, the packet is forwarded according to the session table.

Subtask 9.5.4 Generation process analysis of Server-map

At present, there are several situations in which Server-map entries are generated on FW:

(1) Server-map entries generated when forwarding multichannel protocols such as FTP and RTSP after configuring ASPF. Multichannel protocol dynamically negotiates the data channel by the control channel between the client and the server, that is, the port numbers of both parties are not fixed. After the ASPF function is configured, the device detects the negotiation of the control channel, and dynamically creates a server-map entry according to the address information in the key message payload, which is used for searching when the data channel initiates the connection. This server-map entry contains the information of the negotiated data channel in the multichannel protocol message.

(2) Triple Server-map entries generated when forwarding STUN protocols such as QQ/MSN and TFTP after configuring ASPF. In QQ/MSN and other protocols, when a user logs in, the IP address and port of the user are fixed, but the IP address and port number of the other party who will initiate a conversation with the user are not fixed. By configuring STUN-type ASPF, when

users such as QQ or MSN connect to the server, the device will record the IP address and port information of the users, and dynamically generate STUN-type Server-map. This server-map entry only contains triple information, i.e. IP address, port number and protocol number of the communicating party. In this way, other users can directly communicate with the user through the IP and port.

(3) Static Server-map generated when configuring NAT Server mapping. When using the NAT Server function, the user of the external network initiatively initiates an access request to the internal server. The IP address and port number of the user are uncertain, but the only ones that can be determined are the IP address of the internal server and the port number of the service provided. Therefore, after successfully configuring NAT Server, the device will automatically generate a Server-map table entry, which is used to store the mapping relationship between the Global address and the Inside address. According to this mapping relationship, the device converts and forwards the address of the message. Every effective NAT Server will generate two static Server-map in the forward and reverse directions. In SLB function, because multiple servers in the intranet need to be published to the outside with the same IP address, a Server-map entry similar to NAT Server will also be established, except that one forward entry and n reverse entries need to be established according to the number of intranet servers.

(4) A dynamic Server-map generated when configuring NAT No-PAT. When using NAT function, if the No-PAT parameter is configured, the device will map the IP of the internal network and the IP of the public network one-to-one, without port conversion. At this time, all the port numbers of the intranet IP can be mapped to the corresponding ports of the public network address, and the external network users can initiate connection to any port of the intranet users actively. Therefore, after configuring NAT No-PAT, the device will establish a Server-map table for the data flow with actual traffic, which is used to store the mapping relationship between private IP address and public IP address. According to this mapping relationship, the device converts the address of the message and forwards it.

(5) A dynamic Server-map generated when configuring NAT Full-cone.

(6) A dynamic Server-map generated when configuring PCP(port control protocol).

(7) Static Server-map generated when configuring server load balancing.

(8) A dynamic Server-map generated when configuring NAT Server in DS-Lite scenario.

(9) Static Server-map generated when configuring static NAT64.

Subtask 9.5.5 Port identification support for multichannel protocols

Port identification, also called port mapping, is used by the firewall to identify application-layer protocol packets that use non-standard ports. Application layer protocols supported by port mapping include FTP, HTTP, RTSP, PPTP, MGCP, MMS, SMTP, H323, SIP, and SQLNET. The port mapping function takes effect only for data flows in security zones. Therefore, you must configure security zones and interzones when configuring port mapping.

Port identification is based on ACL. Only packets matching a certain ACL will implement port mapping. Port mapping uses the basic ACL (numbers 2000 to 2999). The destination IP address of packets is used to match the source IP address configured in the basic ACL when the port mapping uses the ACL to filter packets.

As shown in Figure 9-29, the default control channel port number of the FTP server is changed from 21 to 31. In this case, the destination port number of the TCP packet that sent by Host 10.0.0.1 is 31 when Host 10.0.0.1 accesses the FTP server. By default, the intermediate firewall considers only TCP packets whose destination port number is 21 as FTP packets and ASPF processing is performed on these packets. It lead to the data channel fails to be established when the Host attempts to access the FTP server. In this case, you can configure port identification on the firewall. First, defining the IP address which the basic ACL match the FTP server, and then the port identification function is configured to identify TCP packets whose destination IP address matches the ACL and destination port is 31 as FTP packets.

- Port identification maps non-standard protocol ports to identifiable application protocol ports.

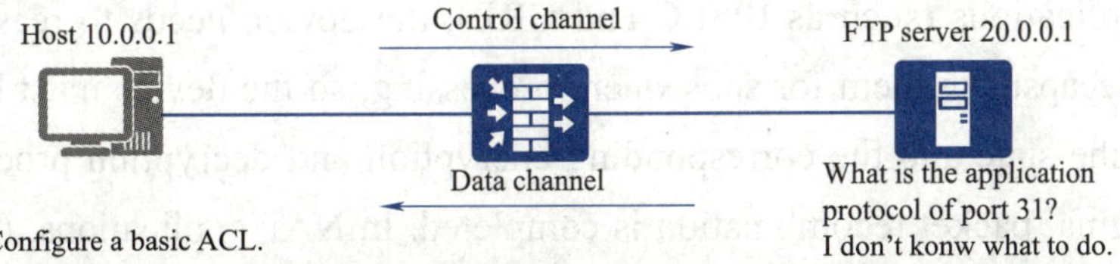

- Configure a basic ACL.

```
[NGFW]acl 2000
[NGFW-acl-basic-2000]rule permit source 20.0.0.1 0
```

- Configure port identification (or port mapping.)

```
[NGFW]port-mapping FTP port 31 acl 2000
```

Figure 9-29 Port identification support for multichannel protocols

Subtask 9.5.6 Understand fragmented caching

The fragment cache function is used to cache subsequent fragments that arrive before the first fragment, preventing fragments from being discarded by the firewall.

If the maximum transfer unit (MTU) configured on the device is smaller than the packet length when the network devices transmit packets, the network device fragments the packet and sends the packet again. As shown in Figure 9-30.

Ideally, fragments are transmitted on the network in a fixed sequence. In the actual transmission process, there may be a phenomenon that the first fragmented packet is not the first to reach the firewall. In this case, the firewall discards the fragmented packets. To ensure normal session operation, the firewall supports fragment caching by default. The device caches non-initial fragment packets to the fragment hash table. After the first fragment arrives and a session is established, all fragments are forwarded. If the first fragment does not arrive within the specified period, the firewall discards the fragment in the fragment cache.

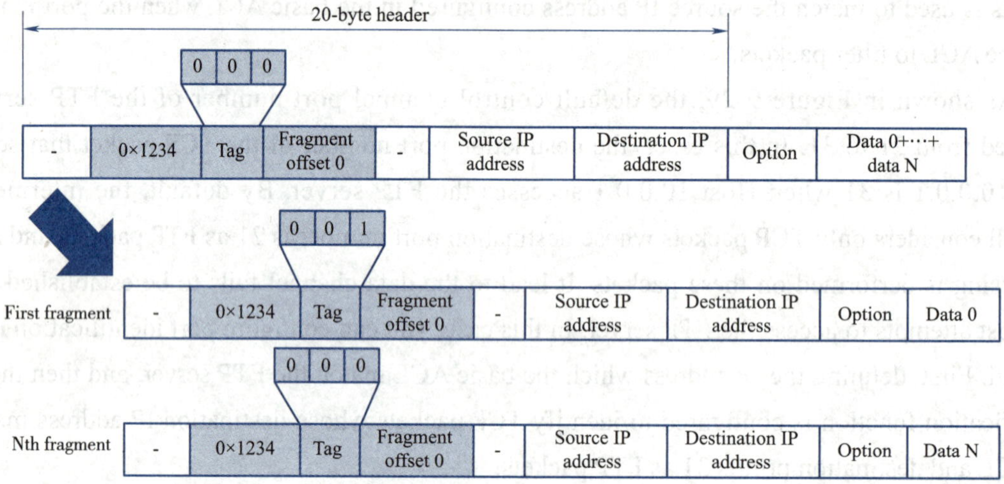

Figure 9-30 IP fragmentation

In VPN applications (such as IPSEC and GRE), the device needs to reassemble fragments and decrypt or decapsulate them for subsequent processing, so the device must be configured with the fragment cache state and the corresponding encryption and decryption process can be carried out after the original packet recombination is completed. In NAT applications, the device needs to reassemble fragment packets to properly parse and translate IP addresses in the packets. Therefore, the device must be configured with the fragment cache state to properly implement NAT.

The direct forwarding of fragment packets is generally used when NAT is not implemented. After this function is enabled, the firewall directly forwards received fragmented packets without creating a session table.

(1) Configuration fragment cache aging time:

```
Firewall session aging-time fragment interval (1-40000)
```

(2) Enable/Disable firewall fragment-forward:

```
Firewall fragment-forward enable/disable
```

Subtask 9.5.7 Understand the background and configuration of long connections

When the time interval between two consecutive messages of a TCP session arriving at the firewall is longer than the aging time of the session, the firewall will delete the corresponding session information from the session table to ensure the security of the network. In this way, after the subsequent message arrives at the firewall, the firewall will discard the message, resulting in connection interruption. In the actual network environment, the session information of some special service data streams needs to be kept from aging for a long time. In order to solve this problem, the firewall supports the function of configuring long connections between security domains, defines data flow rules by referencing ACL, and sets super-long aging time for sessions of specific messages matching ACL rules, so as to ensure normal sessions. By default, the aging time for long connections is 168 hours (7×24 hours). The firewall only supports the function of configuring long connections between domains for TCP protocol messages.

When the state detection mechanism is turned off, the session table can also be established for the non-first packet, so the normal operation of the service can be maintained without using the long connection function.

(1) Configure long connection aging time:

```
Firewall long-link aging-time time
```

(2) Open the long connection function:

```
Firewall interzone zone-name1  zone-name2
lonk-link acl-number { inbound | outbound }
```

Project 10 Network address translation technology

NAT is an address translation technology that translates the addresses in the IPv4 packet header. In most cases, the NAT technology is used to translate the private address in the IPv4 packet header into a public IP address, enabling the communication between the intranet and the public network.

In order to solve the problem of IPv4 public address shortage caused by the expansion of Internet, the temporary solution NAT technology came into being.

Task 10.1 Analysis of network address translation principle

Subtask 10.1.1 Understand the background of NAT

Back in the early 1990s, RFC documentation raised the possibility of IP address exhaustion. Web applications based on TCP/IP rapidly expand the Internet, and the number of IPv4 address applications is increasing. The sustainable development of the Internet is becoming more and more serious. IPv6 is proposed to fundamentally solve the problem of insufficient IPv4 addresses. The

IPv6 address set extends the address number from IPv4's 32 bits to 128 bits. For web applications, the address space is almost infinite. Therefore, IPv6 technology can fundamentally solve the problem of address shortage. However, IPv6 is faced with some sharp problems such as immature technology and huge cost of update, it still has a long way to go before it can replace the mature and widely used IPv4 network.

Since there is no immediate transition to IPv6 networks, some technical means must be used to extend the life of IPv4. The widely used techniques include classless inter-domain routing (CIDR), variable length subnet mask (VLSM), and network address translation (NAT).

The purpose of private network addresses is to realize address reuse and improve the utilization of IP address resources. In order to meet the requirements of private networks which are independent from the Internet for some laboratories, companies, or other organizations, RFC (requests for comment) 1918 reserved three IP address segments For private use. Details are as follows:

(1) Class A IP address: 10.0.0.0-10.255.255.255(10.0.0.0/8).

(2) Class B IP address: 172.16.0.0-172.31.255.255(172.16.0.0/12).

(3) Class C IP address: 192.168.0.0-192.168.255.255(192.168.0.0/16).

The addresses in the above three ranges cannot be allocated on the Internet, so they can be used freely without applying. The internal network uses a private IP address, and the external network uses a public IP address. If NAT is not used to translate the private IP address into a public IP address, it can mess up communications. This is because private routes cannot be configured on the public network (you do not need to apply for private network addresses. For example, the 10.0.0.0/8 network segment may be used by company A, but also by other companies, and these private routes cannot be configured on the public network). After the traffic from the private network to the public network reaches the destination host, the destination host or the intermediate public network equipment will eventually discard the message because there is no private network route, that is, it can send out messages but cannot receive the returned information. Therefore, when using a private network address to communicate with an external network, NAT technology must be used for address translation to ensure normal communication.

Subtask 10.1.2 Analysis of the basic principle of NAT technology

NAT is the process of converting the IP address in the IP data message header into another IP address. It is mainly used to realize the function of internal network (private IP address) to access external network (public IP address). In terms of implementation, the general NAT translation device (network device that implements the NAT function) maintains an address translation table, and all packets that pass the NAT translation device and need address translation will be modified accordingly through this table. The mechanism of address translation is divided into the following two parts:

(1) The IP address and port of the internal network host are converted to the external network address and port of the NAT translation device.

(2) The external network address and port are converted into the IP address and port of the internal network host of the NAT translation device.

That is, the conversion between <private address + port> and <public address + port>.

The NAT translation device is located at the junction of the internal network and the external network. All interactive messages between the internal PC and the external server pass through the NAT translation device. Common NAT translation devices include routers and firewalls.

As shown in Figure 10-1, the private network PC accesses the public network FTP server, the source IP of the constructed IP packet is 10.1.1.1, and the destination IP is 123.3.2.3. Before using the NAT technology, the message will be accurately forwarded to the destination host at 123.3.2.3 (whether it is a private or public network, there will be a public network address route), and the source IP of the IP message that the destination host responds to is 123.3.2.3, the destination IP is 10.1.1.1. Since the public network cannot have private network routes, the IP packets responded by the host will be discarded by the public network equipment. So how to solve this problem of no return? The public network cannot have private network routing, this is a fact that cannot be changed. If you want the response message to be forwarded to the private network egress normally in the public network. It is necessary that the destination IP of the response message is a public network and can be routed to the private network egress IP address. Then it is necessary that the source IP of the IP packet sent by the private network egress device is a public network and can be routed to the IP address of the private network egress.

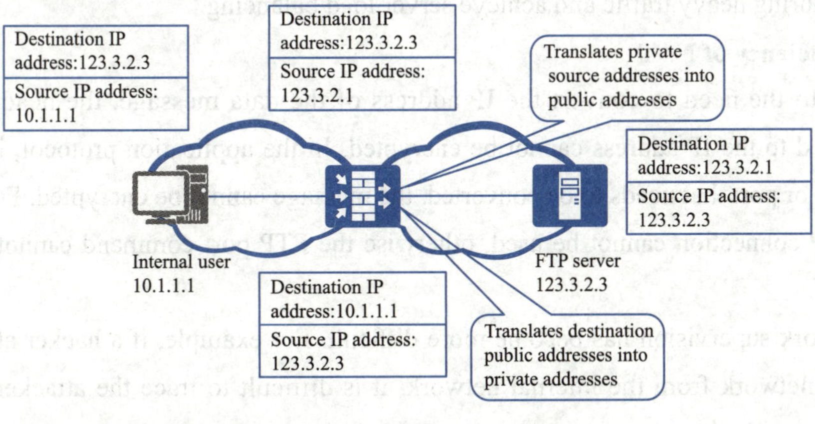

Figure 10-1 NAT technology principle

This requires NAT technology to achieve this. Before the NAT device forwards the IP packets from the private network to the public network, it converts the source IP address 10.1.1.1 in the packet to 123.3.2.1, which is the IP received by the FTP server at this time. The source IP of the message is 123.3.2.1, the destination IP is 123.3.2.3, the source IP of the IP message that the FTP server responds to is 123.3.2.3, the destination IP is 123.3.2.1, and the NAT device receives the destination IP of the message as 123.3.2.1 is converted to 10.1.1.1, and then forwarded to the private network PC according to the routing table, so that the private network can access the public network.

Subtask 10.1.3 Distinguish the advantages and disadvantages of NAT

In addition to the advantages of NAT technology that can realize address reuse and save valuable IP address resources, there are other advantages. The development of NAT technology has also continuously absorbed advanced concepts. In general, the advantages and disadvantages of NAT are as follows:

1. Advantages of NAT

(1) Multiple hosts in a LAN can use a few legal addresses to access external resources, and internal WWW, FTP, Telnet and other services can be set for external networks to use, which solves the problem of increasing shortage of IP addresses.

(2) For internal and external network users, the process of IP address conversion is not felt, and the whole process is transparent to users.

(3) Provide privacy protection for intranet users. Extranet users cannot directly obtain information such as IP addresses and services of intranet users, which has a certain degree of security.

(4) By configuring multiple identical internal servers, it is possible to reduce the pressure on a single server during heavy traffic and achieve server load balancing.

2. Insufficiency of NAT

(1) Due to the need to convert the IP address of the data message, the header of the data message related to the IP address cannot be encrypted. In the application protocol, if the message has an address or port that needs to be converted, the message cannot be encrypted. For example, an encrypted FTP connection cannot be used, otherwise the FTP port command cannot be converted correctly.

(2) Network supervision has become more difficult. For example, if a hacker attacks a server on the public network from the internal network, it is difficult to trace the attacker. Because the address is converted when the message passes through the NAT translation device, it is impossible to determine which one is the hacker's host.

Subtask 10.1.4 Master NAT classification

According to different application scenarios, NAT can be divided into the following categories:

1. Source NAT

Used to enable multiple private network users to access the Internet at the same time.

(1) Address pool method: Use public network addresses in the address pool to perform address translation for private network users, which is suitable for scenarios where a large number of private network users access the Internet.

(2) Outgoing interface address mode (Easy IP): The internal network host directly borrows the IP address of the public network interface to access the Internet, which is especially suitable for the situation where the IP address of the public network interface is obtained dynamically.

(3) Server mapping: Used to enable external network users to access private network servers.

2. Static mapping (NAT Server)

Public network address and private network address are mapped one-to-one, which is used when public network users access the internal server of the private network.

Task 10.2 Principle analysis of source NAT technology

Subtask 10.2.1 Implementation of source NAT without port translation

NAT based on the source IP address refers to the translation of the source address in the IP packet header that initiates the connection. It can achieve the purpose of internal users accessing external networks. By converting the private addresses of internal hosts into public addresses, multiple hosts in a LAN can use a few legal addresses to access external resources, which effectively hides the IP addresses of internal LAN hosts and plays the role of security protection.

The address pool mode without port translation is realized by configuring the NAT address pool, which can contain multiple public network addresses. Only the address is converted during the conversion, and the port is not converted, so that the one-to-one conversion from the private network address to the public network address is realized. If all the addresses in the address pool have been allocated, the remaining intranet hosts will not perform NAT translation when accessing the external network, and will not perform NAT translation until there are free addresses in the address pool.

As shown in Figure 10-2, the NAT address pool defines three public network addresses. When the first three private network users access the public network, they can all provide the public network address for address translation. When the fourth private network user accesses the public network, there is no public network address in the NAT address pool. At this time, address translation cannot be performed. Until one of the first three users does not need to access the public network, the public network address is released.

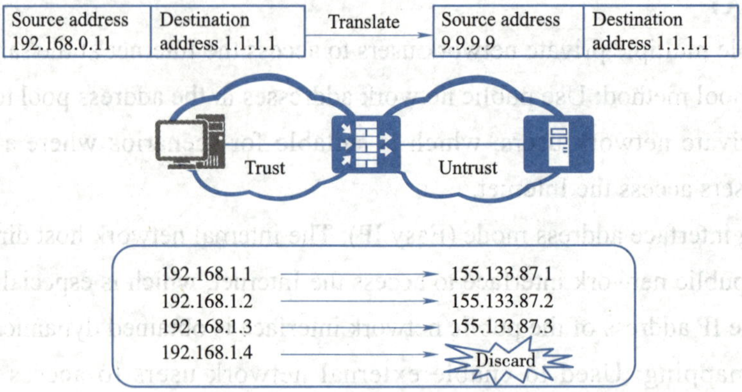

Figure 10-2 Source NAT without port translation

Subtask 10.2.2 Implementation of source NAT with port translation

The address pool mode with port translation is realized by configuring the NAT address pool, which can contain one or more public network addresses. With port conversion, the address and port can be converted at the same time, the requirement that multiple private network addresses share one or more public network addresses can be realized.

In this mode, since the address conversion is also performed at the same time as the port conversion, multiple private network users can use the same public IP address to surf the Internet. The firewall distinguishes different users according to the port, so it can support more users online at the same time. This is a technology that uses the fourth-level information to extend the third-level address. An IP address has 65 535 ports that can be used. In theory, one address can provide NAT translation for other 65 535 addresses, and the firewall can also map data packets from different internal addresses to different port numbers of the same public address, so it can still share the same address. Compared with one-to-one or many-to-many address translation, this greatly improves the address space and increases the utilization of IP addresses. Therefore, port conversion is the most commonly used address conversion method.

In most application layer protocols in the network, the destination port number is fixed, while the source port number is generated in a random manner. In this way, for the destination device, the source port number is irrelevant. At the same time, when uniquely identifying a piece of traffic, you need to use the "five-tuple" to achieve. The "five-tuple" refers to the source and destination IP, protocol number, and source and destination port number. So this provides the possibility for NAT with port translation. As shown in Figure 10-3, when a private network host accesses an application on the public network 1.1.1.1, such as HTTP, the source IP of the constructed message is 192.168.0.11, the source port number is randomly generated as X, and the destination port number is 80, the destination IP is 1.1.1.1. When the message is forwarded by the NAT device, the source IP of the message is converted to 2.2.2.2, and the source port number is converted to Y, and this information is recorded in the form of a NAT session. After the message is forwarded to the destination host, the destination host receives and responds. The destination IP of the response message is 2.2.2.2, the destination port number is Y, the source IP is 1.1.1.1, and the source port number is 80. Such IP packets will be forwarded on the public network to the NAT device at the egress of the private network. After receiving the message, the NAT device queries the NAT session, and finally converts the destination IP of the message to 192.168.0.11 and the destination port number to X, and queries the routing table and forwards it to the private network host. In this way, NAT with port translation is realized. In this way, only a small amount of public network IP can be provided for a large number of private network users to provide access to the external network.

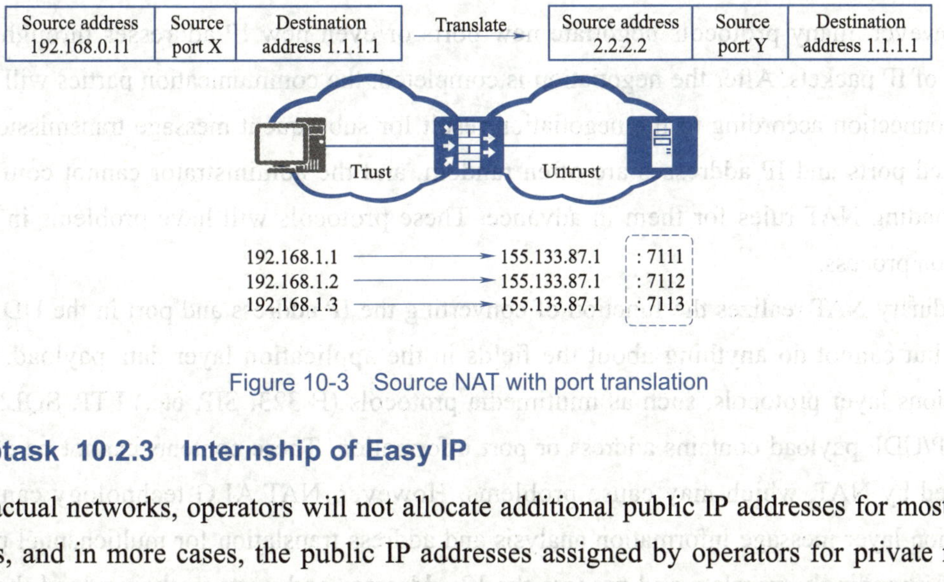

Figure 10-3 Source NAT with port translation

Subtask 10.2.3 Internship of Easy IP

In actual networks, operators will not allocate additional public IP addresses for most private networks, and in more cases, the public IP addresses assigned by operators for private network egress are not fixed. Therefore, in this scenario, you cannot manually configure the NAT address pool. In this case, you can directly use the outgoing interface address as the NAT address pool.

The outgoing interface address method is also called Easy IP, that is, the public network address of the interface is directly used as the translated address, and there is no need to configure a NAT address pool. The address and port are converted at the same time during conversion, it can realize the requirement that multiple private network addresses share the public network address of the external network interface. It can be seen that Easy IP is only a special form of NAT with port translation, and its working process is the same.

Subtask 10.2.4 Understand NAT ALG

NAT ALG (application level gateway) is a conversion agent of a specific application protocol, which can complete the conversion of the address and port number information carried in the application layer data.

As shown in Figure 10-4, in the data message structure, the IP header contains a 32-bit source IP address and a 32-bit destination IP address, and the TCP header contains a 16-bit source port number and a 16-bit destination port number.

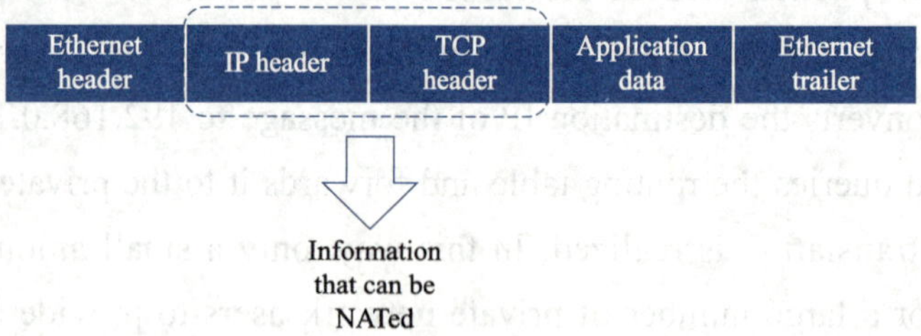

Figure 10-4 Data message structure

However, many protocols negotiate new ports or even new IP addresses through the data payload of IP packets. After the negotiation is completed, the communication parties will establish a new connection according to the negotiation result for subsequent message transmission. These negotiated ports and IP addresses are often random, and the administrator cannot configure the corresponding NAT rules for them in advance. These protocols will have problems in the NAT translation process.

Ordinary NAT realizes the function of converting the IP address and port in the UDP or TCP header, but cannot do anything about the fields in the application layer data payload. In many applications layer protocols, such as multimedia protocols (H.323, SIP, etc.) FTP, SQLNET, etc. The TCP/UDP payload contains address or port information. These contents cannot be effectively translated by NAT, which may cause problems. However, NAT ALG technology can perform application layer message information analysis and address translation for multichannel protocols, and correspondingly translate and process the IP addresses and ports in the payload that require address translation or fields that require special processing. So as to ensure the correctness of application layer communication.

For example, the FTP application is completed by the data connection and the control connection, and the establishment of the data connection is dynamically determined by the load field information in the control connection. This requires ALG to complete the conversion of the payload field information to ensure the correct establishment of subsequent data connections.

In order to realize the forwarding strategy of the application layer protocol, the ASPF function is proposed. The main purpose of the ASPF function is to open the corresponding packet filtering rules for itself through the analysis of the application layer protocol packets. The main purpose of NAT ALG is to open corresponding NAT rules for it. Since the two are usually used in combination, you can turn them on at the same time by using the same command.

Subtask 10.2.5 Implementation process analysis of NAT ALG

The host on the private network side in Figure 10-5 needs to access the FTP server on the public network. The mapping from the private network address 192.168.1.2 to the public network address 8.8.8.11 is configured on the NAT device to realize the NAT translation of the address to support the private network host's access to the public network. In the networking, if there is no ALG processing the message load, after the PORT message sent by the private network host arrives at the server, the server cannot address according to the private network address, and the correct data connection cannot be established. The entire communication process includes the following four stages:

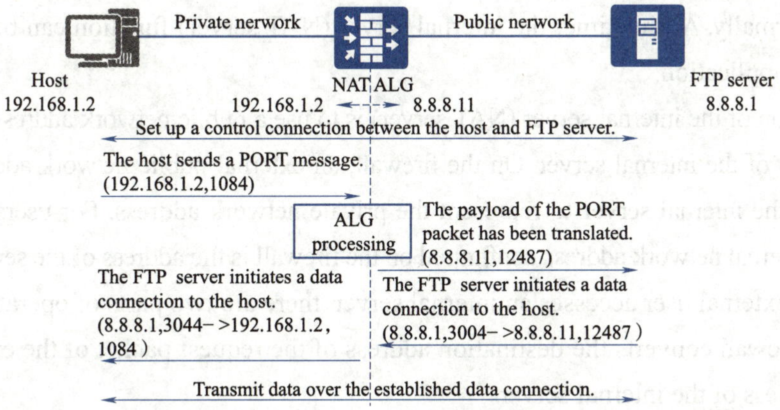

Figure 10-5 Principles of NAT ALG Implementation

(1) Successfully establish a control connection between the private network host and the public network FTP server through the TCP three-way handshake.

(2) After the control connection is established, the private network host sends a PORT message to the FTP server. The message carries the destination address and port of the data connection specified by the private network host to notify the server to use the address and port to connect to itself.

(3) When a PORT message passes through a NAT device that supports the ALG feature, the private network address and port in the message payload will be converted into the corresponding public network address and port. That is, the device converts the private network address 192.168.1.2 in the received PORT packet load into the public network address 8.8.8.11, and the port 1084 into 12487.

(4) After receiving the PORT message, the FTP server on the public network parses its content and initiates a data connection to the host on the private network. The destination address of the data connection is 8.8.8.11, and the destination port is 12487 (Note: Generally, the source port of the message is 20. However, because the FTP protocol is not strictly regulated, the source port of the data connection sent by some servers is a random port greater than 1024. For example, the wftpd server is used in this example, and the source port used is 3004). Since the destination address is a public network address, subsequent data connections can be successfully established, thereby realizing the access of the private network host to the public network server.

Task 10.3 Implementation of server mapping

NAT server, i.e. internal server. NAT hides the structure of the internal network and has the function of "shielding" internal hosts. However, in actual applications, it may be necessary to provide an opportunity for the outside to access the internal host. For example, it is provided to an external WWW server, and the external host has no route to the internal address at all, so it cannot be accessed normally. At this time, the internal server (NAT server) function can be used to realize this functional application.

The function of the internal server (NAT server) is to use a public network address to represent the external address of the internal server. On the firewall, an external public network address is specially configured for the internal server to represent the private network address. For users on the external network, the external network address configured on the firewall is the address of the server.

When an external user accesses an internal server, there are two parts of operations as follows:

(1) The firewall converts the destination address of the request packet of the external user into the private address of the internal server.

(2) The firewall converts the source address (private network address) of the internal server's response message into a public network address.

As shown in Figure 10-6, a public network address such as 202.202.1.1 can be used as the external address of the Web server, or even an IP address plus port number such as 202.202.1.1:8080 can be used as the external address of the web. When an external network user accesses the content server, an IP packet with a destination IP of 202.202.1.1 is sent, and the packet is forwarded by the public network device to the NAT device. The NAT server function needs to be configured on the NAT device in advance to map the public network address of 202.202.1.1 to the private network address 192.168.1.1, convert the destination IP of the message to 192.168.1.1, and find the route to forward the message to the intranet server. The source IP of the response message from the intranet server is 192.168.1.1, and the source IP is converted to 202.202.1.1 when forwarded by the NAT device.

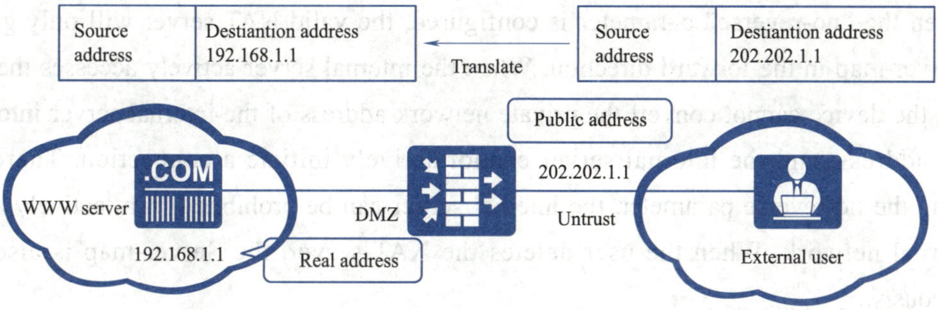

Figure 10-6 NAT Server

Subtask Understand the relationship between NAT server and Server-map table

Generally, if strict packet filtering is configured on the device, the device will only allow intranet users to actively access the external network in one direction. But in practical applications, for example, when transferring files using the port method of the FTP protocol, both the client and the server need to initiate a control connection to the server, and the server also needs to initiate a server data connection to the client. If the packet filtering configured on the device allows unidirectional packets to pass actively, the FTP file transfer cannot be successful.

In order to solve this type of problem, the USG equipment introduced the Server-map table, Server-map is based on triples and is used to store a mapping relationship. This mapping relationship can be the data connection relationship negotiated by the control data, or the address mapping relationship in the configuration NAT. Enable the external network to actively access the internal network through the device.

After the Server-map table is generated, if a data connection matches the Server-map table entry, it can be forwarded normally by the device without checking the session table, which ensures the normal forwarding of some special applications.

As shown in Figure 10-7, after the NAT server is successfully configured, the device will automatically generate a Server-map entry to store the mapping relationship between the Global address and the Inside address. When the "no-reverse" parameter is not configured, each valid NAT server will generate two static server-maps in the forward and reverse directions.

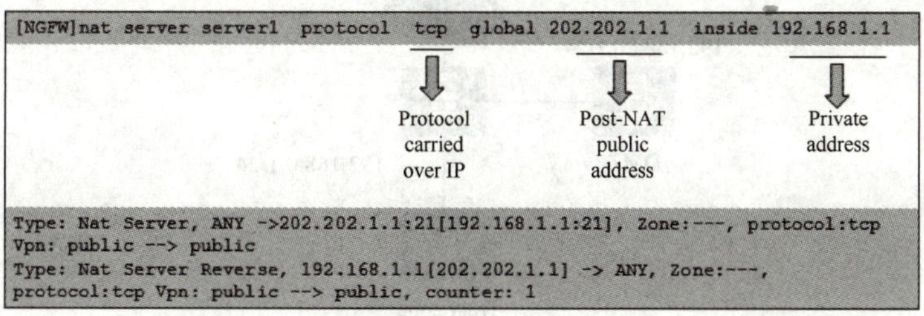

Figure 10-7 NAT server and Server-map

When the "no-reverse" parameter is configured, the valid NAT server will only generate a static Server-map in the forward direction. When the internal server actively accesses the external network, the device cannot convert the private network address of the internal server into a public network address, and the internal server cannot actively initiate a connection. Therefore, by specifying the no-reverse parameter, the internal server can be prohibited from actively accessing the external network. When the user deletes the NAT server, the Server-map is also deleted synchronously.

```
    [NGFW] nat server server protocol tcp global 202.202.1.1 inside 192.168.1.1
no-reverse
    Type: Nat Server, ANY -> 202.202.1.1[192.168.1.1], Zone:---, protocol: tcp Vpn: public -> pubic
```

When the no-reverse NAT server is configured to implement the requirement that an internal server advertise multiple public IP addresses for external networks to access. Because the no-reverse parameter is configured, the internal server will not be able to actively access the external network. At this time, if the internal server wants to access the external network, it is necessary to configure a source NAT policy between the area where the internal server is located and the area where the external network is located to convert the server's private network address to a public network address. The address pool referenced in the source NAT policy can be a global address or other public network addresses.

After NAT No-PAT is configured, the device will create a Server-map table for the data flow with actual traffic generated by the configured multichannel protocol.

Task 10.4 NAT application scenario and configuration

As shown in Figure 10-8, a company deployed a firewall at the exit. It is now required that the company's intranet 192.168.0.0/24 network segment can access the external network through the firewall, and the server with the IP address of 192.168.20.1 needs to provide external services.

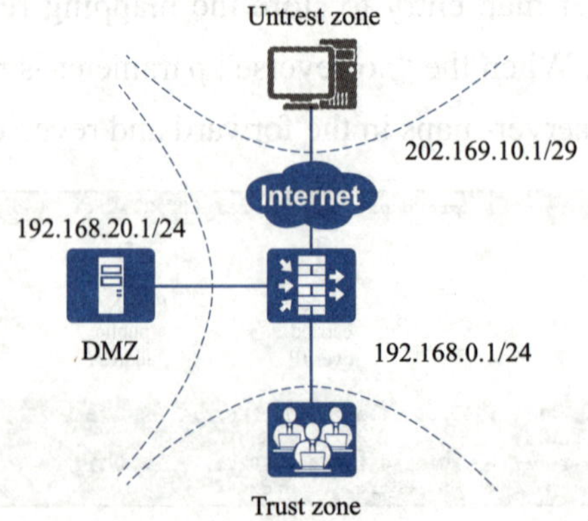

Figure 10-8 NAT application scenario

This requires the deployment of source NAT on the egress firewall to meet the needs of the internal network host to access the external network, and the deployment of a NAT server to enable the internal network server to provide services for the external network.

Subtask 10.4.1 Source NAT configuration of firewall

Source NAT is configured to implement NAT address translation when internal network employees access external networks. The data flow is from high security level to low security level, so the source address should be the network address segment of the internal network. The address pool allocated for intranet users should be the external network address segment, so that intranet users can access internet resources.

1. CLI configuration

Step 1: Configure interface IP and associated security zone, the reference command is omitted.

Step 2: Configure a security policy to allow traffic on the 192.168.0.0/24 network segment.

```
[NGFW]security-policy
[NGFW-policy-security]rule name natpolicy
[NGFW-policy-security-rule-natpolicy]source-address 192.168.0.0 24
[NGFW-policy-security-rule-natpolicy]action permit
```

Step 3: Configure the NAT address pool.

```
[NGFW] nat address-group 1
[NGFW -nat-address-group-1] section 202.169.10.2 202.169.10.6
```

Step 4: Configure NAT policy.

```
[NGFW] nat-policy
[NGFW-policy-nat] rule name nat1
[NGFW-policy-nat-rule-nat1] source-zone trust
[NGFW-policy-nat-rule-nat1] destination-zone untrust
[NGFW-policy-nat-rule-nat1] source-address 192.168.0.0 24
[NGFW-policy-nat-rule-nat1] action nat address-group 1
```

2. Web configuration

Step 1: Configure interface IP and associated security zone, the reference command is omitted.

Step 2: Configure a security policy to allow traffic on the 192.168.0.0/24 network segment, as shown in Figure 10-9.

Step 3: Configure the NAT address pool, as shown in Figure 10-10.

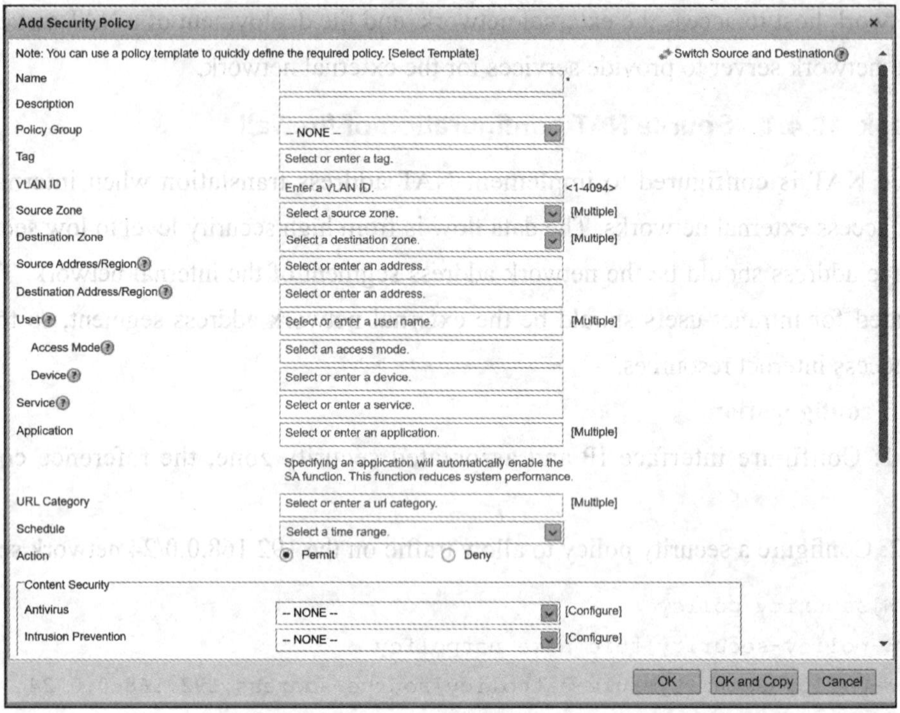

Figure 10-9 Security policy configuration

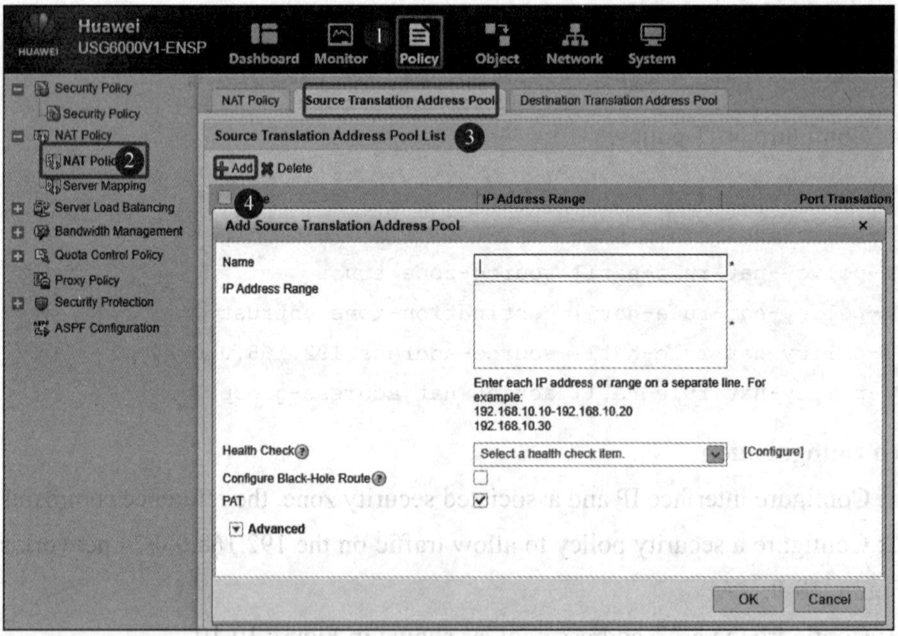

Figure 10-10 Configuring the NAT address pool

Step 4: Configure NAT policy, as shown in Figure 10-11.

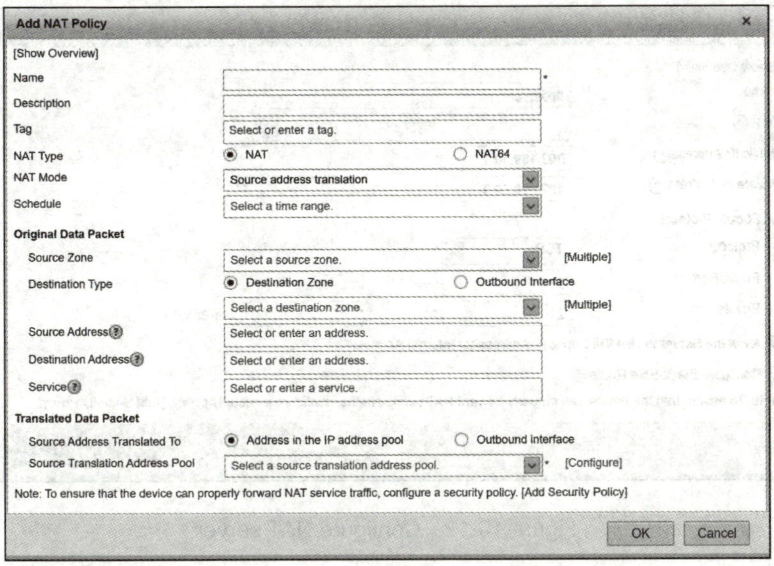

Figure 10-11　Configure NAT policy

Subtask 10.4.2　Firewall NAT server configuration

1. CLI configuration

Step 1: Configure internal Web and FTP servers.

```
[NGFW] nat server wwwserver protocol tcp global 202.169.10.1 80 inside 192.168.20.2 8080
[NGFW] nat server ftpserver protocol tcp global 202.169.10.1 ftp inside 192.168.20.3 ftp
```

Step 2: Configure inter-domain packet filtering rules.

```
[NGFW] security-policy
[NGFW -policy-security] rule name p1
[NGFW -policy-security-rule-p1] source-zone untrust
[NGFW -policy-security-rule-p1] destination-zone dmz
[NGFW -policy-security-rule-p1] destination-address 192.168.20.2 32
[NGFW -policy-security-rule-p1] service http
[NGFW -policy-security-rule-p1] action permit
[NGFW -policy-security] rule name p2
[NGFW -policy-security-rule-p2] source-zone untrust
[NGFW -policy-security-rule-p2] destination-zone dmz
[NGFW -policy-security-rule-p2] destination-address 192.168.20.3 32
[NGFW -policy-security-rule-p2] service ftp
[NGFW -policy-security-rule-p2] action permit
```

2. Web configuration

Step 1: Configure internal Web and FTP servers, as shown in Figure 10-12.

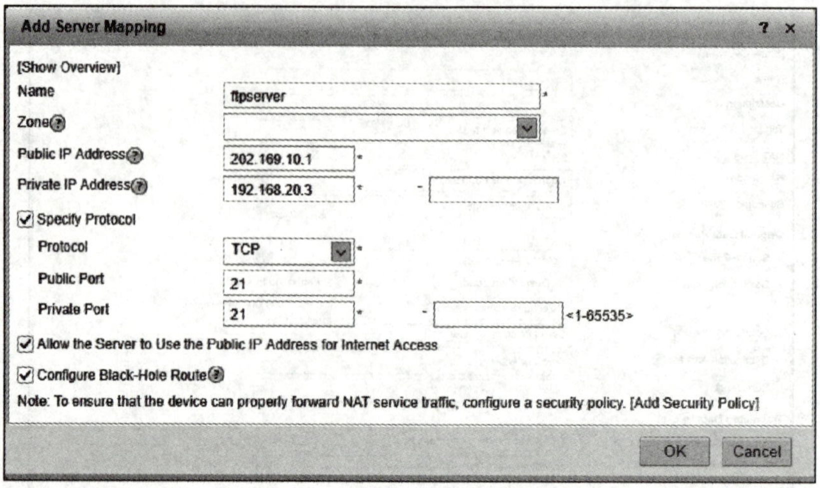

Figure 10-12 Configure NAT server

Step 2: Configure inter-domain packet filtering rules, as shown in Figure 10-13.

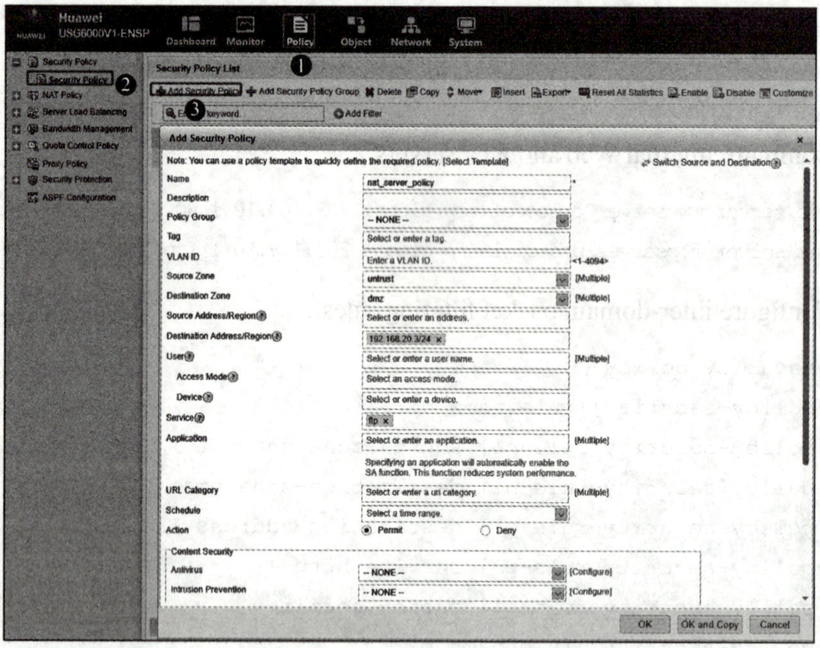

Figure 10-13 Configure filtering rules between domains

Subtask 10.4.3 Two-way NAT technology

Two-way NAT refers to the conversion of the source information and destination information of a message at the same time during the conversion process. Two-way NAT is not a separate function, but a combination of source NAT and destination NAT. Two-way NAT refers to the same flow, and simultaneously translates the source and destination addresses of packets when it passes through the firewall. Two application scenarios of bidirectional NAT:

(1) Inter-domain bidirectional NAT (NAT server + Source NAT).

(2) Intra-domain bidirectional NAT.

When the two communication parties in the two-way NAT application scenario visit each other, the destination address is not the real address, but the NAT-translated address. Generally speaking, the internal network is of high priority, and the external network is of low priority. When an external network user in a low-priority security zone accesses the public network address of the internal server, the destination address of the message is converted to the private network address of the internal server, but the internal server needs to be configured with a route to the public network address.

If you want to avoid configuring routes to public network addresses, you can configure NAT from a low-priority security zone to a high-priority security zone. The access in the same security zone needs to be NAT, and the intra-zone NAT function needs to be configured.

1. Two-way NAT between domains

In order to simplify the configuration of the route from the server to the public network, source NAT configuration can be added on the basis of NAT server.

When configuring the NAT server, the server needs to be configured with a route to the public network address before it can send response packets normally. If you want to simplify the configuration and avoid configuring the route to the public network address, you can also convert the source IP address of the external network user, and the converted source IP address is in the same network segment as the server's private network address. In this way, the internal server will send the response message to the gateway by default, that is, the device itself, and the device will forward the response message.

As shown in Figure 10-14, when the Host accesses the Server, the FW processing process is as follows:

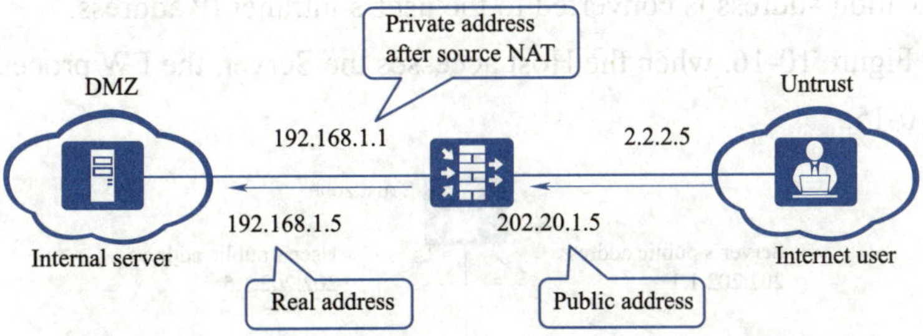

Figure 10-14 Two-way NAT between domains

(1) FW performs address translation on packets matching the two-way NAT processing strategy.

(2) FW selects a private IP address from the destination NAT address pool to replace the destination IP address of the message, and at the same time replaces the destination port number of the message

with the new port number.

(3) Determine whether the requirements of the security policy are met, and after passing the security policy, select a private network IP address from the source NAT address pool to replace the source IP address of the packet. At the same time, replace the source port number of the message with the new port number, establish a session table, and then send the message to the intranet.

(4) After the FW receives the message from the Server in response to the Host, it searches the session table to match the created entry, replaces the source and destination addresses of the message with the original IP address, and replaces the source and destination port numbers of the message with the original. And then send the message to the Internet.

2. Two-way NAT within the domain

If both parties that need address translation are in the same security domain, then the case of intra-domain NAT is involved. As shown in Figure 10-16, when the FTP server and the user are in the trust zone, the user accesses the external public IP address of the FTP server, so that all interactive packets between the user and the FTP server must pass through the firewall. At this time, the internal server and intra-domain NAT need to be configured at the same time.

Intra-domain NAT refers to a scenario where intranet users and servers are deployed in the same security zone, but it is still hoped that intranet users can only access the server's public network address. In the process of implementing intra-domain NAT, it is necessary to convert the destination address of the message that accesses the internal server from a public network address to a private network address, and it is also necessary to convert the source address from a private network address to a public network address.

The firewall converts the destination address of the user's request message into the internal IP address of the FTP server, and the source address into the IP address announced by the user. The firewall converts the source address of the FTP server's response message to the public announced address. The destination address is converted to the user's intranet IP address.

As shown in Figure 10-16, when the Host accesses the Server, the FW processing process is as shown in Figure 10-15.

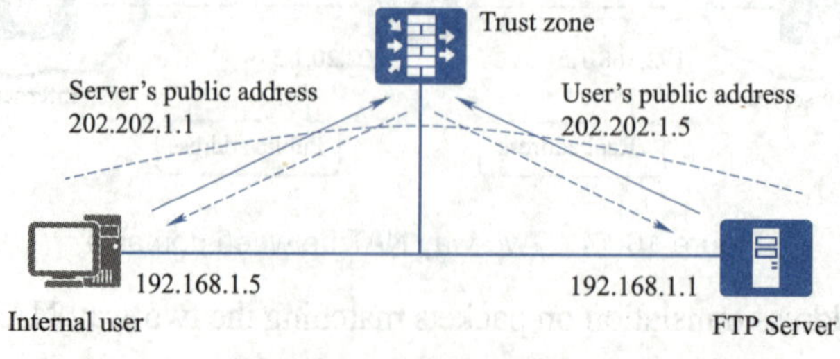

Figure 10-15 Intra-domain bidirectional NAT

Project 11 Dual-system hot backup technology of firewall

With the vigorous development of mobile office, online shopping, instant messaging, Internet finance, Internet education, and other businesses, more and more businesses are carried on the Internet and becoming more and more important. Therefore, how to ensure the uninterrupted transmission of the network has become an urgent problem in the process of network development.

The emergence of dual-system hot backup technology has changed the awkward state of reliability that is difficult to guarantee. By deploying two firewalls at the network exit location, the communication between the internal network and the external network is guaranteed to be smooth.

Task 11.1 Principle analysis of dual-system hot standby technology

Subtask 11.1.1 Understand the causes of hot standby technology

The traditional networking mode is shown in Figure 11-1. All interaction messages between internal users and external users pass through firewall A. If firewall A fails, the communication between all hosts with firewall A as the default gateway in the internal network and the external network will be interrupted, and communication reliability cannot be guaranteed.

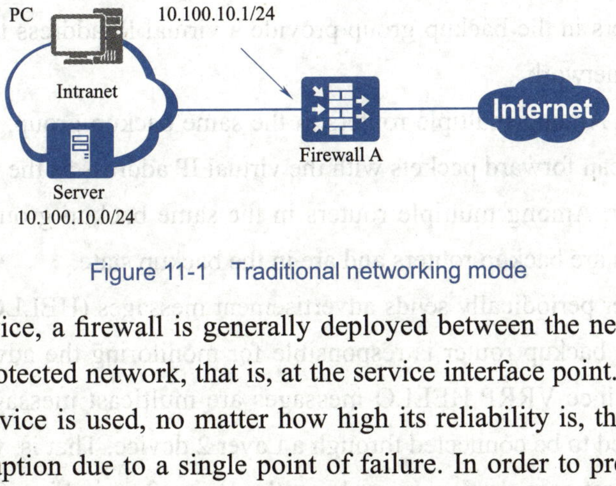

Figure 11-1 Traditional networking mode

As a security device, a firewall is generally deployed between the network that needs to be protected and the unprotected network, that is, at the service interface point. At this business point, if only one firewall device is used, no matter how high its reliability is, the system may bear the risk of network interruption due to a single point of failure. In order to prevent network services from being interrupted by accidental failure of one device, two firewalls can be used to form a dual-system backup.

Subtask 11.1.2 Understand router redundancy deployment scheme

In order to avoid the occurrence of single-point failures caused by the traditional networking of routers, the protection mechanism of multiple links can usually be adopted, and link switching can be carried out by relying on dynamic routing protocols. However, the use of routing protocols for handover protection has certain limitations. When the dynamic routing protocol cannot be used, it will still cause the problem of link interruption. Therefore, another protection mechanism, VRRP (virtual routing redundancy protocol) is introduced. The link protection mechanism of VRRP takes shorter time than relying on the broadcast message of the dynamic routing protocol to switch the link. At the same time, it makes up for the link protection in the case that dynamic routing cannot be used.

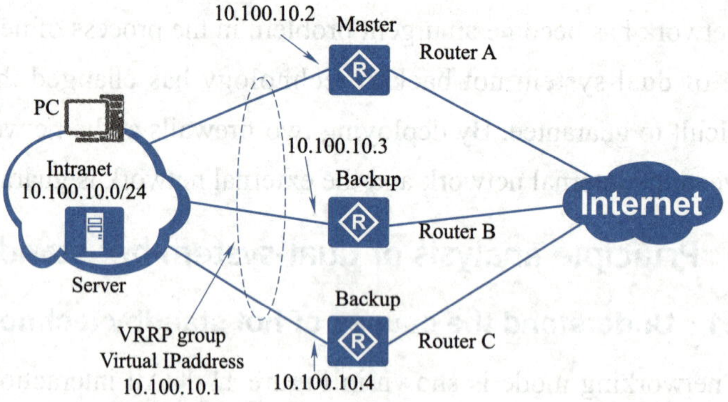

Figure 11-2 The VRRP protocol is used to implement device redundancy in the router network

VRRP is a basic fault-tolerant protocol.

(1) Backup group: A group of routers in the same broadcast domain are organized into a virtual router. All routers in the backup group provide a virtual IP address together as the gateway address of the internal network.

(2) Master router: Among multiple routers in the same backup group, only one is active, and only the master router can forward packets with the virtual IP address as the next hop.

(3) Backup router: Among multiple routers in the same backup group, except for the main router, the other routers are backup routers and are in the backup state.

The primary router periodically sends advertisement messages (HELLO) to the backup router through multicast. The backup router is responsible for monitoring the advertisement message to determine its status. Since VRRP HELLO messages are multicast messages, each router in the backup group is required to be connected through a Layer 2 device. That is, when VRRP is enabled, the upstream and downstream devices must have the Layer 2 switching function, otherwise the backup router cannot receive the HELLO message sent by the primary router. If the networking conditions are not met, VRRP cannot be used.

Subtask 11.1.3 The application of VRRP in multi-zone firewall networking

When providing dual-system backup for multiple zones on the firewall, you need to configure multiple VRRP backup groups on each firewall, as shown in Figure 11-3.

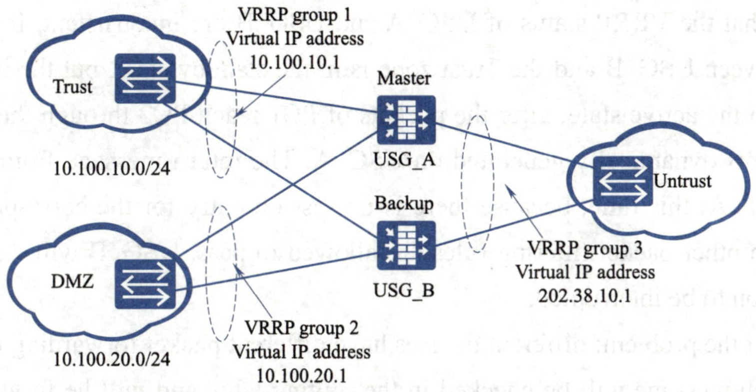

Figure 11-3 Application of VRRP in multi-zone firewall networking

Since the firewall is a stateful firewall, it requires the back and forth path of packets to pass through the same firewall. In order to meet this restriction, the state of all VRRP backup groups on the same firewall is required to be consistent. That is, it is necessary to ensure that all VRRP backup groups on the primary firewall are in the primary state, so that all packets will pass through this firewall. Another firewall acts as a backup device.

Subtask 11.1.4 Understand the defects of VRRP in firewall application

As shown in Figure 11-4, assume that the VRRP states of USG_A and USG_B are the same, that is, all interfaces of USG_A are in the active state, and all interfaces of USG_B are in the standby state.

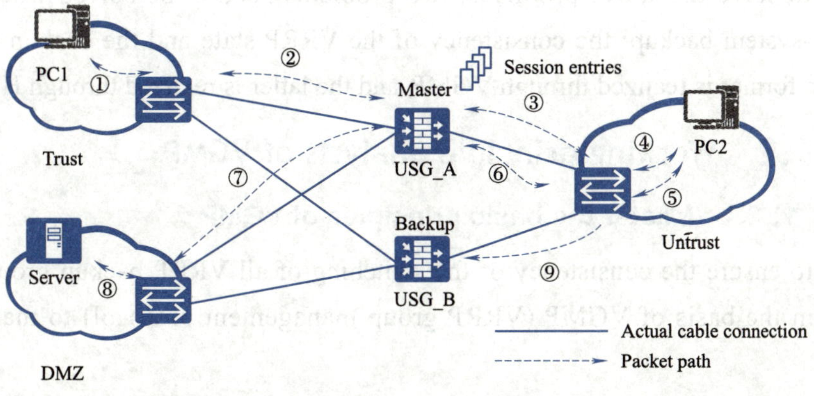

Figure 11-4 Defects of VRRP in firewall applications

At this time, PC1 in the trust zone accesses PC2 in the untrust zone, and the packet forwarding route is ① - ② - ③ - ④ . When USG_A forwards access packets, it dynamically generates session entries. When the return message from PC2 arrives at USG_A through ④ - ③ , it can hit the session entry before it reaches PC1 through ② - ① and returns smoothly. In the same way, the server of PC2 and the server of DMZ can also visit each other.

Assuming that the VRRP status of USG_A and USG_B are inconsistent, for example, when the interface between USG_B and the Trust zone is in the standby state, but the interface with the untrust zone is in the active state, after the packets of PC1 reach PC2 through the USG_A device, the session entry is dynamically generated on USG_A. The return message from PC2 is returned via route ④ - ⑨ . At this time, because there is no session entry for the corresponding data flow on USG_B, if no other packet filtering rules are allowed to pass, USG_B will discard the packet, causing the session to be interrupted.

The cause of the problem: different devices have different packet forwarding mechanisms.

Router: Each message will be checked in the routing table and will be forwarded only when it matches. After the link is switched, subsequent packets will not be affected and continue to be forwarded.

Stateful inspection of the firewall: If the first packet is allowed to pass, a five-tuple session connection will be established, and only subsequent packets (including return packets) that hit the session entry can pass through the firewall; After the link is switched, subsequent packets cannot find the correct entries, which will cause service interruption.

Note:

The same problem exists when the router is configured with NAT, because a NAT-translated entry will be formed after NAT is performed.

In order to solve the above problems, two problems need to be solved when the firewall performs dual-system backup: the consistency of the VRRP state and the backup of the session table state. The former is realized through VGMP, and the latter is realized through HRP.

Task 11.2 Working principle analysis of VGMP

Subtask 11.2.1 Master the basic principles of VGMP

In order to ensure the consistency of the switching of all VRRP backup groups, the VRRP is expanded on the basis of VGMP (VRRP group management protocol) to make up for this limitation.

VGMP puts forward the concept of VRRP management group. Multiple VRRP backup groups on the same firewall are added to a VRRP management group, and the management group manages all VRRP backup groups in a unified manner. By uniformly controlling the switching of the status of each VRRP backup group, the status of all VRRP backup groups in the management group is guaranteed to be consistent. As shown in Figure 11-5, when the VGMP on the firewall is in the Active/Standby state, the state of all VRRP backup groups in the group is unified to the Active/Standby state.

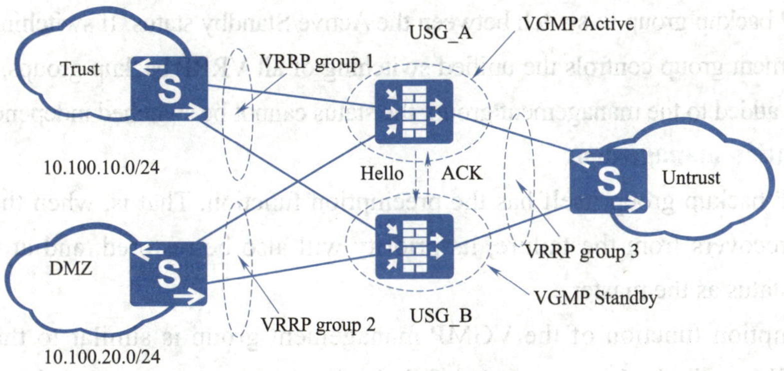

Figure 11-5 Basic principles of VGMP

When the VGMP on the firewall is in the Active state, the state of all VRRP backup groups in the group is unified into the Active state, all packets will pass through the firewall, and the firewall becomes the main firewall. At this time, the corresponding VGMP on the other firewall is in the standby state, and this firewall becomes the standby firewall.

Whether the firewall is the active firewall or the backup firewall is determined by specifying the state of the VGMP group. The VGMP priority of the firewall has an initial priority. When the interface or board of the firewall fails, a certain lower value will be subtracted from the initial priority. The initial priority of the USG6000 and NGFW Module is 45000. The initial priority of the VGMP group of the USG9500 is related to the number of cards on the LPU board (interface board) and the number of CPUs on the SPU board (service board).

The VGMP of the VRRP backup group in the group whose status is Active will also periodically send HELLO messages to the peer to notify the Standby end of its own running status (including priority, VRRP member status, etc.). The status of the members is dynamically adjusted to complete the Active/Standby switchover of the two firewalls, which is similar to VRRP. But unlike VRRP, after the Standby end receives the HELLO message, it will respond with an ACK message, which will also carry its own priority and VRRP member status.

The VGMP HELLO message sending cycle defaults to 1 s. When the Standby end does not receive the HELLO message sent by the opposite end for three HELLO message cycles, it will consider the opposite end to be faulty and switch itself to the Active state.

Subtask 11.2.2 Understand VGMP group management

In order to ensure that the status of all VRRPs in the VGMP group is consistent, the following two points need to be achieved for consistent preemption:

1. State consistency management

The change of the Active/Standby status of each backup group needs to notify the VGMP management group to which it belongs, and the VGMP management group decides whether to allow the VRRP backup group to switch between the Active/Standby status. If switching is required, the VGMP management group controls the unified switching of all VRRP backup groups. After the VRRP backup group is added to the management group, the status cannot be switched independently.

2. Preemption management

The VRRP backup group itself has the preemption function. That is, when the original failed master device recovers from the failure, its priority will also be restored, and at this time, it can regain its own status as the master.

The preemption function of the VGMP management group is similar to that of the VRRP backup group. When the backup group that fails in the management group fails and recovers, the priority of the management group will also be restored. At this time, VGMP can decide whether it needs to re-occupy and is called the master device.

When the VRRP backup group is added to the VGMP management group, the original preemption function on the backup group will become invalid. Whether or not the preemption occurs must be determined by the VGMP management group.

Task 11.3 Analysis of HRP working principle

Subtask 11.3.1 Understand the basic concepts of HRP

The HRP (Huawei redundancy protocol) is used to implement the backup of dynamic state data and key configuration commands between two firewalls.

As shown in Figure 11-6, in a dual-system hot backup network, when the main firewall fails, all traffic will be switched to the backup firewall. Because the USG firewall is a stateful firewall, if the backup firewall does not have status data such as the session table of the original primary firewall, the traffic switched to the backup firewall will not pass through the firewall, causing the existing connection to be interrupted. At this time, the user must re-initiate the connection.

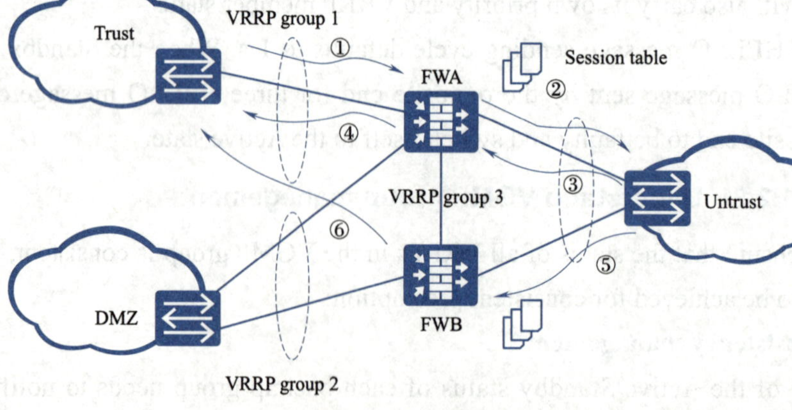

Figure 11-6 Traffic model in firewall dual-system hot backup

The HRP module provides a basic data backup mechanism and transmission function. Each application module collects the data that needs to be backed up by this module and submits it to the HRP module. The HRP module is responsible for sending the data to the corresponding module of the peer firewall. The application module needs to analyze the data submitted by the HRP module and add it to the dynamic running data pool of the firewall.

Backup content: The connection state data to be backed up includes TCP/UDP session table, Server-map table entry, dynamic blacklist, NO-PAT table entry, ARP table entry, etc.

Backup direction: There is a state-based VGMP management group on the firewall, and it is backed up to the opposite end.

Backup channel: Generally, the directly connected ports on two devices are used as backup channels, sometimes called "heartbeat lines" (VGMP also communicates through this channel).

Subtask 11.3.2 Master the working principle of HRP

As shown in Figure 11-7, the backup data between two FWs is sent and received through the heartbeat port, and is transmitted through the heartbeat link (backup channel). The heartbeat port must be a state-independent interface with an IP address. It can be a physical interface (GE interface), or it can be used to increase bandwidth. A logical interface Eth-Trunk is formed by bundling multiple physical interfaces. Normally, the backup data traffic is about 20% to 30% of the business traffic. Please select the number of bundled physical interfaces according to the size of the backup data.

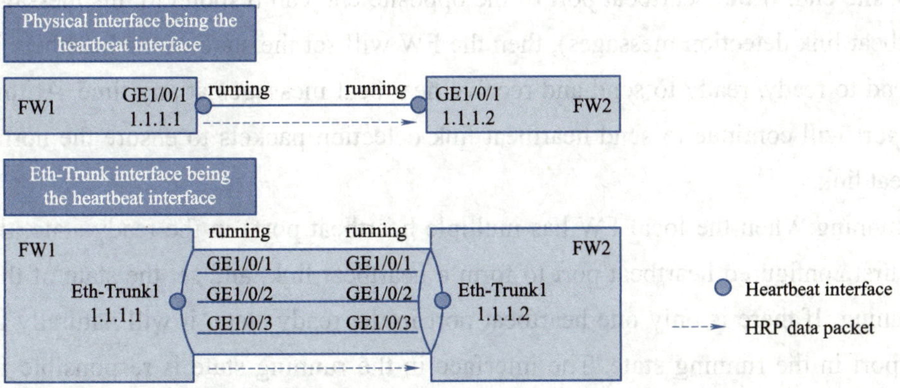

Figure 11-7 HRP heartbeat interface

Subtask 11.3.3 Understand the status of the heartbeat interface

The HRP heartbeat interface has five states, as shown in Figure 11-8.

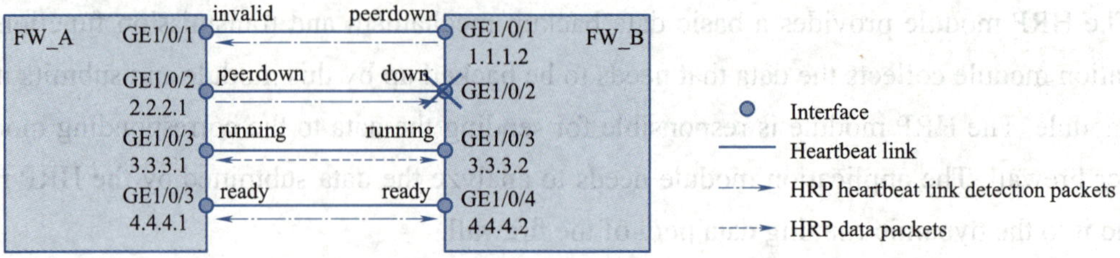

Figure 11-8 HRP heartbeat interface status

(1) Invalid:This status is displayed when the heartbeat port on the local FW is configured incorrectly (physical status up, protocol status down), for example, the specified heartbeat port is a Layer 2 interface or the IP address of the heartbeat interface is not configured.

(2) Down:When the physical and protocol status of the heartbeat port on the local FW are both down, this status will be displayed.

(3) Peerdown:When the physical and protocol status of the heartbeat port on the local FW are both up, the heartbeat port will send a heartbeat link detection packet to the corresponding heartbeat port of the opposite end. If no response packet from the peer is received, the FW will set the state of the heartbeat interface to peerdown. However, the heartbeat port will continue to send heartbeat link detection packets so that when the corresponding heartbeat port of the opposite end is up, the heartbeat link can be in a connected state.

(4) Ready:When the physical and protocol status of the heartbeat port on the local FW are both up, the heartbeat port will send a heartbeat link detection packet to the corresponding heartbeat port of the opposite end. If the heartbeat port of the opposite end can respond to this message (and also send heartbeat link detection messages), then the FW will set the state of the heartbeat interface at the local end to ready, ready to send and receive heartbeat messages at any time. At this time, the heartbeat port will continue to send heartbeat link detection packets to ensure the normal state of the heartbeat link.

(5) Running:When the local FW has multiple heartbeat ports in the ready state, the FW will select the first configured heartbeat port to form a heartbeat link, and set the state of the heartbeat port to running. If there is only one heartbeat port in the ready state, it will naturally become the heartbeat port in the running state. The interface in the running state is responsible for sending HRP heartbeat messages, HRP data messages, HRP link detection messages, VGMP messages, and consistency check messages.

At this time, the rest of the heartbeat ports in the ready state are in the backup state. When the heartbeat ports in the running state or the heartbeat link fail, the other heartbeat ports in the ready state will take over the current heartbeat ports in order (in order of configuration) to process services.

In summary, the function of the heartbeat link detection message is to detect whether the heartbeat port of the peer device can normally receive the message of the local device to determine whether the heartbeat link is available. As long as the physical and protocol status of the local heartbeat interface is up, it will send a heartbeat link detection message to the heartbeat interface of the opposite end for detection. However, the HRP heartbeat message we mentioned earlier is used to detect and perceive whether the peer device (VGMP group) is working properly. HRP heartbeat messages are only sent by the VGMP group of the master device through the heartbeat port in the running state.

Subtask 11.3.4 Master the backup mode of dual-system hot standby

There are four backup methods for firewall dual-system hot backup: automatic backup, manual batch backup, fast session backup, and automatic synchronization of the configuration of the Active/Standby FW after the device restarts.

1. Automatic backup

The automatic backup function is turned on by default, which can automatically back up configuration commands in real time and periodically back up status information, which is suitable for various dual-system hot backup networking.

After the automatic backup function is enabled, every time a command that can be backed up is executed on one FW, the configuration command will be synchronized and backed up to another FW immediately.

After the automatic backup function is enabled, the main device will periodically back up the state information that can be backed up to the backup device. That is, the status information of the primary device will not be backed up immediately after it is established, but will be backed up to the backup device after a period of time (about 10 seconds).

Automatic backup will not back up the following types of sessions:

(1) The session to the firewall itself, such as the session generated when the administrator logs in to the firewall.

(2) TCP semi-connected sessions that have not completed the 3-way handshake (quick backup supports backing up this session).

(3) The session created only for the first UDP packet and not matched by subsequent packets (quick backup supports this session to be backed up).

2. Manual batch backup

Manual batch backup needs to be manually triggered by the administrator. Every time a manual batch backup command is executed, the active device will immediately synchronize configuration commands and status information to the standby device. Therefore, manual batch backup is mainly applicable to scenarios where the configuration between the active and standby devices is not synchronized and manual synchronization is required.

After the manual batch backup command is executed, the active (configuration master) device will immediately synchronize the configuration commands that can be backed up to the backup (configuration backup) device.

After the manual batch backup command is executed, the active device will immediately synchronize the state information that can be backed up to the standby device without waiting for the arrival of the automatic backup cycle.

3. Fast session backup

The fast session backup function is suitable for the work of load sharing to cope with scenarios where the back and forth paths of packets are inconsistent. In a load-sharing network, since both firewalls are active devices and can forward packets, there may be inconsistencies in the back and forth paths of packets. That is, packets in both directions pass through different firewalls. At this time, if the sessions of the two firewalls are not backed up in time, the backhaul packets will be discarded because they do not match the session entries, which will cause service interruption. Therefore, in order to prevent the above phenomenon from happening, it is necessary to configure the fast session backup function under the load-sharing network, so that the two firewalls can back up sessions with each other in real time, so that the return packets can find the corresponding session entries, so as to ensure the internal and external users' business is not interrupted.

After the fast session backup function is enabled, the active device will synchronize all the sessions that can be backed up (including the sessions that are not supported by the automatic backup mentioned above) to the backup device in real time. That is, when the active device session is established, it will be backed up to the backup device in real time.

4. The configuration of the active and standby FW is automatically synchronized after the device restarts

In a dual-system hot backup network, if one FW restarts, the other FW will carry services during the restart. During this period, configurations may be added, deleted, or modified on the FW carrying services. In order to ensure that the configuration of the Active and Standby FWs is consistent, after the FW restarts, it will automatically perform a configuration synchronization with the FW currently carrying the service.

Configuration synchronization will only synchronize configurations that support backup, such as security policies, NAT policies, etc. Configurations that do not support backup, such as OSPF, BGP, etc., continue to use the original configuration.

The configuration synchronization takes a certain amount of time. The synchronization time is related to the amount of configuration. The larger the amount of configuration, the longer the time required to synchronize the configuration. It may take about 1 hour at the longest. During configuration synchronization, configuration commands that support backup cannot be executed on the FW.

Task 11.4 Basic networking and configuration of dual-system hot standby

The most common dual-system hot standby network is that the firewall adopts the routing mode. The downstream switch is connected to the firewall in two lines, and the upstream and downstream service interfaces work in the three-layer mode. When connecting Layer 2 devices, you need to configure a VRRP backup group on the upstream and downstream service interfaces so that the VGMP management group can monitor the Layer 3 service interface through the VRRP backup group. As shown in Figure 11-9, under normal circumstances, firewall A acts as the master. When firewall A's uplink or downlink is down, firewall B automatically switches to the master device, and the switch traffic flows to firewall B.

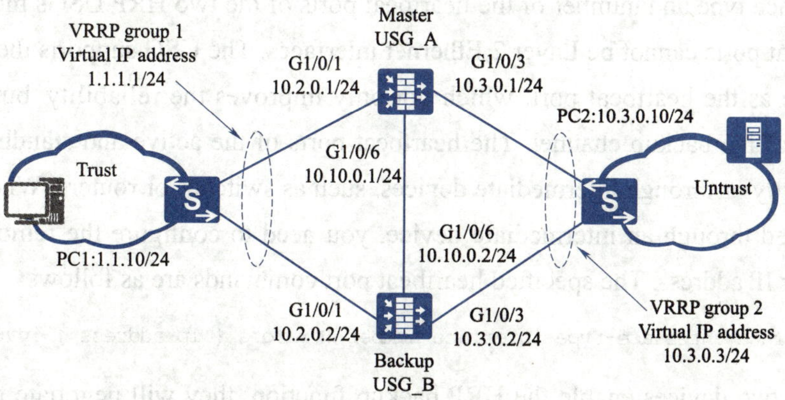

Figure 11-9 Basic networking of dual-system backup

Subtask 11.4.1 Dual system hot standby configuration command-CLI

1. VRRP configuration group commands

Configure VRRP in interface view:

```
vrrp vrid virtual-router-ID virtual-ip virtual-address [ ip-mask | ip-mask-
length ] { active | standby }
```

(1) When executing this command, after specifying the active or standby parameters. That is, the VRRP group is added to the Active or Standby management group of the VGMP management group.

(2) A maximum of 255 VRRP groups can be configured under each ordinary physical interface (GigabitEthernet interface).

By default, the Master management group will send VRRP packets every 1 second. You can adjust the VRRP packet sending interval in the interface view. Modify the VRRP packet sending time in the interface view:

```
vrrp vrid virtual-router-ID timer advertise adver-interval
```

VRRP can also cooperate with ip-link to enable VRRP to switch between active and standby when the uplink is broken. Configure ip-link in interface view:

```
vrrp vrid virtual-router-id ip-link link-id
```

By default, the preemption function of the VGMP management group is enabled, and the preemption delay time is 60 s. The command to configure the preemption delay time of the VGMP management group is as follows:

```
hrp preempt [ delay interval ]
```

2. HRP configuration commands

The interface type and number of the heartbeat ports of the two HRP USGs must be the same, and the heartbeat ports cannot be Layer 2 Ethernet interfaces. The USG supports the use of the Eth-Trunk interface as the heartbeat port, which not only improves the reliability, but also increases the bandwidth of the backup channel. The heartbeat ports of the active and standby USGs can be connected directly or through intermediate devices, such as switches or routers. When the heartbeat port is connected through an intermediate device, you need to configure the remote parameter to specify the peer IP address. The specified heartbeat port commands are as follows:

```
hrp interface interface-type interface-number [ remote { ip-address | ipv6-address } ]
```

When the two devices enable the HRP backup function, they will negotiate the status of the master and backup, and finally get a master device (represented by HRP_A when displayed) and a backup device (represented by HRP_S when displayed). After the first negotiation between the two ends, the master device is obtained, and the active device will back up information such as configuration and connection status to the standby device. The command to enable the HRP backup function is as follows:

```
hrp enable
```

After the function that allows the configuration of the backup device is enabled, all the information that can be backed up can be configured directly on the backup device, and the configuration on the backup device can be synchronized to the main device. If a certain configuration is performed on both the active and standby devices, in terms of time, the information configured later will overwrite the information configured first. The command to enable the configuration of the standby device is as follows:

```
hrp standby config enable
```

To enable automatic backup of commands and status information, the commands are as follows:

```
hrp auto-sync [ config | connection-status]
```

When the USG works in a load-sharing network, the back and forth paths of packets may be inconsistent. Be sure to enable the fast session backup function to synchronize the session information of one USG to another USG immediately to ensure that the services of internal and external users are not interrupted. The command to enable fast session backup is as follows:

```
hrp mirror session enable
```

Subtask 11.4.2 Example of firewall dual-system hot standby configuration -CLI

As shown in Figure 11-10, USG_A and USG_B form a firewall dual-system backup network. Normally, firewall A acts as the master. When firewall A's uplink or downlink is down, firewall B automatically switches to the master device, and the switch traffic flows to firewall B.

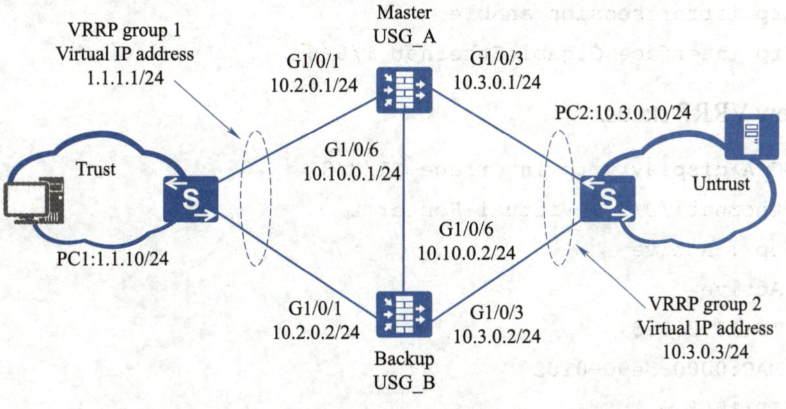

Figure 11-10 Basic networking of dual-system backup

Step 1: Configure basic firewall information, such as interface IP, security zone, etc., which are omitted here.

Step 2: Configure the USG_A VRRP function and set both VRRP groups to active.

```
[USG_A]interface GigabitEthernet 1/0/1
[USG_A-GigabitEthernet 1/0/1 ]ip address 10.2.0.1 24
[USG_A-GigabitEthernet 1/0/1 ]vrrp vrid 1 virtual-ip 1.1.1.1 255.255.255.0 active
```

```
[USG_A]interface GigabitEthernet 1/0/3
[USG_A-GigabitEthernet 1/0/3 ]ip address 10.3.0.1 24
[USG_A-GigabitEthernet 1/0/3 ]vrrp vrid 2 virtual-ip 10.3.0.3 active
```

Step 3: Configure the USG_B VRRP function and set both VRRP groups to standby.

```
[USG_B]interface GigabitEthernet 1/0/1
[USG_B-GigabitEthernet1/0/1 ]ip address 10.2.0.2 24
[USG_B-GigabitEthernet 1/0/1 ]vrrp vrid 1 virtual-ip 1.1.1.1 255.255.255.0 standby
[USG_B]interface GigabitEthernet 1/0/3
[USG_B-GigabitEthernet 1/0/3 ]ip address 10.3.0.2 24
[USG_B-GigabitEthernet 1/0/3 ]vrrp vrid 2 virtual-ip 10.3.0.3 standby
```

Step 4: Configure the USG_A HRP function, enable fast session backup, and specify the HRP heartbeat port as G1/0/6.

```
[USG_A]hrp enable
[USG_A]hrp mirror session enable
[USG_A]hrp interface GigabitEthernet 1/0/6
```

Step 5: Configure the USG_B HRP function, enable fast session backup, and specify the HRP heartbeat port as G1/0/6.

```
[USG_B]hrp enable
[USG_B]hrp mirror session enable
[USG_B]hrp interface GigabitEthernet 1/0/6
```

Step 6: View VRRP status.

```
HRP_A<USG_A>display vrrp interface G1/0/3
GigabitEthernet1/0/3 | Virtual Router 2
VRRP Group : Active
state : Active
Virtual IP:10.3.0.3
Virtual MAC:0000-5e00-0102
Primary IP:10.3.0.1
PriorityRun:120
PriorityConfig:100
MasterPriority: 120
Preempt: YES   Delay Time:0
Advertisement Timer:1
Auth Type:NONE
Check TTL:YES
```

Step 7: View the status information of the firewall in the Master state.

```
HRP_A<USG_A>dis hrp state
The firewall's config state is: ACTIVE
Current state of virtual routers configured as active:
GigabitEthernet1/0/1    vrid   1 : active
GigabitEthernet1/0/3    vrid   2 : active
```

Step 8: View the status information of the firewall in the Slave state.

```
HRP_S[USG_B] display hrp state
The firewall's config state is: Standby
Current state of virtual routers configured as standby:
GigabitEthernet1/0/1    vrid   1 : standby
GigabitEthernet1/0/3    vrid   2 : standby
```

Subtask 11.4.3 Example of firewall dual-system hot standby configuration-Web

Step 1: Enter the dual-system hot standby configuration interface, as shown in Figure 11-11.

(1) Click "System"→"High Reliability"→"Dual-System Hot Backup"→"Configuration" to configure related dual-system hot backup.

(2) Click "Configure" to enter the dual-system hot backup configuration interface. In the configuration interface, you can configure the basic parameters of HRP and monitor the interface, VLAN, IP-Link, and BFD.

(3) Click the "Details" button to view the historical switching information of HRP.

(4) Click the "Consistency Check" button to check the consistency of the configuration of the primary and backup firewalls.

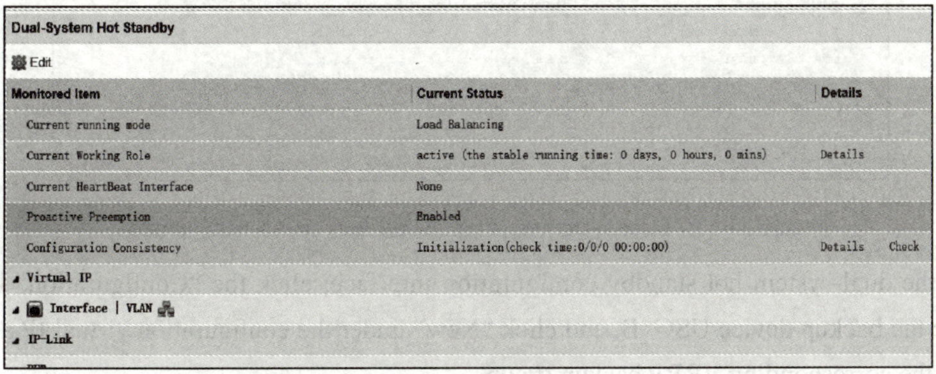

Figure 11-11 Entering the dual-system hot standby configuration interface

Step 2: Dual-system hot standby master device configuration, as shown in Figure 11-12.

On the dual-system hot standby configuration interface, click the "Configuration" button to configure the main device USG_A, and click "New" under the configuration virtual IP address to establish the corresponding VRRP backup group.

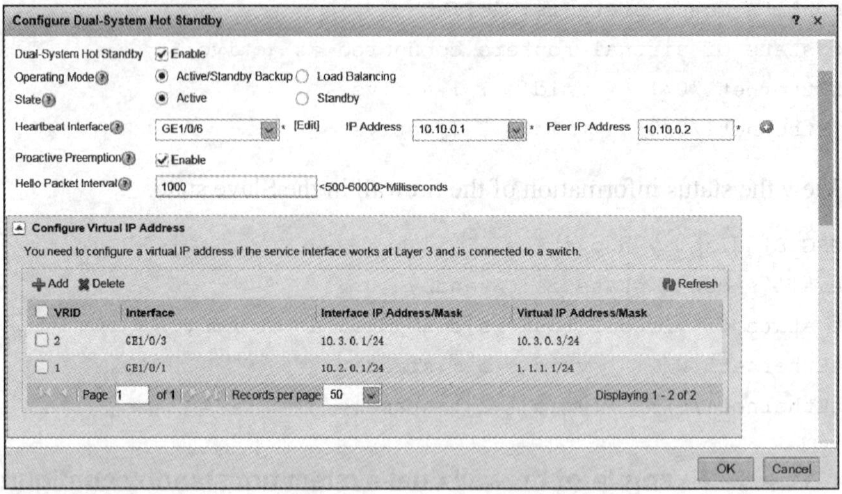

Figure 11-12 Dual-system hot standby master device configuration

Step 3: The dual-system hot standby backup device configuration, as shown in Figure 11-13.

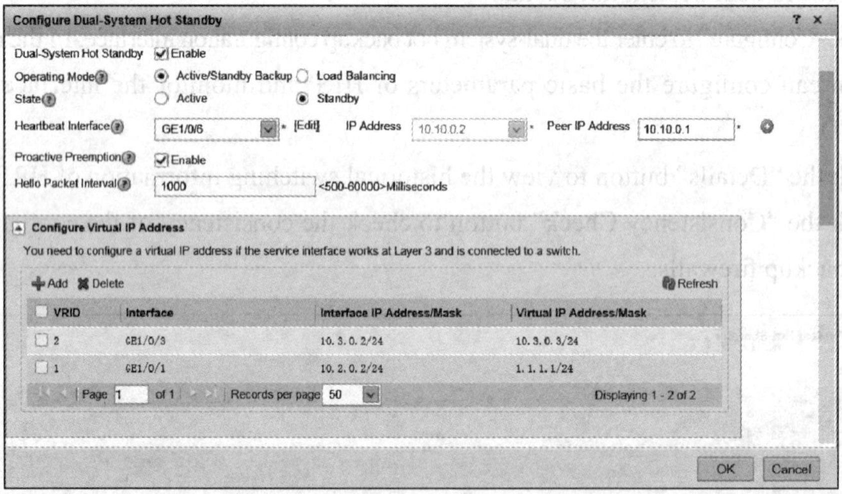

Figure 11-13 Dual-system hot standby backup device configuration

On the dual-system hot standby configuration interface, click the "Configuration" button to configure the backup device USG_B, and click "New" under the configuration virtual IP address to establish the corresponding VRRP backup group.

Step 4: View the historical switching information of dual-system hot backup, as shown in Figure 11-14.

On the dual-system hot standby interface, click "Details" to view the dual-system hot backup master/backup switch information.

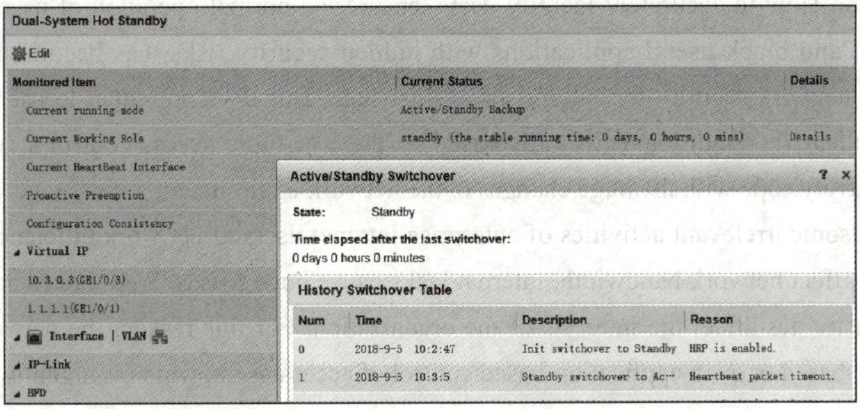

Figure 11-14 Viewing the historical switching information of dual-system hot standby

Step 5: View the status information of dual-system hot standby, as shown in Figure 11-15.

Confirm the operating mode, role, and status information of the VRRP backup group on the dual-system hot standby interface.

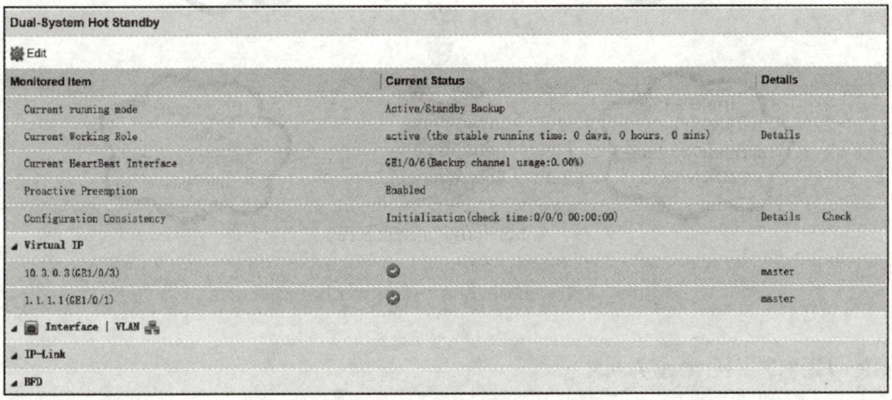

Figure 11-15 Viewing the status information of dual-system hot standby

Project 12 Firewall user management

Task 12.1 Principle analysis of user authentication and AAA technology

Subtask 12.1.1 Understand the background of user authentication

In the current network environment, the threat of network security comes more from the application layer, which also makes enterprises put forward higher requirements for network access control. How to accurately identify users, ensure the normal operation of users' legitimate applications, and block users' applications with hidden security risks, has become the focus of attention on network security by enterprises at this stage. But IP is not equal to user, port is not equal to application. Traditional firewalls based on the five-tuple access control strategy can no longer effectively cope with the huge changes in the network environment at this stage. As shown in Figure 12-1, some irrelevant activities of enterprise intranet users affect work efficiency, irrelevant applications affect network bandwidth, internal leaks cause huge losses, Web threats are pervasive, and illegal activities affect the interests of the organization. For this reason, the network needs to perform user-based or even application-based control of access users, which is urgently needed.

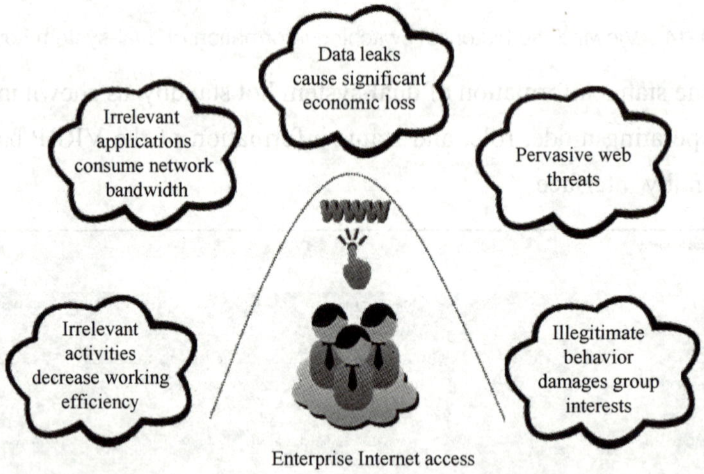

Figure 12-1 User authentication background

Subtask 12.1.2 Know AAA

AAA is the abbreviation for authentication, authorization, and accounting. It provides three security functions: authentication, authorization, and accounting. This technology can be used to verify the legitimacy of a user account, to authorize the services that the user can access, and to record the user's use of network resources. As shown in Figure 12-2.

(1) When the user wants to access the Internet to access resources. First, using authentication technology, the user enters the user name and password.

(2) After passing the authentication, through authorization, different users can access different resources, such as Baidu or Google.

(3) During the client's visit, billing is performed through accounting, and the operations and duration are recorded.

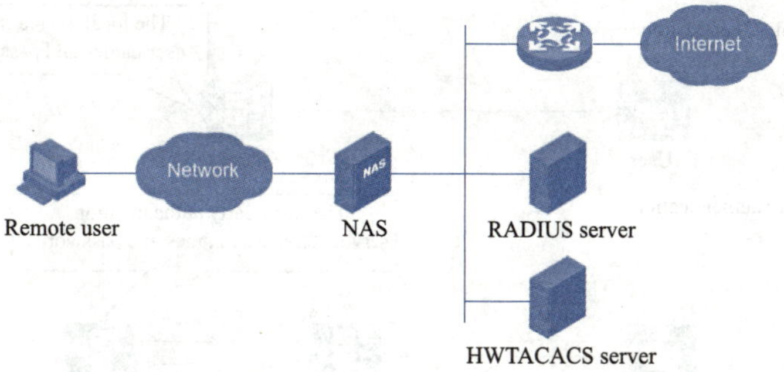

Figure 12-2 AAA

1. Authentication

Authentication refers to the confirmation of the user's identity through certain means. Authentication can be achieved through the information you know, what you have, and unique physical characteristics.

As shown in Figure 12-3, the objects of authentication include:

(1) I know: Information that the user knows [such as password, personal identification number (PIN), etc.].

(2) I own: Information owned by the user (such as token card, smart card or bank card).

(3) I have: The biological characteristics of the user (such as fingerprints, voice, retina, DNA).

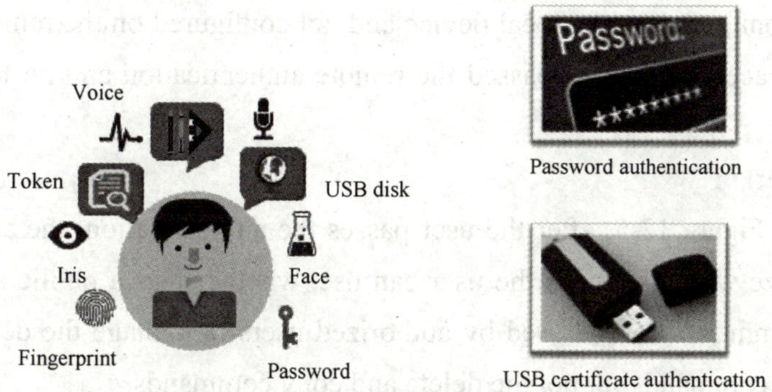

Figure 12-3 Authentication object

AAA supports some authentication methods, as shown in Figure 12-4.

(1) Non-authentication: Trust users very much and do not conduct legal checks on them. Generally, this method is not used.

(2) Local authentication: Configure user information (including the user name, password, and various attributes of local users) on the network access server. The advantage of local authentication is that it is fast and can reduce operating costs. The disadvantage is that the amount of stored information is limited by the hardware conditions of the device.

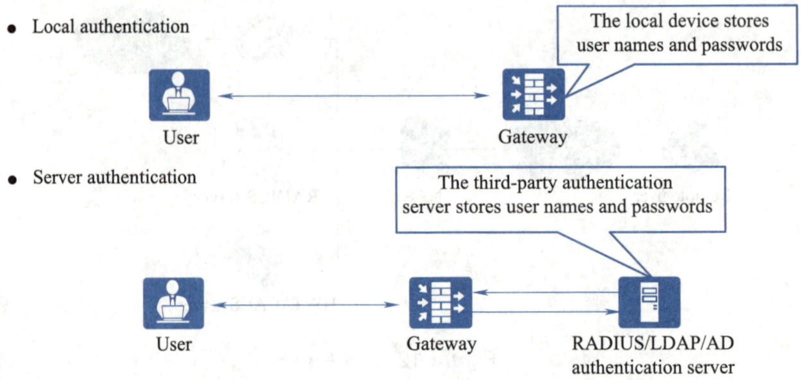

Figure 12-4 Authentication method

(3) Server authentication: Configure user information (including the user name, password, and various attributes of local users) on the authentication server. AAA supports remote authentication through RADIUS (remote authentication dial in user service) protocol or HWTACACS (Huawei terminal access controller access control system) protocol.

If an authentication scheme uses multiple authentication methods, these authentication methods take effect in the order in which they are configured. For example, if the remote authentication is configured first, and then the local authentication is configured, when the remote authentication server does not respond, it will switch to the local authentication mode. If only the login account is configured on the local device and not configured on the remote server, the device considers that the account has not passed the remote authentication and no longer performs local authentication.

2. Authorization

As shown in Figure 12-5, after the user passes the authentication, the authorization function is used to authorize which services the user can use, whether it is a public service or a sensitive service. Those commands can be used by authorized users to manage the device. For example, it can be the display command, but not the delete and copy commands.

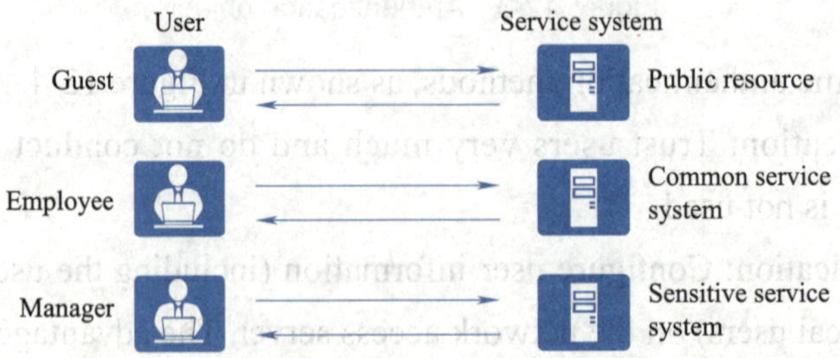

Figure 12-5 Authorization

The AAA authorization function gives the user the authority to access a specific network or device. AAA supports the following authorization methods:

(1) No authorization: No authorization is performed on the user.

(2) Local authorization: Authorize according to the relevant attributes of the local user account configured on the NAS.

(3) Remote authorization: HWTACACS authorization, using the TACACS server to authorize users. RADIUS authorization authorizes users who have passed RADIUS server authentication. The authentication and authorization of the RADIUS protocol are bound together, and RADIUS cannot be used alone for authorization.

If multiple authorization methods are used in an authorization scheme, these authorization methods take effect in the order in which they are configured. The non-authorization method finally takes effect. Authorization and authentication are often unified, and when users complete authentication, they will also be authorized.

3. Accounting

The billing function is used to monitor the network behavior of authorized users and the usage of network resources. There are three main meanings of billing:

(1) How long does the user spend.

(2) How much did the user spend.

(3) What operations did the user do.

The billing function is used to monitor the network behavior of authorized users and the usage of network resources. AAA supports the following two accounting methods:

(1) No billing: Provide users with free Internet services without generating related activity logs.

(2) Server accounting: Remote accounting through RADIUS server or HWTACACS server. The RADIUS server or HWTACACS server has sufficient storage space, can store the network access activity log of each authorized user, and supports the accounting function.

Subtask 12.1.3 Understand the working mechanism of RADIUS protocol

AAA can be implemented with a variety of protocols, the most commonly used is the RADIUS (remote authentication dial-in user service) protocol. RADIUS is widely used in NAS (network access server) systems. The NAS is responsible for passing user authentication and accounting information to the RADIUS server. The RADIUS protocol specifies how to transfer user information, accounting information, and authentication and accounting results between the NAS and the RADIUS server. The RADIUS server is responsible for receiving the user's connection request, completing the authentication, and returning the result to the NAS.

The RADIUS server authenticates users by establishing a unique user database and storing user names and passwords. RADIUS protocol uses UDP (user datagram protocol) as the transmission protocol, which has good real-time performance. At the same time, it also supports the retransmission mechanism and the backup server mechanism, which has better reliability.

The message flow between RADIUS client and server, as shown in Figure 12-6.

(1) When a user logs in to a network device such as the USG or access server, the user name and password will be sent to the network access server.

(2) The RADIUS client (network access server) in the network device receives the user name and password, and sends an authentication request to the RADIUS server.

(3) After receiving the legal request, the RADIUS server completes the authentication and returns the required user authorization information to the client. For illegal requests, the RADIUS server returns authentication failure information to the client.

Figure 12-6 Message flow between RADIUS client and server

RADIUS message structure, as shown in Figure 12-7.

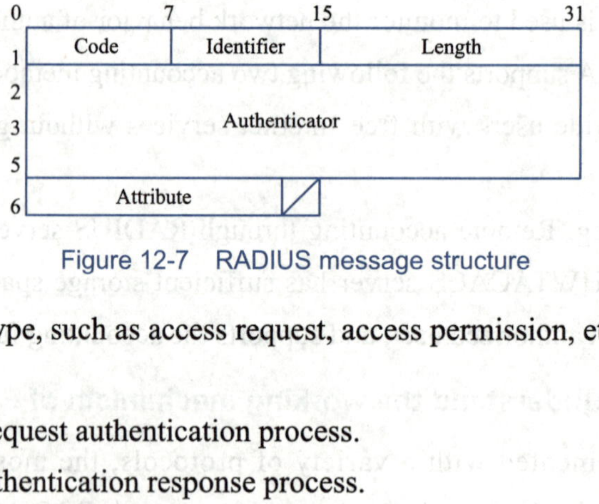

Figure 12-7 RADIUS message structure

(1) Code: Message type, such as access request, access permission, etc., occupies 1 byte and is defined as follows:

Access-Request—Request authentication process.

Access-Accept—Authentication response process.

Access-Reject—Certification rejection process.

Accounting-Request—Request billing process.

Accounting-Response—Billing response process.

Access-Challenge —Access question.

(2) Identifier: Generally, it is a sequentially increasing number, and this field in the request message and the response message must match.

(3) Length: The total length of all fields.

(4) Authenticator: Authentication key, used to verify the legitimacy of RADIUS.

(5) Attribute: The main content of the message is mainly various attributes related to the user.

The RADIUS message interaction process is shown in Figure 12-8.

(1) The user enters user name and password.

(2) Certification request.

(3) Certification acceptance.

(4) Billing start request.

(5) Accounting start request response message.

(6) User access to resources.

(7) Accounting end request message.

(8) Accounting end request response message.

(9) End of visit.

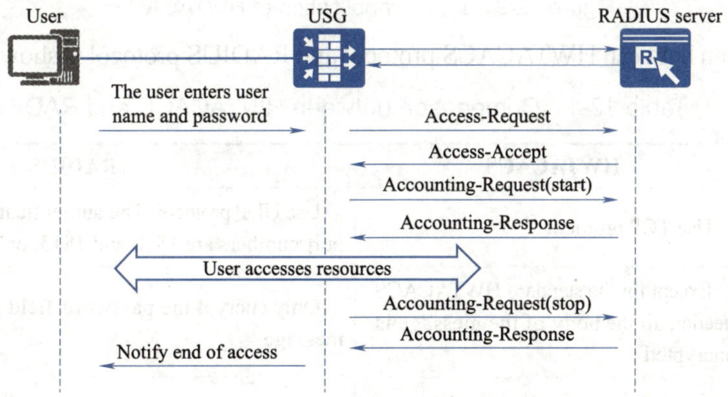

Figure 12-8 RADIUS packet exchange process

Subtask 12.1.4 Understand the working mechanism of HWTACACS protocol

HWTACACS is a security protocol with enhanced functions based on the TACACS protocol. The function of this protocol is similar to that of the RADIUS protocol, which uses the client/server mode to realize the communication between the NAS and the TACACS+ server.

As shown in Figure 12-9, the typical application of HWTACACS is to authenticate, authorize, and charge terminal users who need to log in to the device to perform operations. As the HWTACACS client, the device sends the user name and password to the HWTACACS server for verification. After the user is authenticated and authorized, he can log in to the device to perform operations.

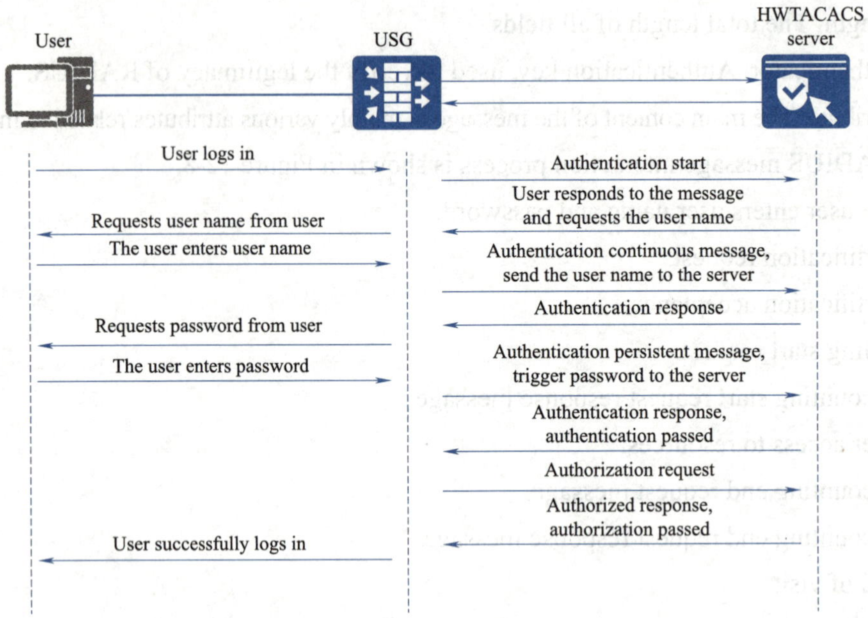

Figure 12-9 Typical application of HWTACACS

The comparison between HWTACACS protocol and RADIUS protocol is shown in Table 12-1.

Table 12-1 Comparison between HWTACACS and RADIUS

Items	HWTACACS	RADIUS
Port usage	Use TCP protocol	Use UDP protocol. The authentication and authorization port numbers are 1812 and 1813, or 1645 and 1646
Encryption situation	Except for the standard HWTACACS header, all the body of the message is encrypted	Only encrypt the password field in the authentication message
Authentication and authorization	Separation of authentication and authorization	Authentication and authorization are handled together
Application	Suitable for safety control	Suitable for billing
Configure command authorization	Support for authorization of configuration commands	Does not support authorization of configuration commands

Subtask 12.1.5 Understand how LDAP works

LDAP (lightweight directory access protocol) is developed from the X.500 directory access protocol. X.500 is hierarchical, and all objects are organized into a tree structure, as shown in Figure 12-10. X.500 directory service realizes identity authentication and access control. LDAP is to simplify the complexity of the X.500 directory while

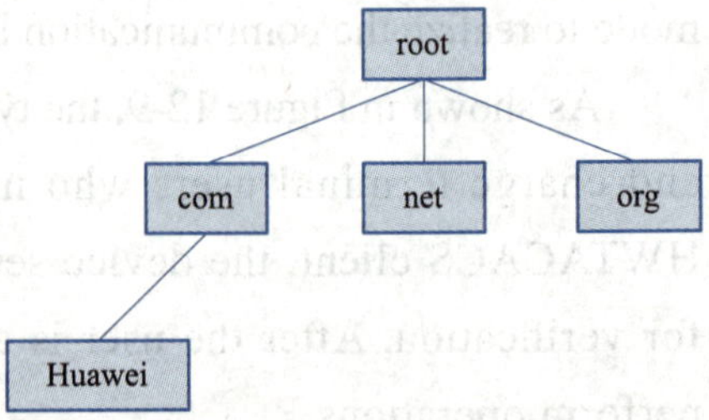

Figure 12-10 Tree structure

adapting to the needs of the Internet.

Directory service is a system composed of a directory database and a set of access protocols. Information similar to the following is suitable for storage in a directory.

(1) Enterprise employee information, such as name, e-mail address, mobile phone number, etc.;

(2) The physical information of the company's equipment, such as IP address, placement location, manufacturer, purchase time, etc.;

(3) Public certificates and security keys;

The characteristics of directory services are as follows:

(1) Organize data in a directory;

(2) Provide a unified access point to the outside world;

(3) Support distributed data storage of data;

(4) Optimize query, read data faster.

1. Basic concepts of LDAP

In LDAP, information is organized in a tree-like manner. The basic data unit is an entry, and each entry is composed of attributes, which are composed of a type (Type) and one or more values (Value).

(1) DIT (directory information tree): A collection of directory entries constitutes a directory information tree.

(2) Entry: A node in the directory information tree, which is the most basic unit in the directory information and consists of a series of attributes.

(3) Attribute: The attribute describes the characteristics of the object. An attribute consists of an attribute type and one or more attribute values

(4) RDN（relative distinguished name）: The name of the entry, which uniquely identifies the child entry of the same parent entry.

(5) DN（distinguished name）: A name that uniquely identifies an Item in a directory information tree. DN has three attributes, namely DC (domain component), CN (common name), and OU (organizational unit).

For example,CN=admin, OU=guest, DC=domainname, DC=com. In the above code, CN=admin may represent a user name, and OU=guest represents an organizational unit in the active directory. The meaning of this sentence may be that the admin object is in the guest organizational unit of the domainname.com domain.

2. LDAP authentication process

The LDAP authentication process is shown in Figure 12-11.

(1) The user enters user name and password to initiate a login request, and the firewall establishes a TCP connection with the LDAP server.

(2) The firewall sends a binding request message to the LDAP server with the administrator's DN

and password to obtain query authority.

(3) After the binding is successful, the LDAP server sends a binding response message to the firewall.

(4) The firewall uses the user name entered by the user to send a user DN query request message to the LDAP server.

(5) The LDAP server searches according to the user's DN, and sends a query response message if the query is successful.

(6) The firewall uses the queried user DN and the password entered by the user to send a user DN binding request message to the LDAP server. The LDAP server queries whether the user password is correct.

(7) After the binding is successful, the LDAP server sends a binding response message.

(8) After the authorization is successful, the firewall notifies the user that the login is successful.

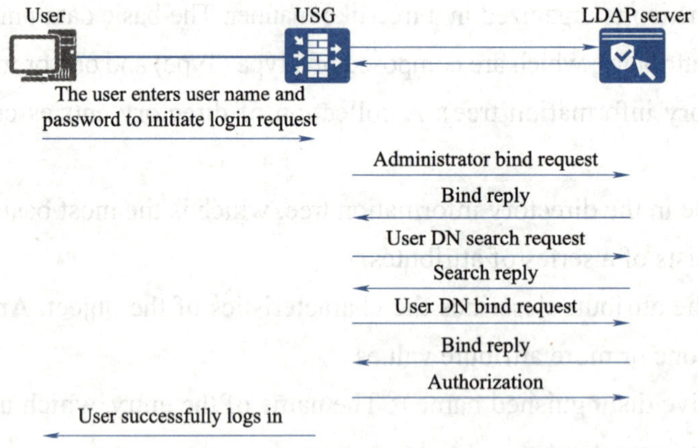

Figure 12-11 LDAP authentication process

Subtask 12.1.6 Master user authentication classification

AAA technology provides a means for user authentication. User authentication categories are:

(1) Local authentication: The visitor sends the user name and password identifying his identity to the FW through the Portal authentication page. The password is stored on the FW, and the verification process is performed on the FW. This method is called local authentication.

(2) Server authentication: The visitor sends the user name and password identifying his identity to the FW through the Portal authentication page. No password is stored on the FW. The FW sends the user name and password to the third-party authentication server. The verification process is performed on the authentication server. This method is called server authentication.

(3) Single sign-on: The visitor sends the user name and password identifying his identity to the third-party authentication server. After the authentication is passed, the third-party authentication

server sends the visitor's identity information to the FW. FW only records the identity information of the visitor and does not participate in the authentication process. This method is called single sign-on.

(4) SMS authentication: The visitor obtains the SMS verification code through the Portal authentication page, and then enters the SMS verification code to pass the authentication. FW authenticates visitors by verifying the SMS verification code.

Task 12.2 User management and application

Subtask 12.2.1 Understand the organizational structure and classification of users

User management will be divided into different user groups, through user authentication, users will be tagged, and different permissions and applications will be assigned to the user groups, so as to achieve the goal of security.

As shown in Figure 12-12, add company employees (users) to user groups, and then control and audit network behaviors for users or user groups. Visually formulate strategies based on users or user groups to improve the ease of use of strategy formulation. User information is reflected in reports. Analyze the online behavior of users to achieve tracking and auditing of the behavior of users (rather than pure IP addresses), and solve the problem of application behavior policy control caused by frequent changes in the corresponding IP of the same user in existing network applications.

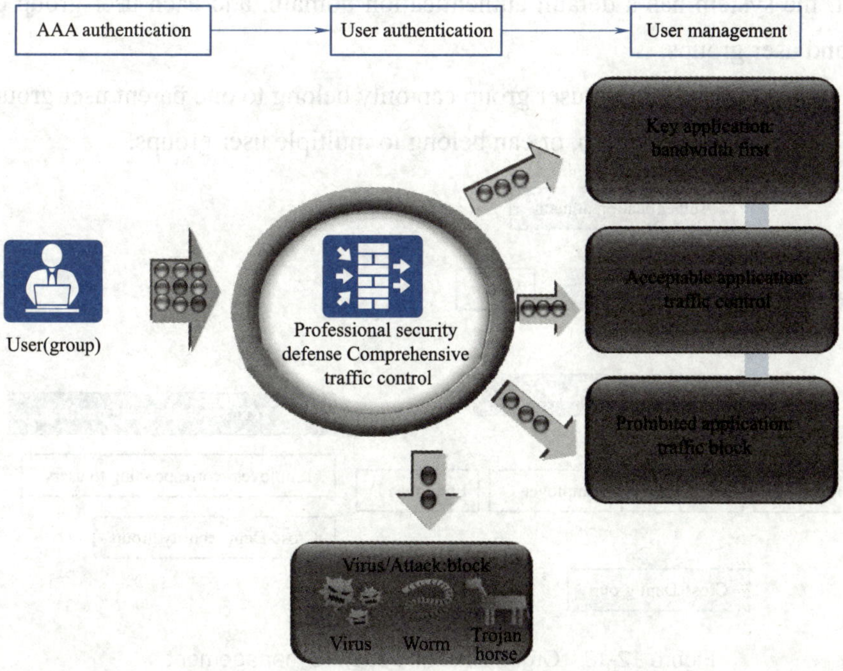

Figure 12-12 User management background

1. User organization structure

The user is the main body of network access and the basic unit of FW for network behavior control and network authority distribution. The following three concepts are involved in the user organization structure:

(1) Authentication domain: The container of the user organization structure. The default authentication domain exists in the firewall by default, and users can create new authentication domains according to their needs.

(2) User groups/Users: Users are organized in a tree structure, and users belong to groups (departments). The administrator can create departments and users according to the organization structure of the enterprise.

(3) Security group: A cross-departmental group with a horizontal organizational structure. When you need to manage users based on dimensions other than departments, you can create cross-departmental security groups. For example, groups established across departments in an enterprise.

When an enterprise stores its organizational structure through a third-party authentication server, there are similar horizontal groups on the server. In order to configure policies based on these groups, security groups need to be created on the FW to be consistent with the organizational structure on the server.

2. Organizational structure management

By default, the system has a default authentication domain, and each user group can include multiple users and user groups.

As shown in Figure 12-13, each user group can only belong to one parent user group, and each user belongs to at least one user group, or can belong to multiple user groups.

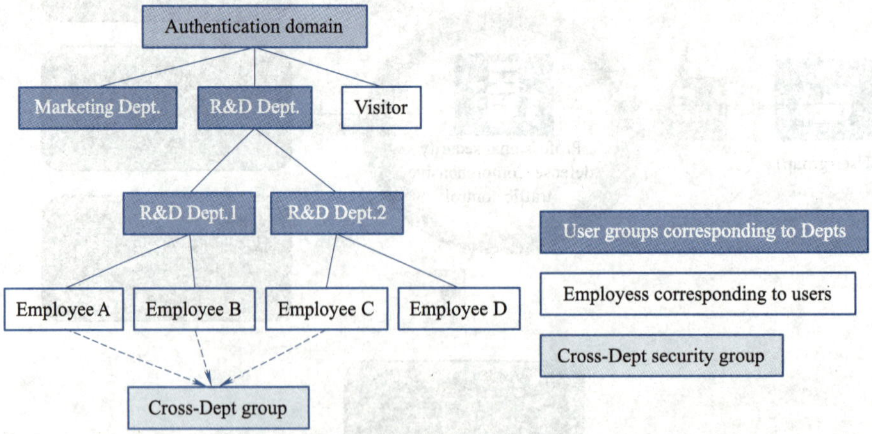

Figure 12-13 Organizational structure management

The authentication domain is an important part of the authentication process. The configuration of the authentication domain determines the authentication method for users and the user's organizational structure. For users with different authentication methods, the role of the authentication domain is different.

FW "distributes" all users to be authenticated to the corresponding authentication domain by identifying the authentication domain contained in the user name. The user is authenticated according to the configuration on the authentication domain.

In order to perform differentiated management and assign different permissions to different users or departments, it is necessary to plan and manage the organizational structure. The firewall supports the creation of a tree-like organizational structure, which is similar to the usual administrative structure and is very convenient for planning and management.

Each user (group) can be referenced by security policies, current limiting policies, etc., so as to achieve user-based permissions and bandwidth resource control.

If the administrator uses the default authentication domain to authenticate the user, the user only needs to enter the user name when logging in. If the administrator uses the newly created authentication domain to authenticate users, the user needs to enter "Username@Authentication Domain Name" when logging in.

3. User classification

For the purpose of users accessing the network, users can be divided into the following three categories:

(1) Administrator: An administrator user refers to a user who accesses the device and configures or operates the device through Telnet, SSH, Web, FTP and other protocols or through the console interface.

(2) Internet users: Internet users are the identification subject of network access and the basic unit of network authority management of equipment. The device obtains the user's identity by authenticating the user who accesses the network, and performs corresponding policy control on the user's identity. For example, internal employees in the corporate headquarters, Internet users can directly access network resources through the firewall.

(3) Access users: The subjects who access network resources in the external network, such as branch employees and employees on business trips. Access users need to access the firewall through SSL VPN, L2TP VPN, IPSec VPN, or PPPoE before they can access the network resources of the corporate headquarters.

Subtask 12.2.2 Administrator certification process and configuration

1. Administrator login method

The administrator is mainly to realize the management, configuration and maintenance of the equipment, the login methods can be divided into:

(1) Console: The Console interface provides a command line method to manage the device. It is usually used for the first configuration of the device, or the device configuration file is lost and there is no configuration. When the device system cannot be started, you can perform diagnosis through the Console port or enter BootRom to upgrade.

(2) Web: The terminal logs in to the device through HTTP/HTTPS for remote configuration and management.

(3) Telnet: Telnet is a traditional login method, usually used to configure and manage the device through command lines.

(4) FTP: The FTP administrator mainly uploads and downloads files in the storage space of the device.

(5) SSH: SSH provides secure information protection and powerful authentication functions, and provides a secure "channel" on an insecure network. At this time, the device acts as an SSH server.

2. Console/Telnet/FTP device management

Console/Telnet/FTP device management can be configured through the Web, as shown in Figure 12-14.

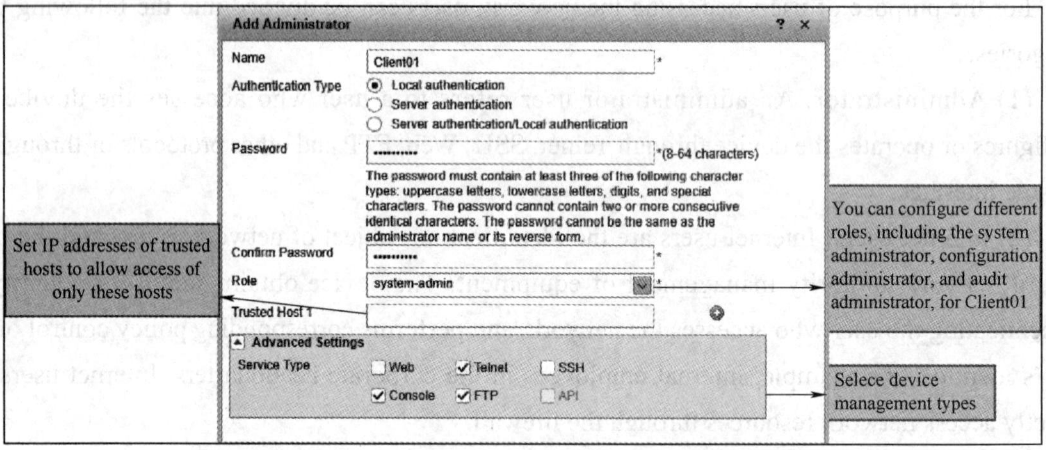

Figure 12-14 Console/Telnet/FTP device management

Console/Telnet/FTP device management can also be configured through CLI:

Step 1: User-interface.

```
Console:
[USG] user-interface console 0
[USG-ui-con0] authentication-mode aaa
```

```
Telnet:
[USG] user-interface vty 0 4
[USG-ui-vty0] authentication-mode aaa
```

Step 2: AAA view.

```
[USG] aaa
[USG-aaa]manager-user client001
[USG-aaa-manager-user-client001]password cipher Admin@123
[USG-aaa-manager-user-client001]service-type terminal telnet ftp
[USG-aaa-manager-user-client001]level 3
[USG-aaa-manager-user-client001]ftp-directory hda1:
```

3. SSH device management

SSH device management can be configured through the Web, as shown in Figure 12-15.

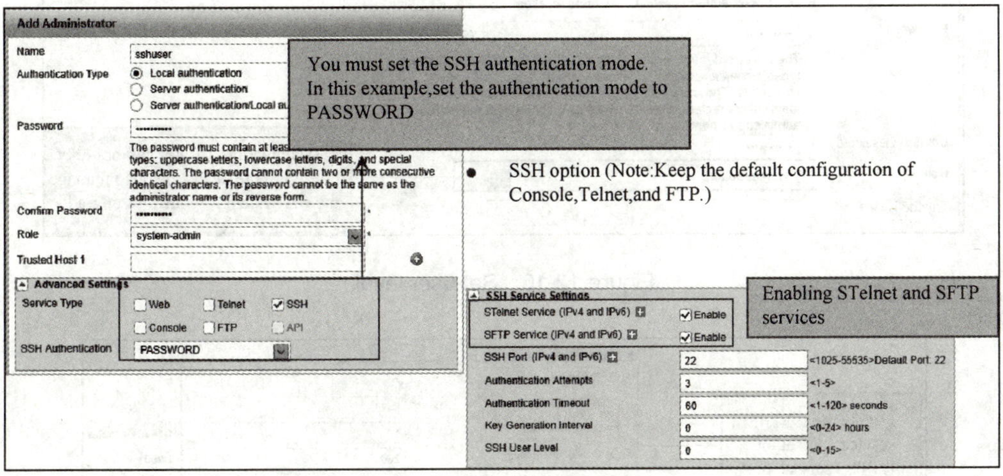

Figure 12-15 SSH device management

SSH device management can also be configured via CLI:

Step 1: Start the SSH service.

```
[USG]stelnet server enable
Info: The Stelnet server is already started.
```

Step 2: Configure the password for the SSH user sshuser as Admin@123.

```
[USG] aaa
[USG-aaa] manager-user sshuser
[USG-aaa-manager-user-client001] ssh authentication-type password
[USG-aaa-manager-user-client001] password cipher Admin@123
[USG-aaa-manager-user-client001] service-type ssh
```

After the above configuration is completed, run the client software that supports SSH to establish an SSH connection

4. Web device management

Web device management can be configured via Web:

Step 1: Create an administrator webuser and set the user level, as shown in Figure 12-16.

Step 2: Set the HTTPS/HTTP service port in "System" → "Administrator" → "Settings". After logging in to the device using HTTP, the HTTP service cannot be closed, and the HTTP service port cannot be modified. After logging in to the device using HTTPS, the HTTPS service cannot be closed, and the HTTPS service port cannot be modified, as shown in Figure 12-17.

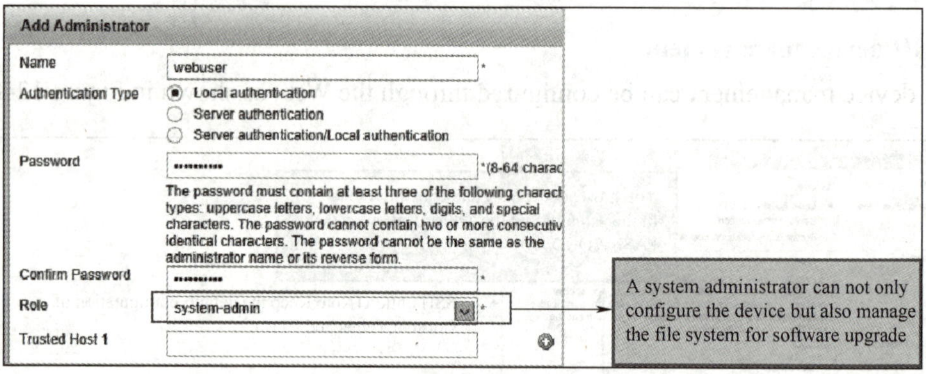

Figure 12-16 Set user level

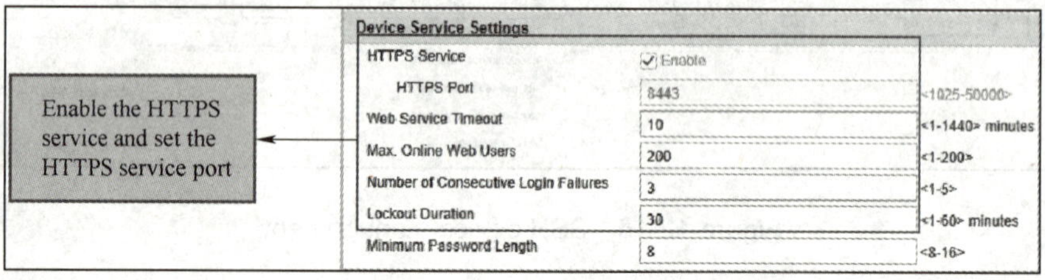

Figure 12-17 HTTP/HTTPS settings

Web device management can also be configured via CLI:

Step 1: Start the Web management function.

```
[USG] web-manager security enable port 6666
```

Step 2: Configure Web users.

```
[USG] aaa
[USG-aaa]manager-user webuser
[USG-aaa-manager-user-webuser]password cipher Admin@123
[USG-aaa-manager-user-webuser]service-type web
[USG-aaa-manager-user-webuser]level 3
```

Subtask 12.2.3 Authentication process of Internet users and access users

The authentication process for Internet users and access users is shown in Figure 12-18.

In the single sign-on scenario of Internet users, as long as the user has passed the authentication of other authentication systems, it is equivalent to have passed the FW authentication. After the user is authenticated, the FW can learn the corresponding relationship between the user and the IP, so as to perform policy management based on the user. This method is suitable for scenarios where the authentication system has been deployed before the firewall user authentication function is deployed.

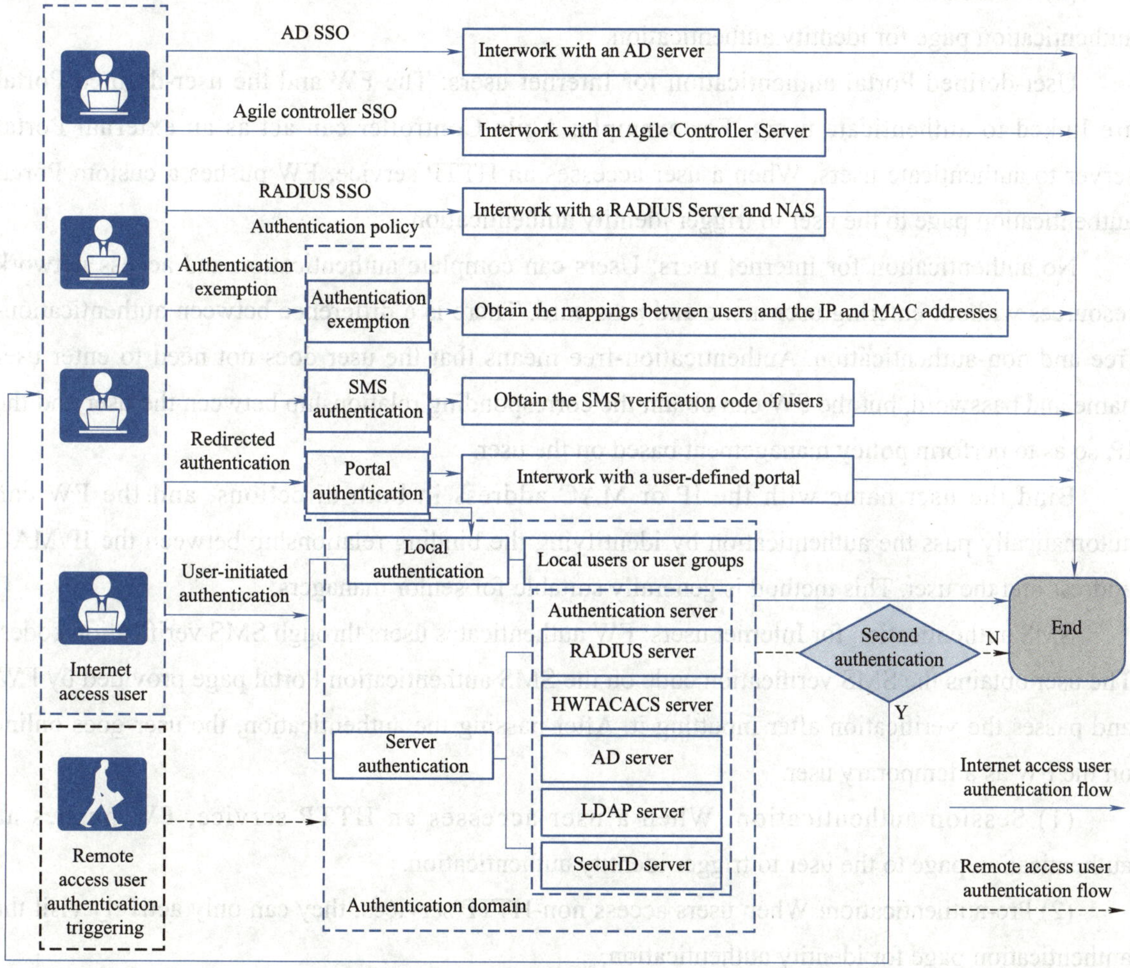

Figure 12-18 Authentication process

AD single sign-on: The user logs in to the AD domain and is authenticated by the AD server.

(1) Agile controller single sign-on: Users are authenticated by Huawei's Agile Controller system (Policy Center or Agile Controller).

(2) RADIUS single sign-on: The user accesses the NAS device, and the NAS device forwards the authentication request to the RADIUS server for authentication.

Built-in Portal authentication for Internet users: FW provides a built-in Portal authentication page (default is https://interface IP address:8887) to authenticate users. FW can forward the authentication request to the local user database and authentication server. This method is suitable for scenarios where users are authenticated through FW.

(1) Session authentication: When a user accesses an HTTP service, FW pushes an authentication page to the user to trigger identity authentication.

(2) Pre-authentication: When users access non-HTTP services, they can only actively visit the authentication page for identity authentication.

User-defined Portal authentication for Internet users: The FW and the user-defined Portal are linked to authenticate users. For example: Agile Controller can act as an external Portal server to authenticate users. When a user accesses an HTTP service, FW pushes a custom Portal authentication page to the user to trigger identity authentication.

No authentication for Internet users: Users can complete authentication and access network resources without entering user name and password. There is a difference between authentication-free and non-authentication. Authentication-free means that the user does not need to enter user name and password, but the FW can obtain the corresponding relationship between the user and the IP, so as to perform policy management based on the user.

Bind the user name with the IP or MAC address in both directions, and the FW can automatically pass the authentication by identifying the binding relationship between the IP/MAC address and the user. This method is generally suitable for senior managers.

SMS authentication for Internet users: FW authenticates users through SMS verification codes. The user obtains the SMS verification code on the SMS authentication Portal page provided by FW, and passes the verification after inputting it. After passing the authentication, the user goes online on the FW as a temporary user.

(1) Session authentication: When a user accesses an HTTP service, FW pushes an authentication page to the user to trigger identity authentication.

(2) Pre-authentication: When users access non-HTTP services, they can only actively visit the authentication page for identity authentication.

Access user authentication: FW authenticates the user during the user VPN access process. If you want to re-authenticate after successful access authentication and before accessing resources, you can configure secondary authentication. Support local authentication and server authentication.

Authentication strategy:

(1) When Internet users use authentication-free or session authentication to trigger the

authentication process, and access users who have already connected to the FW use session authentication to trigger the authentication process, they must go through the authentication policy.

(2) The role of the authentication strategy is to select data streams that need to be authenticated-free or session-authenticated. For data streams without authentication, the FW identifies the user based on the binding relationship between the user and the IP/MAC address; For the data flow of session authentication, FW will push the authentication page. The authentication strategy does not work for single sign-on or pre-authentication.

Local authentication/Server authentication user: Decide whether to use local authentication or server authentication for user authentication. If it is server authentication, it also includes which authentication server to use.

Single sign-on users: FW does not participate in the authentication process of single sign-on users, but only receives user login/logout messages from the authentication server, so the authentication mode configuration in the authentication domain does not work for single sign-on users. However, the new user option configured in the authentication domain with the same name as the user domain name (dc field) is effective for single sign-on users.

Subtask 12.2.4 Single sign-on

Single sign-on means that the firewall is not an authentication point. The firewall obtains user login messages from other authentication systems to enable users to go online on the firewall.

AD single sign-on means that after the user hopes to pass the authentication of the AD server, it will automatically pass the authentication of the FW, and then can access all network resources. Visitors can use AD single sign-on to trigger authentication on the FW. By enabling the single sign-on function, the device can identify users who have passed the authentication of these identity authentication systems, and avoid users who are required to enter a user name and password again when surfing the Internet. In AD single sign-on, FW supports multiple ways to obtain the message of user authentication. There are three main login implementation methods for AD single sign-on:

(1) Accept PC message mode.

(2) Query the security log mode of the AD server.

(3) The firewall monitors AD authentication packets.

1. Accept PC message mode

As shown in Figure 12-19, the administrator needs to deploy the AD single sign-on service on the AD monitor. Set the login script and logout script on the AD server (AD domain controller), and configure the AD single sign-on parameters on the firewall to receive user login/logout messages sent by the AD single sign-on service. The AD monitor can be an AD domain controller or other machines in the AD domain.

The specific login process is as follows:

(1) The visitor logs in to the AD domain, and the AD server returns a login success message to the user and issues a login script.

(2) The visitor PC executes the login script and sends the user login information to the AD monitor.

(3) The AD monitor connects to the AD server to query the login user information, and if the user's information can be queried, it forwards the user login information to the FW.

(4) FW extracts the corresponding relationship between the user and the IP from the login information and adds it to the online user table.

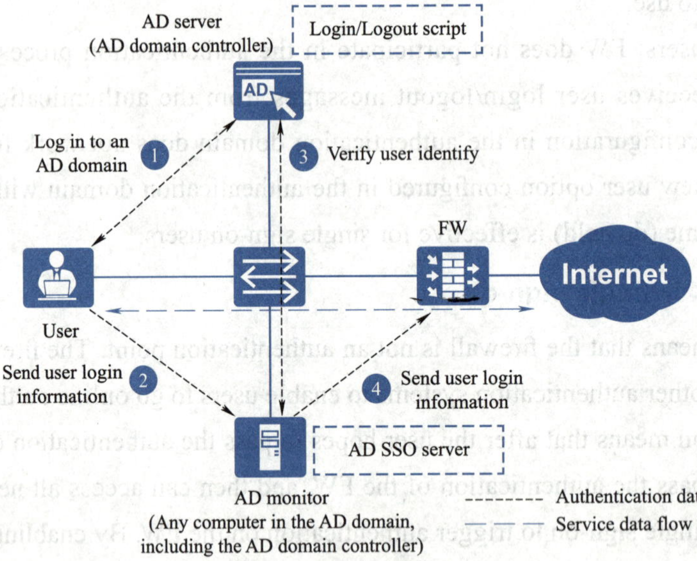

Figure 12-19 Receiving PC message mode

If the interactive packets between the visitor and the AD server, the visitor and the AD monitor, and the AD monitor and the AD server have passed the FW, you need to ensure that the authentication policy on the FW does not authenticate these packets. At the same time, in the security policy, it is ensured that such packets can pass through the FW normally.

2. Query the security log mode of the AD server

As shown in Figure 12-20, the administrator needs to deploy the AD single sign-on service on the AD monitor. At the same time, configure the AD single sign-on parameters on the FW, and receive the user login message sent by the AD single sign-on service.

The specific login process is as follows:

(1) The visitor logs in to the AD domain, and the AD server records the user online information in the security log.

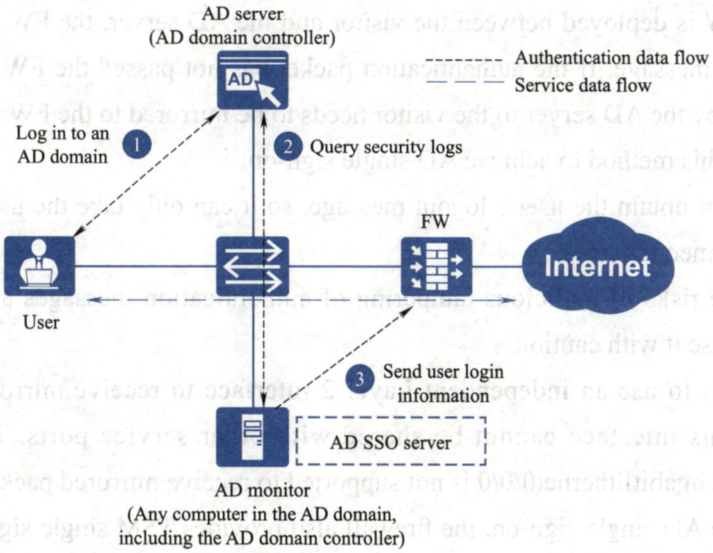

Figure 12-20 Querying the security log mode of the AD server

(2) The AD monitor connects to the AD server through the WMI (Windows management instrumentation) interface provided by the AD server to query security logs and obtain user login messages.

(3) The AD monitor starts from the time when the AD single sign-on service starts, and regularly queries the security logs generated on the AD server.

(4) The AD monitor forwards the user login message to the FW, and the user goes online at the FW.

3. The firewall monitors AD authentication packets

As shown in Figure 12-21, the administrator does not need to install the program on the AD server. FW obtains the authentication result by monitoring the authentication message that the visitor logs in to the AD server. If the authentication is successful, the corresponding relationship between the user and the IP is added to the online user table.

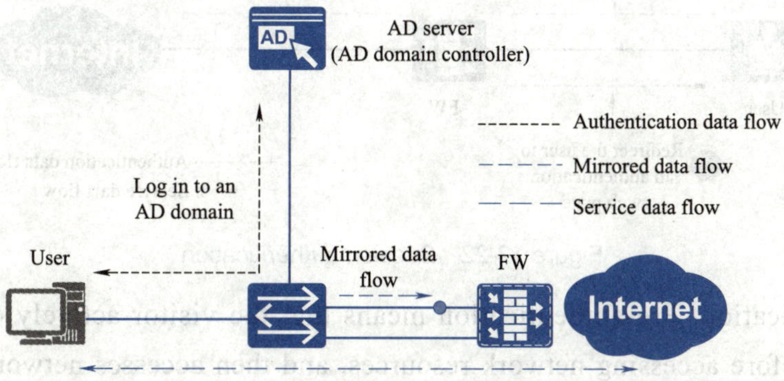

Figure 12-21 The firewall monitors AD authentication packets

When the FW is deployed between the visitor and the AD server, the FW can directly obtain the authentication message. If the authentication packet has not passed the FW, the authentication result packet sent by the AD server to the visitor needs to be mirrored to the FW.

When using this method to achieve AD single sign-on:

(1) FW cannot obtain the user's logout message, so it can only take the user offline based on the online user's timeout period.

(2) There are risks of malicious tampering of authentication messages and forgery of user identities. Please use it with caution.

(3) FW needs to use an independent Layer 2 interface to receive mirrored authentication messages, and this interface cannot be shared with other service ports. The configuration management port GigabitEthernet0/0/0 is not supported to receive mirrored packets.

In addition to AD single sign-on, the firewall also provides TSM single sign-on and RADIUS single sign-on.

Subtask 12.2.5 User Portal authentication

Portal authentication means that a firewall or a third-party server provides a Portal authentication page to authenticate users. The triggering methods of the firewall's built-in Portal authentication include: session authentication and pre-authentication.

Session authentication: Session authentication means that the user does not actively perform identity authentication, first performs HTTP business access, and then authenticates during the access process. After the authentication is passed, business visits can be carried out.

As shown in Figure 12-22, when the firewall receives the user's first HTTP service access data stream, it redirects the HTTP request to the authentication page to trigger the visitor's identity authentication. After passing the authentication, you can access HTTP services and other services.

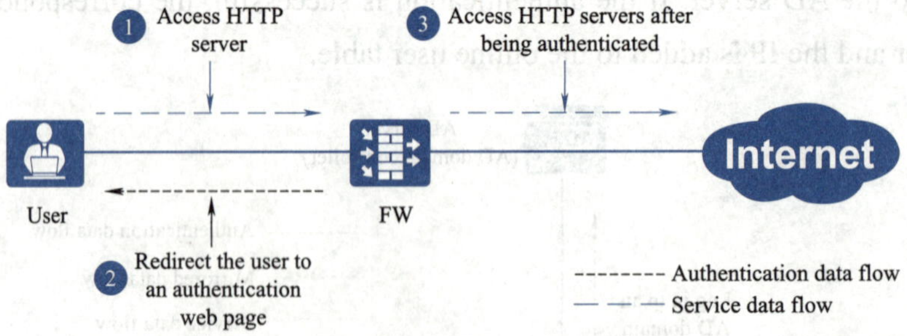

Figure 12-22 Session authentication

Pre-authentication: Pre-authentication means that the visitor actively conducts identity authentication before accessing network resources, and then accesses network resources after passing the authentication.

As shown in Figure 12-23, the user actively initiates an authentication request to the authentication page provided by the firewall. After FW receives the authentication request, it performs identity authentication. After passing the authentication, you can access the Internet.

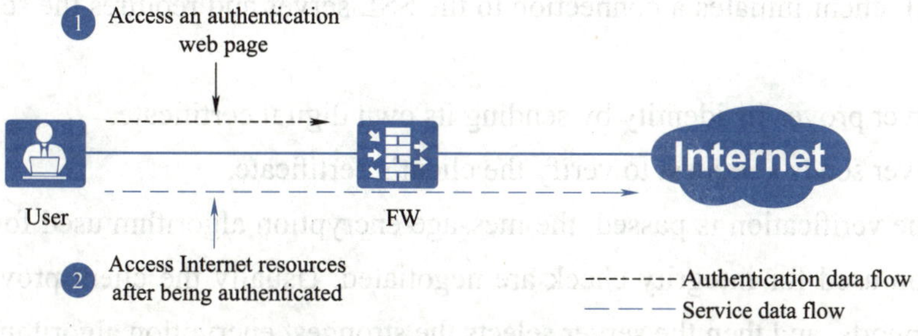

Figure 12-23 Pre-certification

Subtask 12.2.6 Access user authentication

Access user authentication refers to the authentication of various VPN access users. The method of triggering authentication is determined by the access method, including:

(1) SSL VPN access users. The visitor logs in to the authentication page provided by the SSL VPN module to trigger the authentication process. After the authentication is completed, SSL VPN access users can access the network resources of the headquarters.

(2) L2TP VPN access users. For the LAC autonomous dial-up mode, in the access phase, the LAC of the branch triggers the authentication process through the dial-up mode and establishes an L2TP VPN tunnel with the LNS. In the stage of accessing resources, visitors in branches can use pre-authentication, session authentication and other methods to trigger the authentication process. After the authentication is completed, L2TP VPN access users can access the network resources of the headquarters.

For the NAS-Initiated/Client-Initiated method, in the access phase, the visitor triggers the authentication process by dialing, and establishes an L2TP VPN tunnel with the LNS. In the resource access stage, visitors in the branch can directly access the network resources of the headquarters. You can also use pre-authentication, session authentication and other methods to trigger secondary authentication. After passing the secondary authentication, you can access the network resources of the headquarters.

(3) IPSec VPN access users. After the branch and the headquarters establish an IPSec VPN tunnel, visitors in the branch can use pre-authentication, session authentication and other methods to trigger the authentication process. After the authentication is completed, IPSec VPN access users can access the network resources of the headquarters.

SSL VPN is a VPN technology based on HTTPS, which works between the transport layer and the application layer, and provides a confidential security protocol based on the Internet. The main services provided are Web proxy, network extension, file sharing and port forwarding. The handshake steps of SSL protocol communication are as follows:

(1) The SSL client initiates a connection to the SSL server and requires the server to verify its identity.

(2) The server proves its identity by sending its own digital certificate.

(3) The server sends a request to verify the client's certificate.

(4) After the verification is passed, the message encryption algorithm used for encryption and the hash function used for integrity check are negotiated. Usually the client provides a list of all algorithms it supports, and then the server selects the strongest encryption algorithm.

(5) The client and server generate a session key through the following steps:

The client generates a random number, encrypts it with the server's public key (obtained from the server certificate), and sends it to the server.

The server responds with random data (the client key is used when the client's key is available, otherwise the data is sent in plain text).

Use a hash function to generate a key from random data.

As shown in Figure 12-24, an enterprise deploys a firewall at the network boundary as a VPN access gateway to connect the internal network and the Internet. After traveling employees access the FW through SSL VPN, they use the network extension service to access network resources.

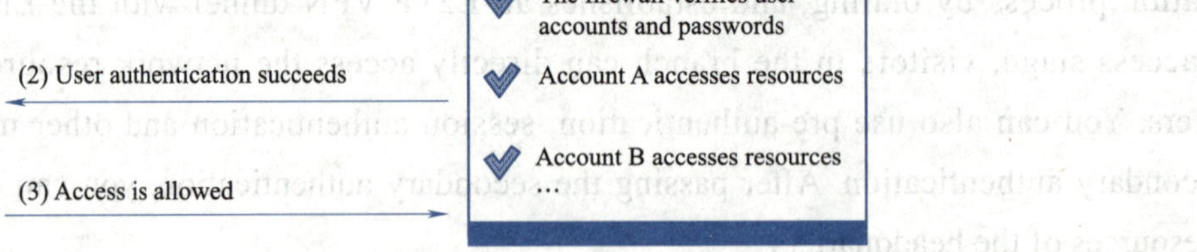

Figure 12-24 SSL access users access internal resources

Task 12.3 Formulation of user authentication strategy

The authentication strategy is used to determine which data flows the firewall needs to authenticate. Data flows matching the authentication strategy must pass the firewall's identity authentication before they can pass. By default, the firewall does not authenticate data flows that pass through itself, and the data stream to be authenticated need to be selected through the authentication policy. If the traffic passing through the firewall matches the authentication policy, the following actions will be triggered:

(1) Session authentication: When a user accesses an HTTP service, if the data flow matches the authentication policy, the firewall will push the authentication page to ask the visitor to authenticate.

(2) Pre-authentication: When accessing non-HTTP services, users must actively visit the authentication page for authentication, otherwise access to service data streams matching the authentication policy will be prohibited by FW.

(3) Authentication-free: When users access services, if they match the authentication-free authentication policy, they do not need to enter a user name and password to directly access network resources. The firewall recognizes the user based on the binding relationship between the user and the IP/MAC address.

(4) Single sign-on: The online of single sign-on users is not controlled by the authentication policies, but user business traffic must match the authentication policies in order to perform policy control based on users.

The following traffic will not trigger authentication even if it matches the authentication policy:

(1) Access to the device or the traffic initiated by the device.

(2) DHCP, BGP, OSPF, and LDP packets.

(3) The DNS packet corresponding to the first HTTP service data flow that triggers authentication is not controlled by the authentication policy, and the DNS packet after the user is authenticated and goes online is controlled by the authentication policy.

An authentication policy is a collection of multiple authentication policy rules. The authentication policy determines whether to authenticate a piece of traffic. Authentication policy rules are composed of conditions and actions. Conditions refer to the basis for FW matching packets, including:

(1) Source/destination security zone.

(2) Source address/region.

(3) Destination address/region.

Action refers to the processing method that FW takes on the matched data stream, including:

(1) Portal authentication. Perform Portal authentication on eligible data streams.

(2) SMS authentication. Perform SMS authentication on eligible data streams and require users to enter an SMS verification code.

(3) Authentication-free. Authentication-free is performed on data streams that meet the

conditions, and the firewall uses other means to identify users. Mainly used in the following situations:

For senior managers of enterprises, on the one hand, they hope to omit the certification process; On the other hand, they can access confidential data and have stricter security requirements. To this end, the administrator can bind such users with IP/MAC addresses in a two-way manner to exempt such data streams from authentication, but require them to only use designated IP or MAC addresses to access network resources. The firewall identifies the user to which the data flow belongs through the binding relationship between the user and the IP/MAC address.

In the scenario of AD/TSM/RADIUS single sign-on, the firewall has already obtained user information from other authentication systems and exempts the service traffic of single sign-on users from being authenticated.

(4) No authentication. No authentication is performed on data streams that meet the conditions, and it is mainly used in the following situations:

Data streams that do not need to be FW certified, such as data streams for mutual visits between intranets.

In the scenario of AD/TSM/RADIUS single sign-on, if the data flow between the unauthenticated visitor and the authentication server passes through the FW, such data flow will not be authenticated.

There is a default authentication policy on the firewall, all matching conditions are any, and the action is no authentication.

Task 12.4 User authentication configuration case

Figure 12-25 shows the authentication configuration process for online users, where:

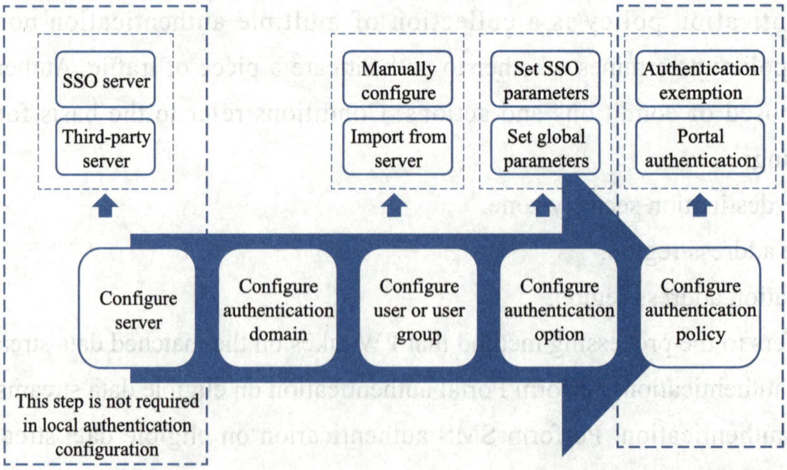

Figure 12-25 Online user authentication configuration process

(1) Configure groups/users: Before the device implements user/user group-based management, users/user groups must be created first. The device supports multiple creation methods of manual configuration by the administrator, local import and server import.

(2) Manually configure groups/users: The default authentication domain exists on the FW by default, and users/groups can be created at the lower level. If you need to plan the organizational structure of other authentication domains, please configure the authentication domain first.

This step is required when you need to create user groups based on the corporate organizational structure, and perform management such as network permission assignment based on user groups.

When performing local password authentication for users, users must be created locally and local password information must be configured.

(3) Local import: Local import supports importing user information in CSV format files and database dbm files to the local device.

(4) Server import: In the network, there are many cases where third-party authentication servers are used. Many companies have authentication servers on their networks, and all users and user group information are stored on the authentication servers. Importing users in batches from the authentication server refers to importing the user (group) information on the authentication server into the device through the server import policy.

(5) The configuration authentication options include the configuration of three parts: global parameters, single sign-on, and customized authentication pages.

(6) The global parameter configuration is mainly for local authentication and server authentication, and its configuration includes:

Set the strength of the user's password, the user must modify the password for the first login, and the password expiration settings;

Set how to deal with current authentication when authentication conflicts;

Define the jump page after user authentication;

Define the protocol and port used by the authentication interface;

Define the limit of the number of user login errors, the lock time after the limit is reached, and the user online timeout period.

(7) The single sign-on part includes the configuration methods of AD single sign-on, TSM single sign-on and RADIUS single sign-on. In this textbook, only the AD single sign-on will be explained in detail.

(8) According to the actual situation, the customized authentication page can be set with a Logo picture, a background picture, a welcome message or a help message, etc., to meet the needs of personalized page customization.

(9) When an Internet user or an access user who has already accessed the FW uses the session authentication method to trigger the authentication process, it must be processed by the authentication policy.

(10) The authentication policy is a collection of multiple authentication policy rules. When the FW matches packets, it is always performed between multiple rules, and the matching is performed from top to bottom. When the attributes of the message match all the conditions of a certain rule, it is considered that the rule is successfully matched, and subsequent rules will not be matched. If all the rules are not matched, it will be processed according to the default authentication policy.

(11) There is a default authentication policy on the FW, all matching conditions are any, and the action is no authentication.

Subtask 12.4.1 Configure servers

Here, only the RADIUS server and AD server are used as examples to introduce the server configuration.

1. Configure RADIUS server

In the scenario where a RADIUS server is used for server authentication of users, the FW acts as a proxy client of the RADIUS server and sends the user name and password to the RADIUS server for authentication. And the parameters configured on the FW must be consistent with the parameters on the RADIUS server.

As shown in Figure 12-26, choose "Object" → "Authentication Server" → "RADIUS" → "New" to create a new RADIUS server.

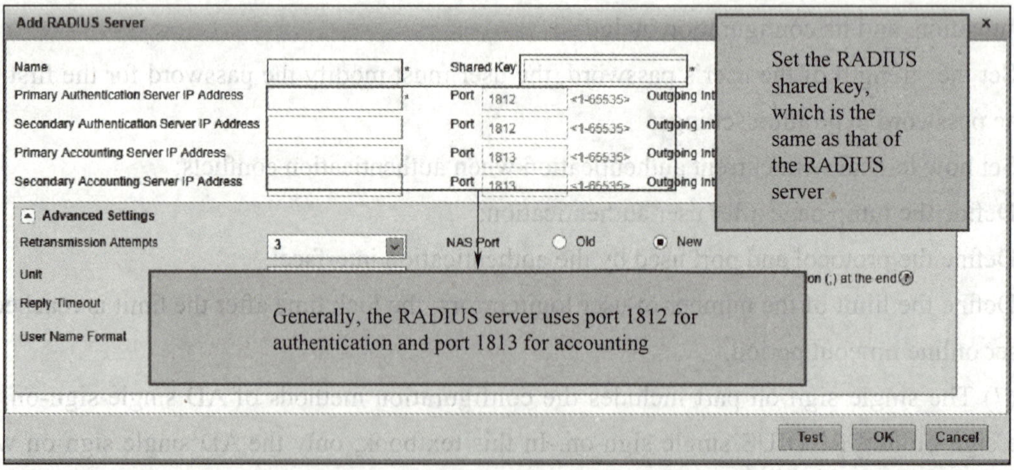

Figure 12-26 Configuring the RADIUS server

2. Configure AD server

When configuring the AD server, make sure that the system time and time zone of the FW are consistent with the time and time zone of the AD server. As shown in Figure 12-27, in "Object" → "Authentication Server" → "AD" → "New", create an AD server.

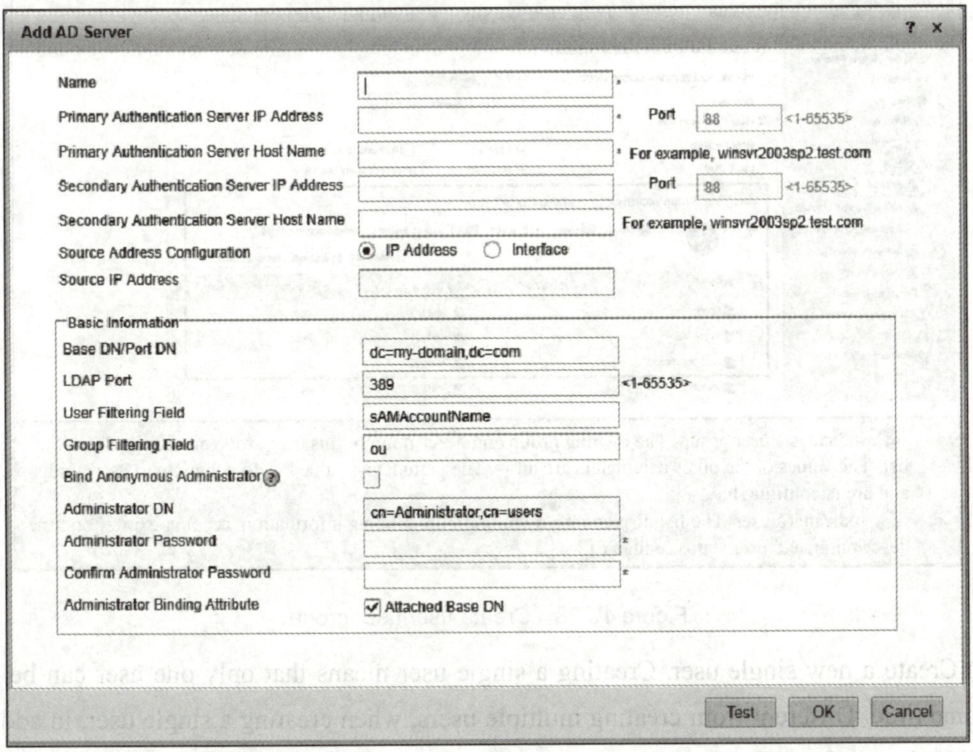

Figure 12-27 Configuring the AD server

Subtask 12.4.2 Users and user groups

1. Create users and user groups

As shown in Figure 12-28, in "Object" → "User" → "default", create a new user/user group.

(1) Group/User. Before the device implements user/user group-based management, the user/user group must exist on the device. Through the group/user node, manually create users/user groups on the device.

(2) Create a new user group. The root group is the root user group that comes with the device by default. It cannot be deleted or created. The group name of the root group cannot be modified, but its description can be configured for easy identification.

(3) All created user groups are subgroups of the root group, or subgroups of subgroups.

Select "Object" → "User" → "User/Group".

Select the authentication domain for which the group needs to be created. By default, there is only the default authentication domain.

In Member Management, click "New" and select "New Group".

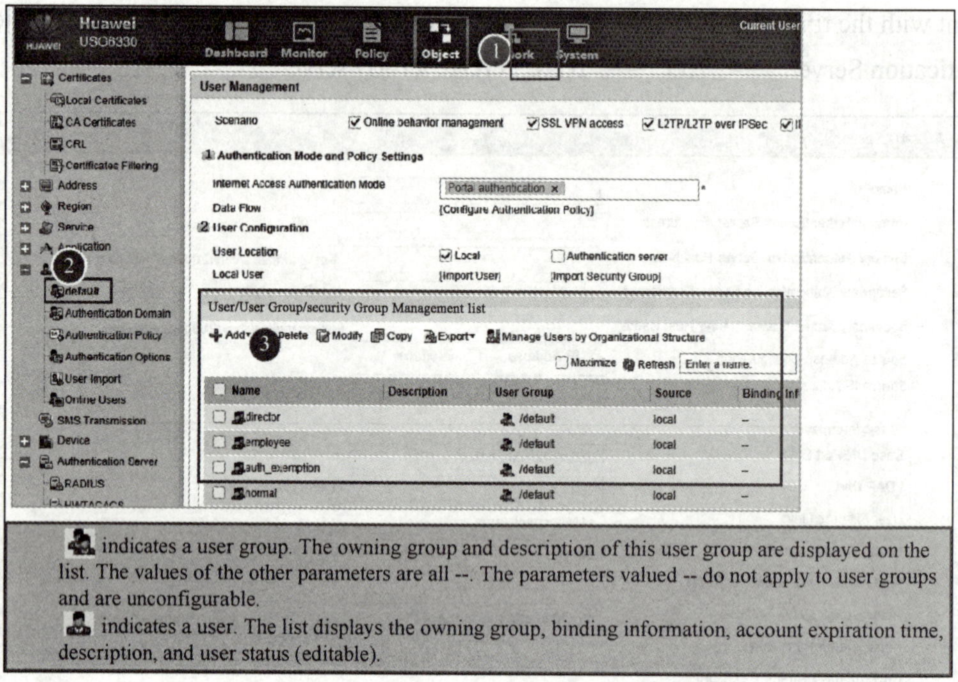

Figure 12-28 Create user/user group

(4) Create a new single user. Creating a single user means that only one user can be created at the same time. Different from creating multiple users, when creating a single user, in addition to completing the configuration items involved in creating multiple users, you can also configure the two-way binding of the user's display name and IP/MAC address.

Select "Object" → "User" → "User/Group".

2. Configure user attributes

As shown in Figure 12-29, for an existing user, you can use "Edit" to modify the user's attributes.

(1) Account expiration time. The expiration time of the user's account.

(2) Allow multiple people to log in with the account at the same time. Selecting this parameter means that multiple people are allowed to log in with the user's login name at the same time, that is, the login name is allowed to log in on multiple computers at the same time.

Unselecting this parameter means that only the login name is allowed to log in on one computer at the same time.

(3) IP/MAC binding method. The binding method of user and IP/MAC address.

Checking "One-way binding" means that users can only use the specified IP/MAC address for authentication, but at the same time allow other users to also use this IP/MAC address for authentication.

Select "Two-way binding", it means that the user can only use the specified IP/MAC address for authentication, and the specified IP/MAC address is only for the user. When an IP/MAC address is two-way bound, other users who are one-way bound to this IP/MAC address will not be able to log in.

(4) IP/MAC address. The IP address, MAC address or IP/MAC address pair bound to the user.

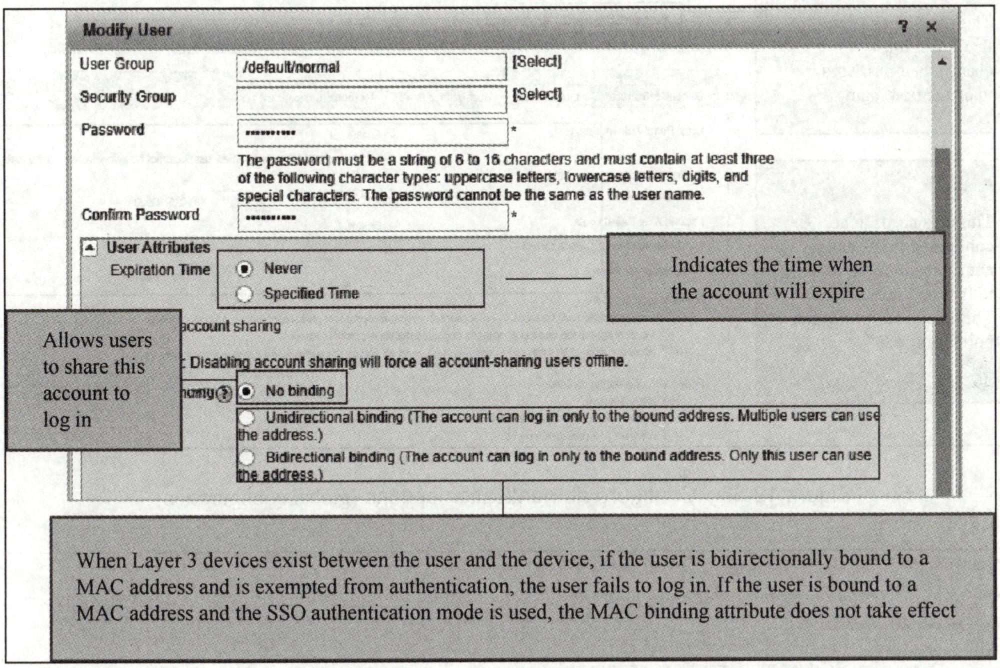

Figure 12-29 Configure user attributes

Subtask 12.4.3 Configure authentication options

1. Global configuration

As shown in Figure 12-30, configure it in "Object" → "User" → "Authentication Options" → "Global Configuration/Local Portal".

Portal authentication related functions need to be completed in cooperation with the Portal server, and the Portal server needs to provide an authentication page to push users. Currently, the Portal server that can be connected to the FW is Huawei's Agile Controller or Policy Center.

If you use web push page authentication, you also need to configure the corresponding security policy to allow the data stream with the port number 8887 to reach the firewall itself.

2. Single sign-on

As shown in Figure 12-31, in "Object" → "User" → "default" → "Internet access and authentication policy configuration", configure single sign-on.

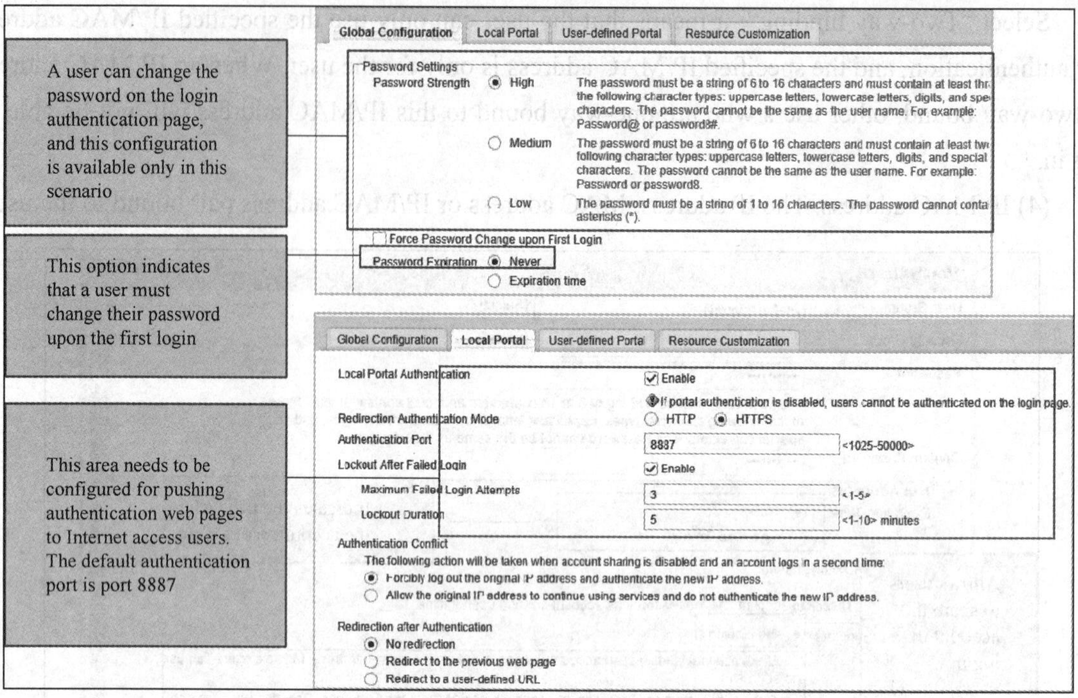

Figure 12-30 Configure authentication options (global configuration)

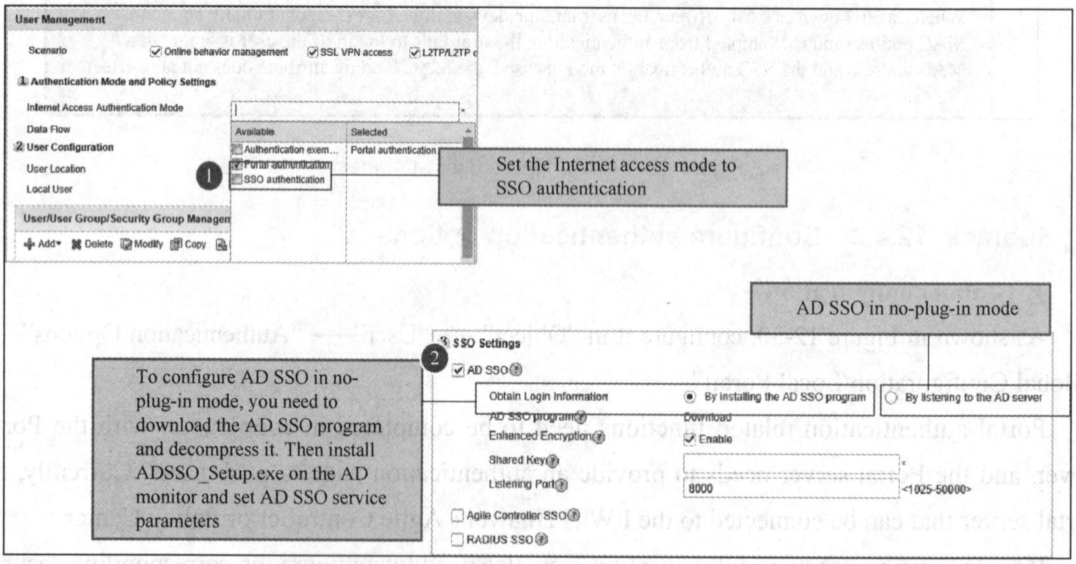

Figure 12-31 Configure authentication options (single sign-on)

If the FW is deployed between the user and the AD domain controller, the user's authentication packet passes through the FW. In order to realize the single sign-on function, the data stream is not authenticated when the authentication policy is configured. In addition, the authentication message also needs to pass the security check of the security policy, that is, the administrator needs to configure the following security policy on the FW:

(1) Source security zone: The security zone where the user's PC is located.

(2) Destination security zone: The security zone where the AD server is located.

(3) Destination address: The IP address of the AD server.

(4) Action: Allowed.

Subtask 12.4.4 Configure authentication policies

As shown in Figure 12-32, in "Object" → "User" → "Authentication Policy" → "New", configure the authentication policy.

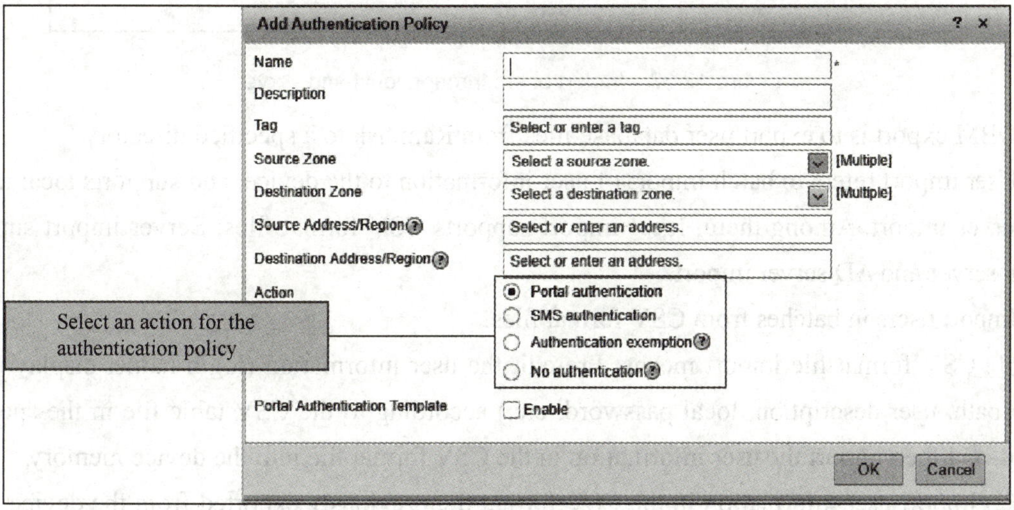

Figure 12-32 Configure authentication policy

When an Internet user or an access user who has already accessed the FW uses the session authentication method to trigger the authentication process, the authentication policy must be processed.

Subtask 12.4.5 Import users through local and server

As shown in Figure 12-33, choose "Object" → "User" → "User Import".

(1) Local import: Local import supports importing user information in CSV format files and database dbm files to the local device.

(2) Server import: Importing users in batches from the authentication server refers to importing the user (group) information on the authentication server into the device through the server import policy.

CSV supports login name, display name, belonging group, description information, password, IP/MAC binding information, binding mode, account status, and validity period information.

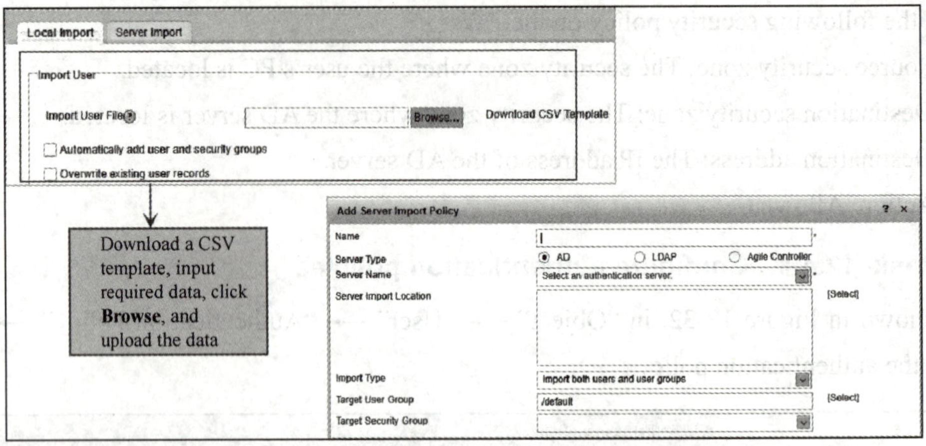

Figure 12-33 Import users through local and server

DBM export is to export user database files from Ramdisk to a specified directory.

User import refers to batch import of user information to the device, and supports local import and server import. Among them, local import supports CSV format files; Server import supports LDAP server and AD server import.

Import users in batches from CSV format files.

(1) CSV format file import means: Pre-edit the user information (login name, display name, group path, user description, local password, etc.) according to the CSV table file in the specified format, and then import the user information in the CSV format file into the device memory.

(2) Import user information in the CSV format file previously exported from the device to the device memory.

Select "Object" → "User" → "User Import".

Select the "Local Import" tab.

Import users in batches from the authentication server.

(1) In the network, there are many cases where third-party authentication servers are used. Many companies have authentication servers on their networks, and all users and user group information are stored on the authentication servers. Importing users in batches from the authentication server refers to importing the user (group) information on the authentication server into the device through the server import policy.

(2) The device supports batch import of users from AD, LDAP, and TSM servers.

Select "User" → "User" → "User Import". Select the "Server Import" tab.

Subtask 12.4.6 Online user management

As shown in Figure 12-34, in "Object" → "User" → "Online User", you can manage online users.

If you need to restrict all online behaviors of certain users within a certain period of time, you can freeze designated online users. If the administrator perceives that some users are not trustworthy, they can forcibly log off the designated online users.

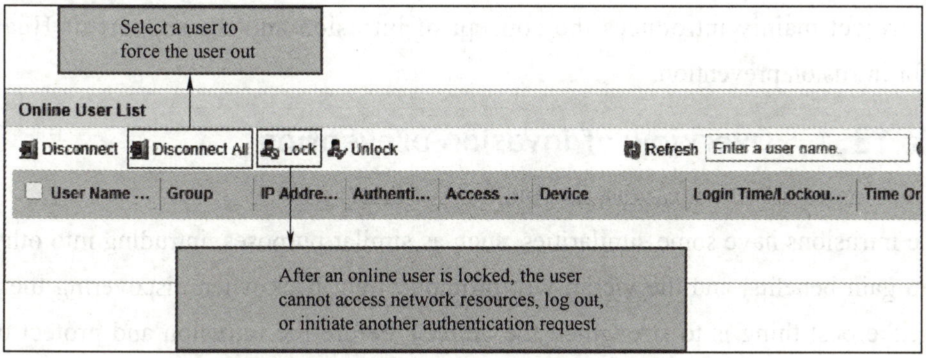

Figure 12-34 Online user management

Through the online user list, you can view online users who have passed device authentication. The administrator can perform related management operations on online users, such as forced logout.

View online users:

(1) Choose "Object" → "User" → "Online User".

(2) Locate online users to be viewed. Administrators can use the following methods to locate online users to be viewed:

① In the "Organization Structure", click the group to which the user to be viewed belongs, and all online users belonging to this group will appear in the "Online User List".

② Use simple query or advanced query function to find the user to be viewed, and the search result will appear in the "online user list".

Force logout of online users:

(1) Choose "Object" → "User" → "Online User".

(2) Locate online users to be logged out.

(3) The administrator can use the following methods to locate online users to be logged out:

① In the "Organization Structure", click the group to which the user to be logged out belongs, and all online users belonging to this group will appear in the "Online User List".

② Use the simple query or advanced query function to find the user to be logged out, and the search result will appear in the "online user list".

③ In the "Online User List", select all users to be logged out, and click "Forced Logout".

④ If the operation is successful, the logged-out user will not appear in the "Online User List".

Project 13 Intrusion prevention

This project mainly introduces the concept of intrusion and the support of Huawei USG products for intrusion prevention.

Task 13.1 Overview of invasion prevention

When it comes to intrusion, what do you think of first?

These intrusions have some similarities, such as similar purposes, intruding into other parties' domains to gain benefits, and the victim will promptly fight back when discovering the intrusion. Of course, the best thing is to strengthen the defense before the intrusion and protect one's own territory from intrusion. So what about intrusion and defense in the network?

Subtask 13.1.1 Understand the current situation of cyber threats

Nowadays, most viruses and other cyber threats no longer simply attack computer systems, but are used by hackers and lawbreakers as a tool for them to gain benefits. Therefore, traditional cyber threats such as computer viruses are developing and changing to profit-driven, comprehensive cyber threats.

As shown in Figure 13-1, among the various security threats currently emerging, malicious programs (viruses and worms, Bot, Rootkit, Trojan horses and backdoor programs, vulnerability attack programs, and mobile device malicious programs) account for a high proportion. The influence of grayware (spyware/adware) has gradually expanded, and security threats related to criminal programs have become an important factor threatening network security.

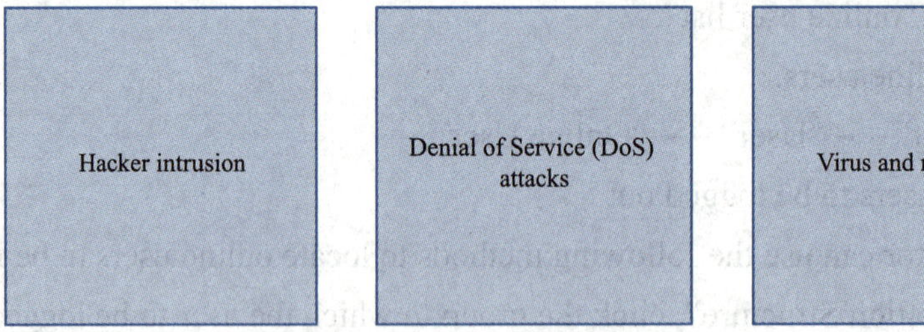

Figure 13-1 Current cyber threats

At present, users are no longer facing traditional virus attacks. "Network threats" are often a mixture of viruses, hacker attacks, Trojan horses, zombies, spies and other harms. Therefore, it is often difficult to resist such threats by relying on past anti-virus or anti-hacking technologies alone.

Subtask 13.1.2 Understand hacking

As shown in Figure 13-2, network hackers and malicious employees within the enterprise use vulnerabilities in the system and software to intrude the server, seriously threaten the security of key business data of the enterprise.

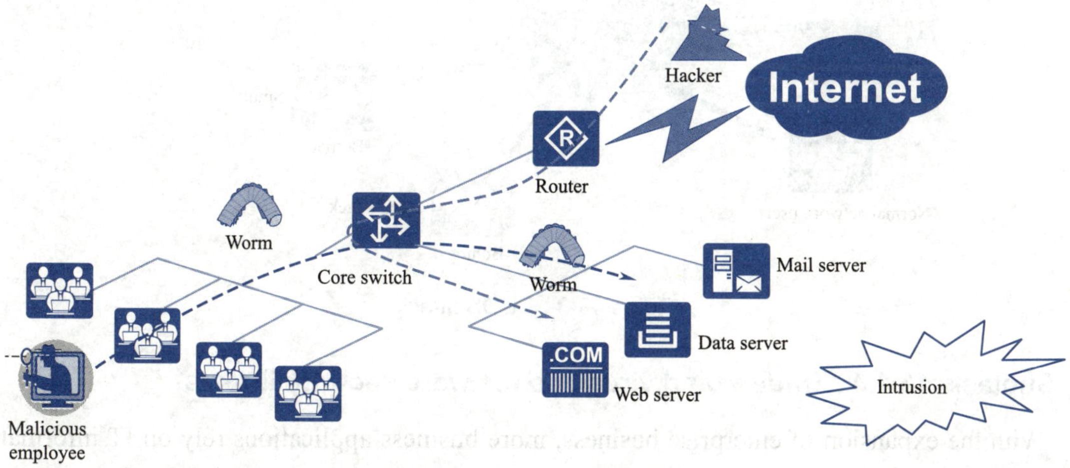

Figure 13-2 Hacking

Vulnerabilities pose serious security threats to enterprises:

(1) There may be vulnerabilities in many application software in the enterprise intranet;

(2) The Internet makes application software vulnerabilities spread rapidly;

(3) Worms use application software vulnerabilities to spread wildly, consume network bandwidth, and destroy important data;

(4) Hackers and malicious employees use vulnerabilities to attack or intrude corporate servers, and business secrets are tampered, destroyed, and stolen.

Subtask 13.1.3 Understand the threat of denial of service attacks

DDOS attacks aimed at economic benefits continue to threaten the normal operation of enterprises, and the damage caused by attacks is becoming more and more serious.

As shown in Figure 13-3, DDOS attack threats:

(1) The formation of a global black industry chain with economic interests as the goal. There are a large number of botnets on the Internet.

(2) Extortion by criminals, malicious competition among peers, etc. may cause enterprises to suffer DDoS attacks.

(3) When suffering from a DDoS attack, the network bandwidth is occupied a lot, and the network is paralyzed. The resources of the attacked server are exhausted and cannot respond to normal user requests. In severe cases, the system will crash and corporate services will not operate normally.

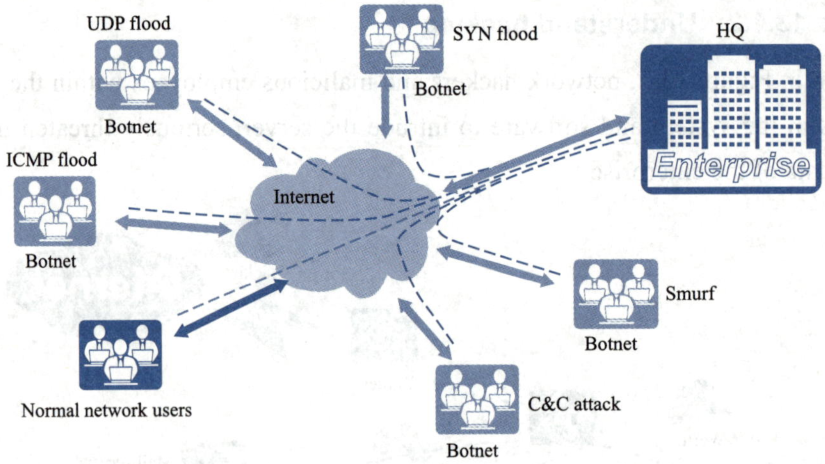

Figure 13-3 DDOS attack

Subtask 13.1.4 Understand virus and malware security threats

With the expansion of enterprise business, more business applications rely on IT information systems to complete. In the course of business operations, they are constantly facing serious threats such as viruses, Trojan horses, and spyware, as shown in Figure 13-4.

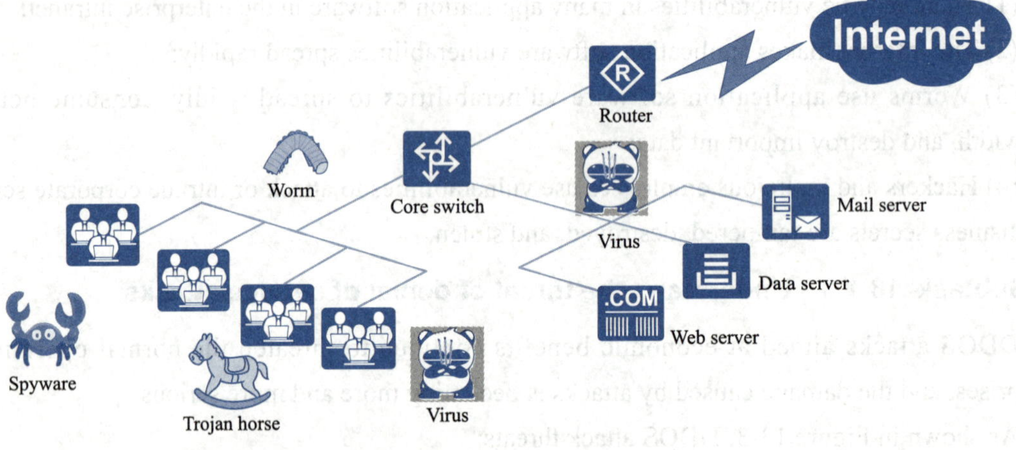

Figure 13-4 Virus and malware

A virus is a malicious code that can infect or attach to applications or files. It is generally spread through protocols such as e-mail or file sharing, threatening the security of user hosts and networks.

Some viruses will exhaust host resources and occupy network bandwidth. Some viruses will control host permissions and steal user data. Some viruses may even cause damage to host hardware.

Virus and malware security threats:

(1) Web browsing and e-mail transmission are the main ways for viruses, Trojan horses, and spyware to enter the intranet.

(2) Viruses can damage computer systems, tamper with, and damage business data.

(3) Trojan horse enables hackers not only to steal important information on the computer, but also to damage the intranet computer. Spyware collects, uses, and disseminates sensitive information of enterprise employees, which seriously interferes with the normal business of the enterprise.

(4) Desktop antivirus software is difficult to prevent the spread of viruses from a global perspective.

Subtask 13.1.5 Understand the definition of intrusion

What are the similar characteristics of the aforementioned cyber threats? As shown in Table 13-1.

Table 13-1 Network threat characteristics

Threat feature	Unauthorized access	Unauthorized tampering	Unauthorized destruction
Hacking	√	√	√
Denial of service attack threat	√		√
Virus and malware	√	√	√

Intrusion refers to unauthorized attempts to access information system resources, tamper with the data in the information system, and make the information system unreliable or unusable. Intrusion attempts to destroy the integrity, confidentiality, availability and controllability of the information system.

Typical invasion behavior:

(1) Tampering with web pages;

(2) Cracking the system password;

(3) Copy/view sensitive data;

(4) Use network sniffing tools to obtain user passwords;

(5) Access unauthorized servers;

(6) Other special hardware obtains the original network package;

(7) Implant a Trojan horse program into the host.

Task 13.2 Understanding intrusion prevention system

Subtask 13.2.1 Understand the position of safety equipment in the safety system

As shown in Figure 13-5, in the construction of information security, the intrusion detection system plays the role of a monitor. By monitoring the flow of key nodes in the information system, in-depth analysis is carried out to discover the security incidents that are occurring. A vivid analogy is: IDS is like a camera in a security surveillance system. Through IDS, system administrators can capture the flow of key nodes and do intelligent analysis, discover abnormal and suspicious network behaviors, and report them to the administrator.

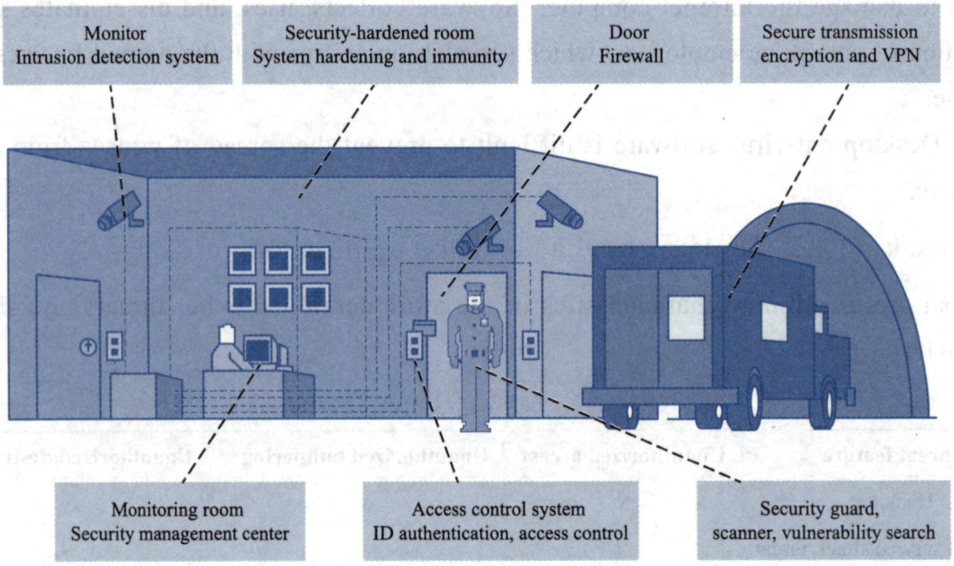

Figure 13-5 The position of safety equipment in the safety system

ID (intrusion detection) is a process of real-time detection of intrusions by monitoring various operations, analyzing and auditing various data and phenomena. It is an active and dynamic security defense technology. The content of intrusion detection covers a variety of authorized and unauthorized intrusions.

IDS (intrusion detection system) can immediately activate the relevant security mechanism to respond when it discovers that there is a violation of security policy or the system has traces of being attacked.

The difference between firewall and IDS:

(1) The firewall is a serial device and needs to be forwarded quickly and cannot be used for in-depth inspection.

(2) The firewall cannot correctly analyze the malicious code doped in the allowed application data stream, and cannot detect malicious operations or misoperations from insiders.

(3) The firewall is a coarse-grained access control, and IDS is a fine-grained detection device. The existing network can be monitored more accurately through IDS.

(4) IDS can be linked with firewalls and switches to become a powerful "assistant" of

firewalls, which can better and more accurately control access between external domains.

(5) IDS can be upgraded flexibly and in a timely manner, and it is convenient and flexible to configure and operate strategically.

Subtask 13.2.2 Master the intrusion prevention system

IPS (intrusion prevention system) is an intelligent intrusion detection and defense product. It can not only detect the occurrence of intrusions, but also stop the occurrence and development of intrusions in real time through certain response methods. Protect information systems from substantial attacks.

IPS is generally deployed in two ways in the network:

(1) Bypass mode: Bypass it on the switch, and perform port mirroring through the switch.

(2) Straight way: Connect in series at the network boundary, deploy online, and block online.

As shown in Figure 13-6, the basic technical characteristics of the intrusion prevention system mainly include:

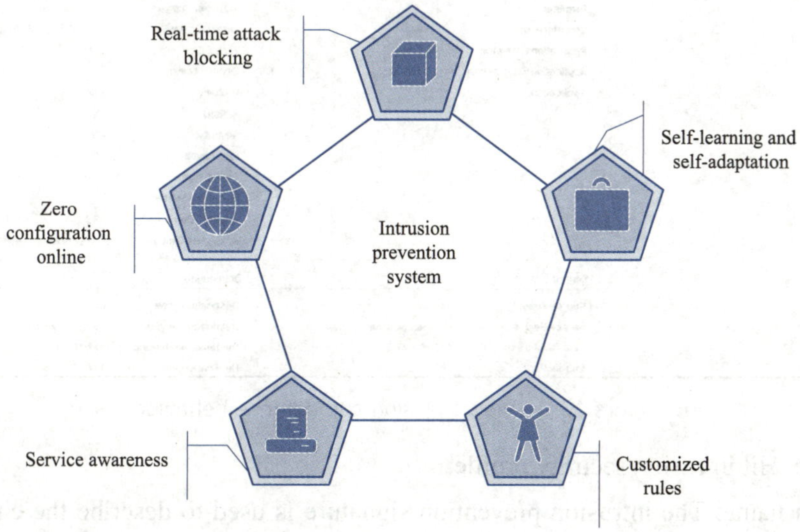

Figure 13-6 Technical characteristics of the intrusion prevention system

(1) Real-time attack blocking: IPS can block discovered network attack behaviors in real time, avoid IDS discovering attacks, can not stop the defects caused by the attack in real time, and maximize the security of the system.

(2) Self-learning and self-adaptation: IPS can minimize the system's false negatives and false positives through self-learning and self-adaptation, and reduce the impact on the business.

(3) Customized rules: IPS can customize intrusion prevention rules to maximize the response to the latest threats.

(4) Service awareness: Allows IPS to detect anomalies and attacks based on the application layer.

(5) Zero configuration online: The system provides a default intrusion prevention security configuration file, which can be directly quoted.

Subtask 13.2.3 Check intrusion and defense behavior

As shown in Figure 13-7. Intrusion log information: virtual system/hit security policy/source and destination address/source and destination port/source and destination security domain/user/protocol/application/hit intrusion security profile/signature name/signature serial number/event count/intrusion target/intrusion Severity/operating system/signature classification/signature action, the key information is as follows:

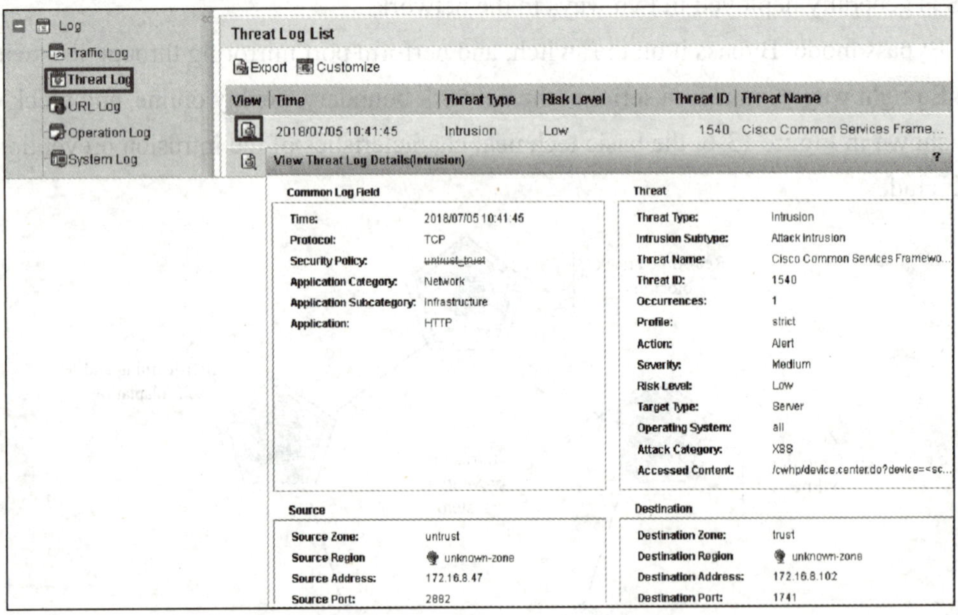

Figure 13-7 View intrusion and defense behavior

(1) Profile: Hit intrusion security profile.

(2) Threat name: The intrusion prevention signature is used to describe the characteristics of the attack behavior in the network, and the data flow is compared with the intrusion prevention signature to detect and prevent attacks.

(3) Event count: The log merge introduces a field. Whether to merge or not must be determined according to the merge frequency and log merge conditions. If no merge occurs, it is 1.

(4) Intrusion target: The attack object of the message detected by the signature. Details are as follows:

server: The target of the attack is the server.

client: The target of the attack is the client.

both: The target of the attack is the server and the client.

(5) Intrusion severity: The severity of the attack caused by the message detected by the signature. Details are as follows:

information: Indicates serious behavior prompts.
low: Indicates severe behavior is low.
medium: Indicates serious behavior is middle.
high: Indicates serious behavior is high.

(6) Operating system: The operating system attacked by the message detected by the signature. Details are as follows:

all: Represents all systems.
android: Represents the Android system.
ios: Represents the Apple system.
unix-like: Represents the Unix system.
windows: Represents the Windows system.
other: Represents other systems.

(7) Signature classification: The threat classification to which the attack characteristics of the message detected by the signature belong.

(8) Signature action: signature action. Details are as follows:

alert: The signature action is an alarm.

block: The signature action is blocking.

Task 13.3 Understanding network antivirus

Subtask 13.3.1 Understand the basic concepts of computer viruses

A computer virus refers to a computer program that destroys computer functions or destroys data that is programmed or inserted into a computer program. A group of computer instructions or program codes that affect the use of a computer and can replicate itself is called a computer virus.

Malicious code is a program that embeds the code into another program without being noticed. So as to achieve the purpose of destroying the data of the infected computer, running intrusive or destructive programs, and destroying the security and integrity of the data of the infected computer.

Computer virus classification:

(1) Classified according to malicious code function: virus, worm, Trojan.

(2) Classification according to transmission mechanism: removable media, network sharing, network scanning, e-mail, P2P network.

(3) Classified according to the infected objects: operating system, application, equipment.

(4) Classification according to the carrier object: executable files, scripts, macros, and boot areas.

Subtask 13.3.2 Distinguish the relationship between viruses, worms and Trojans

A virus is a piece of malicious code that is parasitic in a normal program. When the user activates this normal program, the virus will also be activated, causing damage to the file system of the system.

A worm is a variant of a virus, it is an independent individual and does not need to be parasitic. It can copy itself, use system loopholes or artificial awareness loopholes on the network to spread, the impact is even worse, mainly affecting the performance of the entire network and the system performance of the computer.

Trojan horse is also a parasitic malicious code, which is very concealed. Hackers often use Trojan horses to control a host, making it a "zombie". In addition, Trojan horses can also be used to monitor and obtain victim key information, such as bank card passwords.

The antivirus we often talk about refers to anti-malware code. The characteristics of the relationship between virus, worm, and Trojan horse are shown in Table 13-2.

Table 13-2 Relationship characteristics of virus, worm, and Trojan horse

Project	Virus	Worm	Trojan horse
Existing form	Parasitic	Independent individual	Parasitic
Replication mechanism	Insert into the host program	Self copy	Not self-replicating
Infectious	Host program running	Vulnerability in the system	According to carrier or function
Contagion target	Mainly for local files	For other computers on the network	Zombie
Trigger mechanism	Computer user	The program itself	Remote control
Impact focus	File system	Network performance, system performance	Information theft or denial of service

Subtask 13.3.3 Analysis of antivirus technology

Antivirus technologies are classified according to the protection objects:

(1) Stand-alone antivirus: Stand-alone antivirus can be achieved by installing antivirus software or professional antivirus tools. Virus detection tools are used to detect viruses, Trojan horses, worms and other malicious codes. Some detection tools also provide repair functions. Common antivirus software includes Symantec, etc., and professional antivirus tools include Process Explorer.

(2) Network antivirus: Network antivirus technology refers to the deployment of antivirus strategies on the security gateway.

In the following situations, antivirus features are usually used to ensure network security:

(1) Intranet users can access the external network and often need to download files from the external network.

(2) The servers deployed on the intranet often receive files uploaded by users on the extranet.

As shown in Figure 13-8, NIP (NIP is a new-generation network intelligent protection product launched by Huawei Technologies Co., Ltd.) acts as a gateway device to isolate the internal and external networks. The internal network includes user PCs and servers. Intranet users can download files from the extranet, and extranet users can upload files to the intranet server. To ensure the security of files received by intranet users and servers, you need to configure the antivirus function on the NIP.

After the antivirus function is configured on the NIP, normal files can enter the internal network smoothly, and files containing viruses will be detected and intervened by means such as blocking or warning.

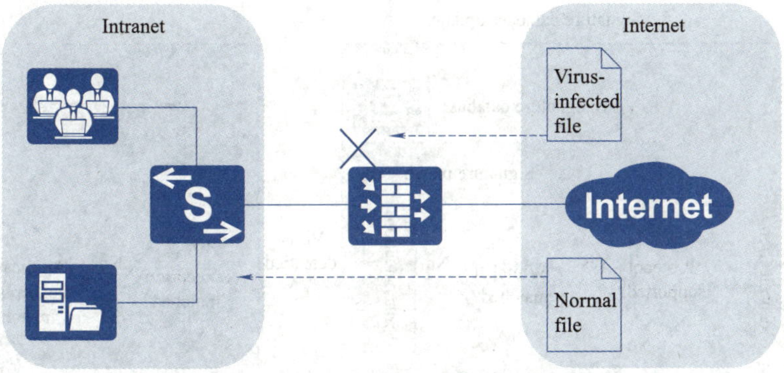

Figure 13-8 Anti-virus application scenario

Subtask 13.3.4 Implementation of gateway antivirus

At present, the scanning methods of equipment manufacturers (including UTM and AVG) are divided into two types: proxy scanning and stream scanning.

(1) Proxy scanning method: Transparently forward all data messages that need to be detected for virus through the gateway to the gateway's own protocol stack. After all the files are cached through the gateway's own protocol stack, they are sent to the virus detection engine for virus detection.

The proxy-based antivirus gateway can perform more advanced operations such as decompression and unpacking, with a high detection rate. However, due to the full file cache, its performance and system overhead will be relatively large.

(2) Stream scanning mode: Rely on state detection technology and protocol analysis technology, simply extract the features of the file and match it with the local signature library.

The antivirus gateway based on stream scanning has high performance and low overhead, but the detection rate of this method is limited, and it cannot deal with files processed by methods such as packing and compression.

Subtask 13.3.5 Analysis of working principles of firewall antivirus

The firewall antivirus workflow is shown in Figure 13-9. After the network traffic enters the intelligent sensing engine, the intelligent sensing engine first conducts an in-depth analysis of the traffic to identify the protocol type corresponding to the traffic and the direction of file transmission.

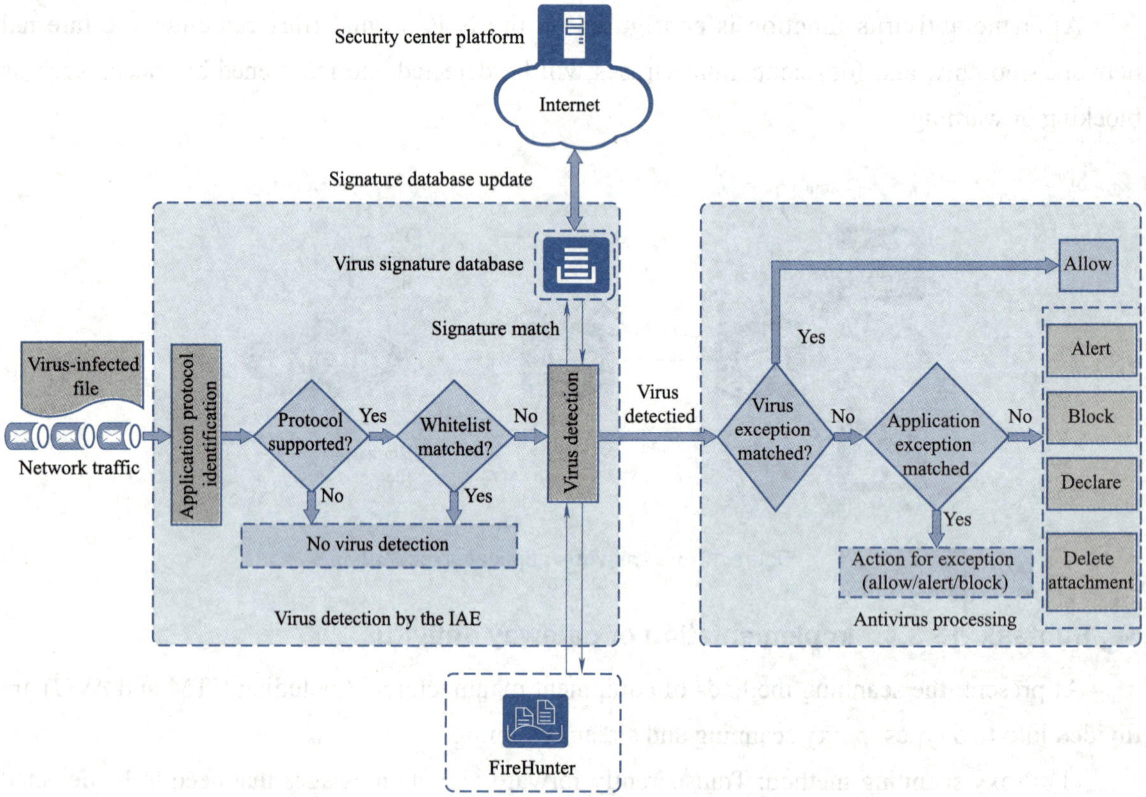

Figure 13-9 Firewall antivirus workflow

1. Determine whether the protocol used for file transfer and the direction of file transfer support virus detection

(1) FW supports virus detection for files transferred using the following protocols.

FTP: file transfer protocol.

HTTP: hypertext transfer protocol.

POPv3: post office protocol version 3.

SMTP: simple mail transfer protocol.

IMAP：Internet message access protocol.

NFS：network file system.

SMB：server message block.

(2) FW supports virus detection for files in different transmission directions.

Upload: Refers to the client sending a file to the server.

Download: Refers to the server sending files to the client.

2. Determine whether to hit the whitelist

After hitting the whitelist, NIP will not perform virus detection on the file.

The whitelist consists of whitelist rules. Administrators can configure whitelist rules for trusted domain names, URLs, IP addresses, or IP address segments to improve the efficiency of antivirus detection. The effective scope of the whitelist rule is limited to the antivirus configuration file in which it is located, and each antivirus configuration file has its own whitelist.

3. Virus detection

(1) The intelligent perception engine extracts the features of the files that meet the virus detection, and the extracted features are matched with the features in the virus signature database. If it matches, the file is considered to be a virus file and processed according to the response action in the configuration file. If it does not match, the file is allowed to pass. When the linkage detection function is enabled, files that miss the virus signature database can also be sent to the sandbox for in-depth detection. If the sandbox detects a malicious file, the file characteristics of the malicious file are sent to the NIP, and the NIP saves the characteristics of the malicious file to the linkage detection cache. When the malicious file is detected next time, it will be processed according to the response action in the configuration file.

(2) The virus signature database is formed by Huawei through analysis of various common virus signatures. The signature database defines various common virus signatures, and assigns a unique virus ID to each virus signature. After the device loads the virus signature database, it can identify the viruses that have been defined in the signature database.

4. When FW detects that the transferred file is a virus file, the following processing is required

(1) Determine whether the virus file hits the virus exception. If it is a virus exception, the file is allowed to pass.

(2) Virus exceptions, that is, virus whitelist. In order to avoid file transmission failures due to system misreports and other reasons, when a user believes that a certain virus has been detected as a misreport, the corresponding virus ID can be added to the virus exception to invalidate the virus rule. If the detection result hits the virus exception, the response to the file is release.

(3) If it is not a virus exception, judge whether the virus file hits the application exception. If

it is an application exception, it will be processed in accordance with the response actions (release, alarm, and block) of the application exception.

(4) Application exceptions can configure response actions different from the protocol for the application. Applications are carried on the protocol, and multiple applications can be carried on the same protocol.

(5) Due to the relationship between the application and the protocol, the following provisions are also made when configuring the response action:

If only the response action of the protocol is configured, all applications carried on the protocol inherit the response action of the protocol.

If both the protocol and the application are configured with a response action, the response action of the application shall prevail.

(6) If the virus file neither hits the virus exception nor the application exception, it will be processed according to the protocol configured in the configuration file and the response action corresponding to the transmission direction.

Subtask 13.3.6 Antivirus response action

The response action after the virus is detected. Include:

(1) Alarm: Allow virus files to pass through and generate virus logs at the same time.

(2) Block: Prevent virus files from passing through, and generate virus logs at the same time.

(3) Announcement: For mail files that carry viruses, the device allows the file to pass, but it will add a message indicating that the virus has been detected in the mail body, and generate a virus log at the same time. The announcement action only takes effect for the SMTP protocol and POPv3 protocol.

(4) Delete attachments: For mail files that carry viruses, the device allows the file to pass, but the device deletes the content of the attachments in the mail and adds an announcement in the mail body, and generates a virus log at the same time. The delete attachment action is only effective for the SMTP protocol and POPv3 protocol.

Subtask 13.3.7 View firewall antivirus results

After the firewall detects the virus, you can view the detailed information of the firewall's anti-virus results through the business log. As shown in Figure 13-10. When the http protocol and mail protocol antivirus are configured, the relevant prompt information can be viewed on the access page or the mail body.

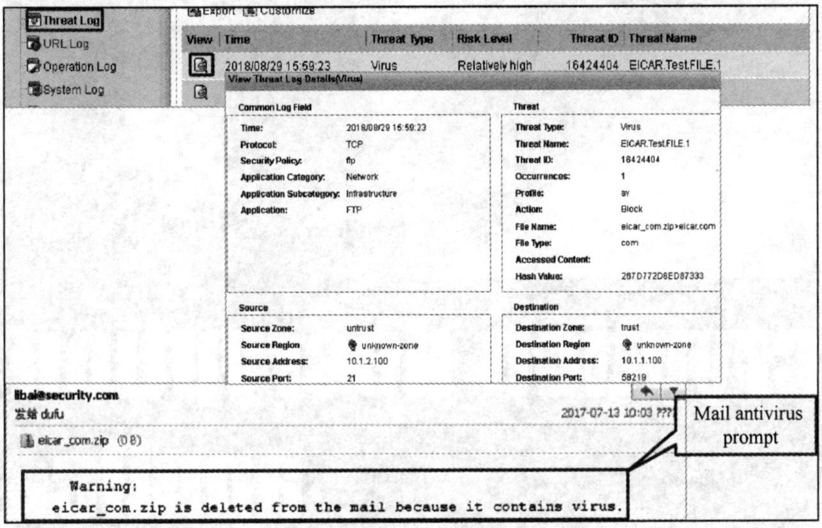

Figure 13-10 View firewall antivirus results

This section will focus on encryption and decryption technology, and focus on how to improve the security of information transmission through three projects.

Encryption and decryption principle: introduces the technical background, development history, common encryption and decryption algorithms.

PKI certificate system: introduces the digital certificate system PKI architecture and certificate application process.

Encryption technology application: introduces the application of encryption technology in VPN, the working principle and configuration of GRE VPN, L2TP VPN, IPsec VPN, SSL VPN and other technologies.

Through these three projects, readers can understand the related technologies and applications of encryption and decryption, and master the working principles and configurations of digital certificates and various VPN technologies.

Understand the principles of encryption and decryption

In the transmission of the Internet, the data encapsulated based on the TCP/IP protocol stack is transmitted in plain text, so there are many potential dangers. For example, passwords and bank account information are stolen or tampered with, users' identities are impersonated, and they are subjected to malicious attacks on the Internet. The application of encryption and decryption

Section 4

Encryption decryption and application

This section will focus on encryption and decryption technology and focus on how to improve the security of information transmission in three projects:

Encryption and decryption principle: Introduce the technical background, development history and common encryption and decryption algorithms.

PKI certificate system: Introduce the digital certificate system, PKI architecture and digital certificate application scenarios.

Encryption technology application: Introduce the application of encryption technology in VPN, the working principle and configuration of L2TP VPN, GRE VPN, IPsec VPN, SSL VPN and other technologies.

Through these three projects, readers can understand the related technologies and applications of encryption and decryption, and master the working principles and configurations of digital certificates and various VPN technologies.

Project 14 Understand the principles of encryption and decryption

In the transmission of the Internet, the data encapsulated based on the TCP/IP protocol stack is transmitted in plain text, so there are many potential dangers. For example, passwords and bank account information are stolen or tampered with, users' identities are impersonated, and they are subjected to malicious attacks on the Internet. The application of encryption and decryption technology in the network can protect the transmitted data and reduce the risk of information leakage.

Task 14.1 Understand the development of encryption technology

Subtask 14.1.1 Understand the encryption technology

Encryption is the use of mathematical methods to convert plain text (data that needs to be concealed) into cipher text (unreadable data) to achieve the purpose of protecting data. As shown in Figure 14-1, in the three elements of mark encryption, the information plain text is represented by the letter P, the key is represented by the letter K, and the information cipher text is represented by the letter C. The relationship between them can be expressed by the formula C=En(K ,P) means. The cipher text is the data calculated in the plain text of the information through the encryption algorithm and the key.

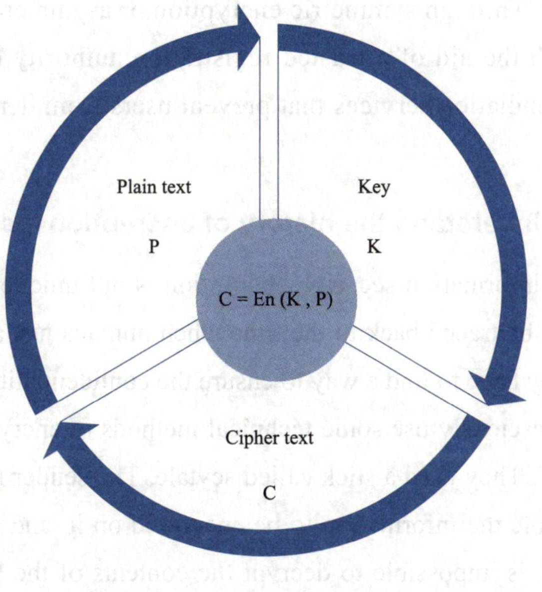

Figure 14-1 Encrypted three elements

Subtask 14.1.2 Understand the function of encryption technology

Encryption is a process that makes the information readable only to the correct recipient, and what other users see is chaotic information, so that the original content can only be displayed after decrypting with the corresponding key. The encryption method is used to protect the data from being stolen and read by unauthorized persons. The role of encryption on the network is to prevent private information from being intercepted and stolen on the network. Encryption can guarantee the confidentiality, integrity, authentication and non-repudiation of information.

(1) Confidentiality: Realized through data encryption. Provide services that only allow specific users to access and read the information, and any unauthorized users cannot understand the information. This is a common reason for using encryption. By carefully using mathematical equations, you can ensure that only the corresponding recipient can view it.

(2) Integrity: Achieved through data encryption, hashing or digital signature. Provide services to ensure that data is not modified (tampered, deleted, inserted, replayed, etc.) by unauthorized users during storage and transmission. For users with higher security requirements, only data encryption is not enough, and the data can still be illegally cracked and modified.

(3) Authentication: It is realized through data encryption, data hashing or digital signature. Provide services related to data and identity recognition, that is, to verify the identity of the sender and recipient of the data.

(4) Non-repudiation: Through symmetric encryption or asymmetric encryption, and digital signatures, etc., and with the aid of a trusted registration authority or certificate authority to achieve. Provide non-repudiation services that prevent users from denying previous statements or actions.

Subtask 14.1.3 Understand the history of encryption technology

As a way to ensure information security, encryption is not unique to modern times. It has a very long history and can be traced back to the time when humans just appeared and tried to learn how to communicate. They have to find a way to ensure the confidentiality of their communications. But the first people to consciously use some technical methods to encrypt information may be the ancient Greeks in 500 BC. They used a stick called scytale. The sender first rolled a piece of paper around the stick, then wrote the information to be encrypted on it, and then opened the paper and gave it to the recipient. It is impossible to decrypt the contents of the letter without knowing the width of the stick (here as the key).

Around 50 BC, Caesar, the ruler of ancient Rome, invented a method for transmitting encrypted information during war, which was later called the "Caesar Code". Its principle is: Arrange the 26 letters in a natural order, and they are connected end to end. Each letter in the plain text is replaced with the third letter after it. For example, HuaweiSymantec becomes KxdzhlvBPdqwhf after encryption.

In the First World War, the Germans used a dictionary to write passwords, such as: 10-4-2, which is the second word in the fourth paragraph on page 10 of a dictionary.

Task 14.2 Master the principle of encryption and decryption technology

Subtask 14.2.1 Master the classification of encryption technology

1. Symmetric encryption

Symmetric encryption algorithm is also called traditional cryptographic algorithm (shared key encryption, secret key algorithm, single key algorithm), which uses the same key to encrypt and decrypt data. The encryption key can be calculated from the decryption key. The sender and recipient share the same key, which is used for both encryption and decryption. Symmetric key encryption is an effective method for encrypting large amounts of data. There are many algorithms for symmetric key encryption, but all of these algorithms have a common purpose: to convert plain text (unencrypted data) into cipher text in a reversible manner. Since symmetric key encryption uses the same key for encryption and decryption, the security of this encryption process depends on whether an unauthorized person obtains the symmetric key. Special attention: Both parties who wish to use symmetric key encryption to communicate must exchange keys securely before exchanging encrypted data.

The workflow of symmetric encryption technology is shown in Figure 14-2.

A and B negotiate a symmetric key in advance, and the specific encryption and decryption process is as follows:

(1) A uses a symmetric key to encrypt the plain text and sends the cipher text to B.

(2) After receiving the cipher text, B uses the symmetric key to decrypt the cipher text to obtain the original plain text.

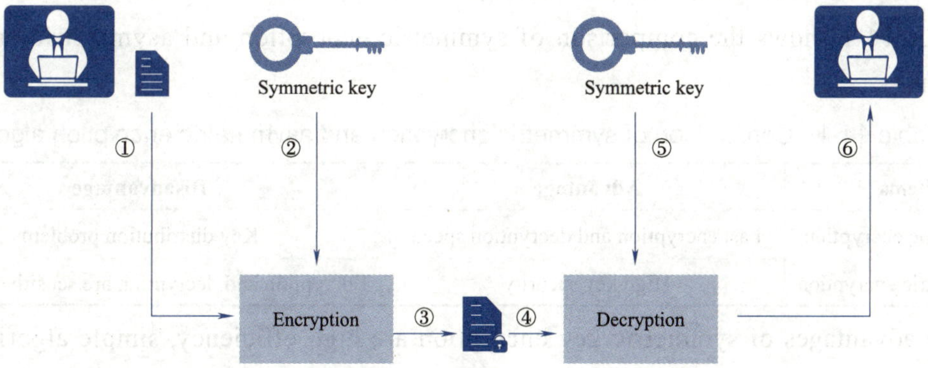

Figure 14-2 The workflow of symmetric encryption technology

2. Asymmetric encryption

The asymmetric algorithm is also called public key encryption. It uses two keys: a public key and a private key. These two keys are mathematically related. In public key encryption, the public key can be publicly transmitted between the communicating parties or published in a public repository, but the related private key is kept secret. Only the private key can decrypt the data encrypted with the public key. Data encrypted with the private key can only be decrypted with the public key.

Encryption and decryption use two different keys, the private key is used to protect the data, and the public key is shared by people in the same system to verify the authenticity and identity of the message and its sender.

The workflow of asymmetric encryption technology is shown in Figure 14-3.

A obtains B's public key in advance, and the specific encryption and decryption process is as follows:

(1) A uses B's public key to encrypt the plain text and sends the cipher text to B.

(2) After receiving the cipher text, B uses his private key to decrypt the cipher text to get the original plain text.

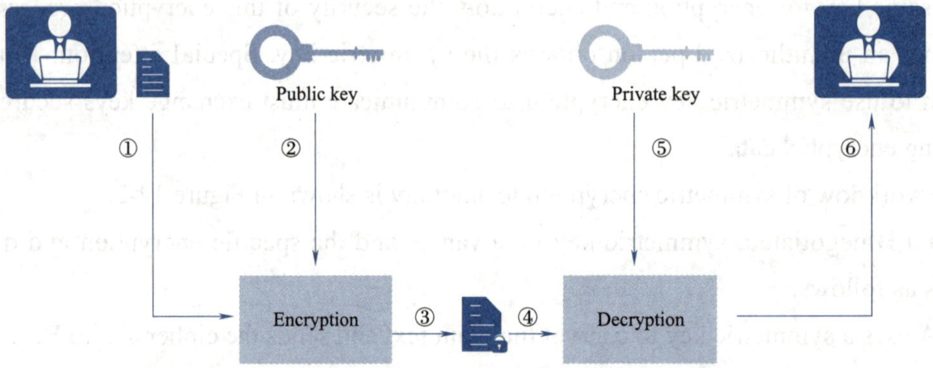

Figure 14-3 The workflow of asymmetric encryption technology

3. Comparison of symmetric and asymmetric encryption

Table 14-1 shows the comparison of symmetric encryption and asymmetric encryption algorithms.

Table 14-1 Comparison of symmetric encryption and asymmetric encryption algorithms

Items	Advantage	Disadvantage
Symmetric encryption	Fast encryption and decryption speed	Key distribution problem
Asymmetric encryption	High key security	Encryption and decryption are sensitive to speed

The advantages of symmetric key encryption are high efficiency, simple algorithm, low system overhead, and it is suitable for encrypting large amounts of data. The disadvantage is that it is difficult to implement and poor scalability. The reason why it is difficult to realize is that the key exchange needs to be performed in a secure manner before secure communication.The poor scalability is manifested in the need to negotiate a key between each pair of communication users, and a community of n users needs to negotiate $n(n-1)/2$ different keys.

The advantage of public key encryption is that one key cannot be derived from another key. Information encrypted by the public key can only be decrypted with the private key. The disadvantage of public key encryption is that the algorithm is very complex, which leads to a long time for encrypting a large amount of data, and the encrypted message is longer, which is not conducive to network transmission.

Based on the advantages and disadvantages of public key encryption, public key encryption is suitable for encrypting sensitive information such as keys or identity information, so as to meet the needs of users in terms of security.

Subtask 14.2.2 Data encryption application scenario—digital envelope

Digital envelope refers to the data obtained after the sender uses the receiver's public key to encrypt the symmetric key. When using a digital envelope, the receiver needs to use its own private key to open the digital envelope to obtain the symmetric key.

A obtains B's public key in advance, and the specific encryption and decryption process is shown in Figure 14-4.

(1) A uses a symmetric key to encrypt the plain text to generate cipher text information.

(2) A uses B's public key to encrypt the symmetric key to generate a digital envelope.

(3) A sends the digital envelope and cipher text information to B together.

(4) After receiving the encrypted information of A, B uses his private key to open the digital envelope and obtain the symmetric key.

(5) B uses the symmetric key to decrypt the cipher text information to obtain the original plain text.

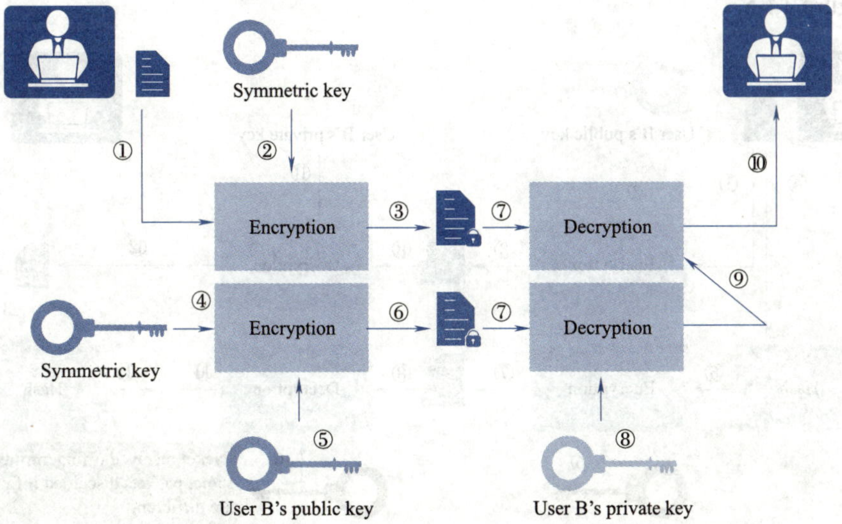

Figure 14-4 Digital envelope

From the encryption and decryption process, it can be seen that the digital envelope technology combines the advantages of symmetric key encryption and public key encryption, solves the problems of symmetric key issuance and slow public key encryption, and improves security, scalability and efficiency, etc.

However, there is still a problem with the digital envelope technology. If an attacker intercepts A's information, encrypts the forged information with his symmetric key, and encrypts his symmetric key with B's public key, and then sends it to B. After receiving the encrypted information, B decrypts the plain text obtained, and B always considers it to be the information sent by A. At this point, a method is needed to ensure that the information received by the receiver is sent by the designated sender.

Subtask 14.2.3 Data verification application scenario—digital signature

Digital signature refers to the data obtained after the sender encrypts the digital fingerprint with its own private key. When using a digital signature, the receiver needs to use the sender's public key to unlock the digital signature and obtain a digital fingerprint.

Digital fingerprint is also called information digest, it refers to the data obtained after the sender calculates the plain text information through the Hash algorithm. When using a digital fingerprint, the sender will send the digital fingerprint and plain text to the receiver, and the receiver uses the same Hash algorithm to calculate the data fingerprint generated by the plain text. Match with the received digital fingerprint. If they are consistent, it can be determined that the plain text information has not been tampered with.

A obtains B's public key in advance, and the specific encryption and decryption process is shown in Figure 14-5.

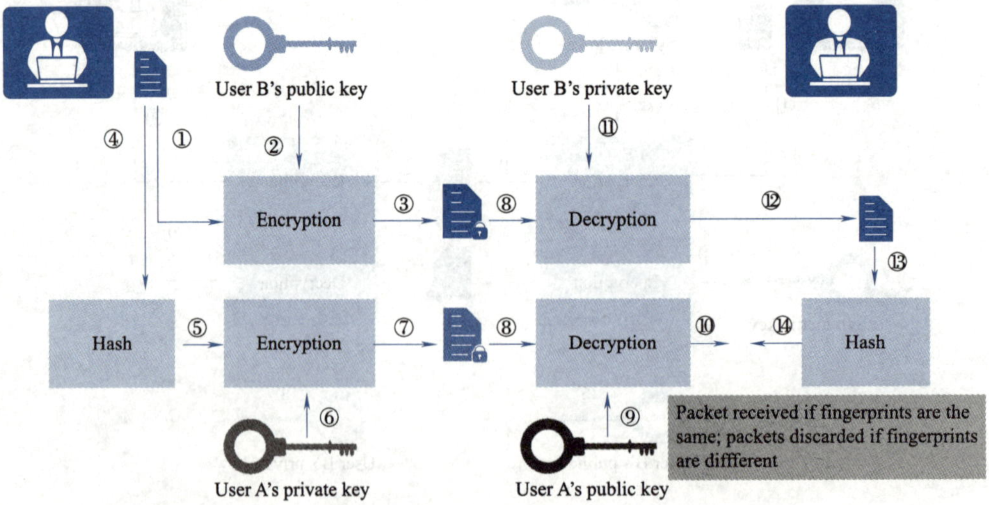

Figure 14-5 Digital signature

(1) A uses B's public key to encrypt the plain text to generate cipher text information.

(2) A uses the Hash algorithm to hash the plain text to generate a digital fingerprint.

(3) A uses his private key to encrypt the digital fingerprint and generate a digital signature.

(4) A sends the cipher text information and the digital signature to B.

(5) B uses A's public key to decrypt the digital signature and obtain a digital fingerprint.

(6) After receiving the encrypted information of A, B uses his private key to decrypt the cipher text information to obtain the original plain text.

(7) B uses the Hash algorithm to hash the plain text to generate a digital fingerprint.

(8) B compares the generated digital fingerprint with the obtained digital fingerprint. If they are consistent, B accepts the plain text. if they are inconsistent, B discards the plain text.

From the encryption and decryption process, it can be seen that the digital signature technology not only proves that the information has not been tampered with, but also proves the identity of the sender. Digital signature and digital envelope technology can also be used in combination.

However, there is still a problem with digital signature technology. If the attacker changes the public key of B, what A obtains is the public key of the attacker. The attacker intercepts the information sent by B to A, digitally signs the forged information with his private key, and then sends it to A together with the encrypted forged information using A's public key. After A receives the encrypted information, decrypts the obtained plain text, and verifies that the plain text has not been tampered with, then A still believes that the message was sent by B. At this point, a method is needed to ensure that a specific public key belongs to a specific owner.

Task 14.3 Understand common encryption and decryption algorithms

Subtask 14.3.1 Understand symmetric encryption algorithms

Symmetric encryption algorithms mainly include two types according to the encrypted objects:

1. Stream algorithm

The stream encryption algorithm continuously inputs elements during the algorithm process, and produces one output element at a time. A typical stream cipher algorithm encrypts one byte of plain text at a time, and the key is input to a pseudo-random byte generator to generate a seemingly random byte stream, called the key stream. Stream encryption algorithms are generally used in data communication channels, browsers or network links.

Common stream encryption algorithm: RC4 is a stream encryption algorithm designed by Ron Rivest for RSA Security in 1987. It is a stream cipher with a variable key size. It uses byte-oriented operations to encrypt information into a whole in real time. The speed of this algorithm can reach about 10 times that of DES encryption.

2. Block algorithm

The input of the block encryption algorithm is the plain text block and key. The plain text is divided into two halves, and the two halves of the data are combined into cipher text packets after n rounds of processing, and the input of each round is the output of the previous round. At the same time, the subkey is also generated by the key. The typical packet length is 64 bits.

Block algorithm include the following:

DES(data encryption standard):DES was developed by the National Institute of Standards and Technology (NIST). DES is the first widely used cryptographic algorithm, using the same key for encryption and decryption. DES is a block encryption algorithm. The input plain text is 64 bits, the key is 56 bits, and the generated cipher text is 64 bits (encrypt the data into a 64-bit block). Because the cipher capacity is only 56 bits, 3DES was later proposed for its weakness of insufficient security.

3DES(Triple DES):3DES uses a 128-bit key. The information is first encrypted with a 56-bit key, then decoded with another 56-bit key, and finally encrypted with the original 56-bit key, so that 3DES uses an effective 128-bit key. The biggest advantage of Triple DES is that it can use existing software and hardware, and the technology on the DES encryption algorithm can easily implement Triple DES.

AES(advanced encryption standard): AES uses a 128-bit packet length, supports key lengths of 128-bit, 192-bit, and 256-bit length, and supports different platforms. The 128-bit key length provides sufficient security and requires less processing time than longer keys. So far, AES has not had any fatal flaws. However, due to the mass production of fast DES chips, DES can still be used. However, AES replaces DES and 3DES to enhance security and efficiency is a general trend.

IDEA(international data encryption algorithm): IDEA is a symmetric block cipher algorithm. The input plain text is 64 bits, the key is 128 bits, and the generated cipher text is 64 bits. There are many applications, among which SSL includes IDEA in its encryption algorithm library.

RC2 is a variable-length key encryption algorithm designed by Ron Rivest for RSA. It is a block mode cipher text, which encrypts information into 64-bit data. Because it can use keys of different lengths, its key length can be from zero to infinite, and the speed of encryption depends on the size of the key.

RC5 is a new type of block cipher algorithm designed by Rivest of RSA in 1994. RC5 is similar to RC2 and is also a block cipher text, but this algorithm uses different block sizes and key sizes. In addition, the round that the data passes through in this algorithm is also different. It is generally recommended to use the RC5 algorithm with a 128-bit key and run 12 to 16 rounds. It is a block iterative cipher algorithm with variable block length, key length and number of iteration rounds.

Unlike some other newer encryption algorithms, RC6 includes the entire family of algorithms. The RC6 series was proposed in 1998 after the RC5 algorithm was proposed. After investigation, it was discovered that there was a theoretical loophole in the encryption of special round.

The national secret algorithm is a commercial cipher block standard symmetric algorithm compiled by the State Cryptography Administration. The block length and key length of the national secret algorithm are both 128 bits. In the case of high security level requirements, the use of SM1 or SM4 national encryption algorithm can fully meet the encryption requirements.

Currently, the more commonly used symmetric key encryption algorithms mainly include DES (data encryption standard), 3DES (triple data encryption standard), and AES (advance encrypt standard) algorithms.

Subtask 14.3.2 Understand asymmetric encryption algorithms

At present, the commonly used public key encryption algorithms mainly include DH (Diffie-Hellman), RSA and DSA (digital signature algorithm) algorithms.

The DH algorithm is generally used for both parties to negotiate a symmetric encryption key, that is, encryption and decryption are all the same key. The essence is that both parties share some parameters, and then each generates a key, and then according to mathematical principles, the keys generated by each are the same. This key does not involve propagation in the link, but the interaction of the previous parameters will involve the link transmission.

The RSA public key encryption algorithm was developed in 1977 by Ron Rivest, Adi Shamirh and LenAdleman (Massachusetts Institute of Technology, USA). The name RSA comes from the development of the names of the three of them. RSA is currently the most influential public key encryption algorithm. It can resist all cryptographic attacks known so far and has been recommended by ISO as a public key data encryption standard. RSA is also the first algorithm that can be used for both encryption and digital signature.

DSA is a variant of Schnorr and ElGamal signature algorithms, and is adopted as DSS (digital signature standard) by NIST in the United States. It plays a very important role in ensuring the integrity, privacy, and non-repudiation of data. DSA is based on the integer finite field discrete logarithm problem, and its security is similar to that of RSA. In DSA digital signature and authentication, the sender uses his own private key to sign a file or message, and the recipient uses the sender's public key to verify the authenticity of the signature after receiving the message. DSA is only an algorithm. The difference from RSA is that it cannot be used for encryption and decryption, and it cannot be used for key exchange. It is only used for signatures. It is much faster than RSA.

Subtask 14.3.3 Understand hashing algorithms

MD5 (information authentication code) is a hash function widely used in the field of computer security to provide integrity protection of messages. Data (such as Chinese characters) will be calculated into another fixed length value. Its function is to allow a large amount of information to be "compressed" into a confidential format before signing the private key with digital signature software. In addition to being used for digital signatures, it can also be used for secure access authentication.

SHA (secure Hash algorithm) was developed by NIST. The original HMAC function was revised in 1994 and was called SHA-1. SHA-1 is described in RFC2404. SHA-1 produces a 160-bit message digest. SHA-1 is slower than MD5, but more secure. Because its signature is longer, it has a stronger anti-breaking function, and it can find the shared key more effectively.

Compared with SHA-1, the encrypted data length of SHA-2 algorithm has increased, and its security performance is much higher than SHA-1. SHA-2 algorithms include SHA2-256, SHA2-384, and SHA2-512. The key lengths are 256 bits, 384 bits, and 512 bits, respectively.

SM3 (national secret algorithm) is a commercial algorithm compiled by the State Cryptography Administration. It is used for digital signature and verification in cryptographic applications, message authentication code generation and verification, and random number generation. It can meet the security requirements of a variety of cryptographic applications.

The above algorithms have their own characteristics. The MD5 algorithm is faster than the SHA-1 algorithm, and the SHA-1 algorithm is more secure than the MD5 algorithm. Compared with SHA-1, the increase in the number of encrypted data bits in SHA-2 and SM3 algorithms increases the difficulty of cracking, making the security performance much higher than SHA-1.

Project 15 PKI certificate system

Digital signature technology not only proves that the information has not been tampered with, but also proves the identity of the sender. However, there are certain problems. If an attacker forges the public key of the receiver, the information sent by the sender will be decrypted by the attacker with the private key and snooped. Therefore, the authenticity of the receiver's public key will directly determine the confidentiality of data transmission.

At this point, a method is needed to ensure that a specific public key belongs to a specific owner.

Task 15.1 Understand digital certificates

Subtask 15.1.1 Understand the definition of digital certificate

A digital certificate is a string of numbers that mark the identity infor mation of all parties in the Internet communication, and provides a way to verify the identity of the communicating entity on the Internet. A digital certificate is not a digital ID card, but a seal or seal (or a signature added to the digital ID card) stamped on the digital ID card by the identity certification agency. It is issued by an authority—CA organization, also known as a Certificate Authority center, and people can use it to identify each other's identity on the Internet.

A digital certificate is a file that is digitally signed by a certificate authority and contains information about the owner of the public key and the public key. The simplest certificate contains a public key, name, and digital signature of the certificate authority. Another important feature of digital certificates is that they are only valid for a certain period of time.

A digital certificate can be said to be a security passport or ID card on the Internet. When people travel to other countries, their identity can be verified with a passport and they are allowed to enter this country. Digital certificates provide proof of identity on the network. The encryption technology (encrypted transmission, digital signature, digital envelope and other security technologies) with digital certificates as the core can encrypt and decrypt the information transmitted on the network, digital signature and signature verification, and ensure the confidentiality, integrity and integrity of the information transmitted on the Internet. The non-repudiation of the transaction. With the use of digital certificates, even if the information you send is intercepted by others on the Internet, or even if you lose your personal account, password and other information, you can still guarantee the safety of your account and funds.

The digital certificate technology solves the problem that the public key cannot be determined as the designated owner in the digital signature technology.

There are four types of digital certificates according to the issuing authority and users of digital certificates:

(1) Self-signed certificate: Also known as the root certificate, it is a certificate issued to itself, that is, the issuer and subject name in the certificate are the same. When the applicant cannot apply for a local certificate from the CA, a self-signed certificate can be generated by the device, which can realize the simple certificate issuance function. The device does not support life cycle management (such as certificate renewal, certificate revocation, etc.) of the self-signed certificate it generates.

(2) CA certificate: CA's own certificate. If there is no multi-level CA in the PKI system, the CA certificate is a self-signed certificate. If there are multiple levels of CA, it will form a CA hierarchy, the top CA is the root CA, it has a CA "self-signed" certificate. The applicant trusts the CA by verifying the digital signature of the CA, and any applicant can obtain the CA's certificate (including the public key) to verify the local certificate issued by it.

(3) Local certificate: The certificate issued by the CA to the applicant.

(4) Device local certificate: The device issues a certificate to itself based on the CA certificate.

The issuer name in the certificate is the name of the CA server. When the applicant cannot apply for a local certificate from the CA, the device can generate a device local certificate through the device, which can realize the simple certificate issuance function.

Subtask 15.1.2 Analysis of digital certificate structure

The simplest certificate contains a public key, user name, and a digital signature from a certificate authority. In general, the certificate also includes the validity period of the key, the name of the issuer (certificate authority), the serial number of the certificate, and other information. The structure of the certificate follows the specifications provided by the X.509 v3 version.

The content of the certificate is shown in Figure 15-1, and the meaning of each field is as follows:

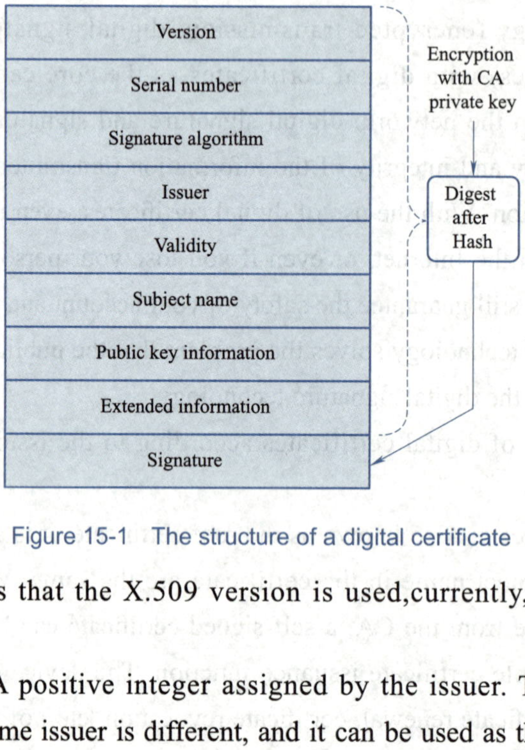

Figure 15-1 The structure of a digital certificate

(1) Version:It means that the X.509 version is used,currently, the v3 version (0x2) is commonly used.

(2) Serial number: A positive integer assigned by the issuer. The serial number of the certificate issued by the same issuer is different, and it can be used as the unique identification of the certificate together with the name of the issuer.

(3) Signature algorithm: The signature algorithm used by the issuer when issuing the certificate.

(4) Issuer: It is the name of the device that issued this certificate. It must be the same as the subject name in the issuer's certificate, usually the name of the CA server.

(5) Validity: Includes the valid start and end dates. Certificates that are not within the validity period are invalid certificates.

(6) Subject name: The name of the certificate owner. If it is the same as the issuer, it means that the certificate is a self-signed certificate.

(7) Public key information: User public key and public key algorithm information.

(8) Extended information: Usually includes optional fields such as the usage of the certificate and the issuing address of the CRL.

(9) Signature:It is the signature of the certificate information by the issuer using the private key.

Subtask 15.1.3 Analysis of digital certificate file format

The main file types and protocols of digital certificates include PEM, DER, PEX, JKS, KDB, CER, KEY, CSR, CRT, CRL, OCSP, SCEP, etc.

Huawei devices mainly support digital certificates using the following 3 file formats:

PKCS#12 (public-key cryptography standards#12) saves the certificate in binary format, which may or may not contain the private key. Commonly used suffixes are .P12 and .PFX.

DER (distinguished encoding rules) saves the certificate in binary format and does not contain the private key. Commonly used suffixes are .DER, .CER and .CRT.

PEM (privacy enhanced mail) saves the certificate in ASCII format, which may or may not contain the private key. Commonly used suffixes are .PEM, .CER and .CRT.

Task 15.2 Understand PKI architecture

As shown in Figure 15-2, with the development of network technology and information technology, e-commerce has gradually been accepted by people and has been continuously popularized. However, when conducting e-commerce transactions through the Internet, there are the following problems:

(1) The two parties of the transaction do not conduct the transaction on-site, and the legal identities of both parties cannot be confirmed.

(2) The information is easily stolen and modified when transmitted through the network, and the security of the information cannot be guaranteed.

(3) There is no evidence to refer to when there is a dispute between two parties of the transaction, and arbitration cannot be provided.

In order to solve the above problems, PKI technology came into being, which uses public key technology to ensure that identity authentication, confidentiality, data integrity and non-repudiation can be achieved in the transaction process. Therefore, PKI technology has been widely used in network communication and network transactions, especially in e-government and e-commerce businesses.

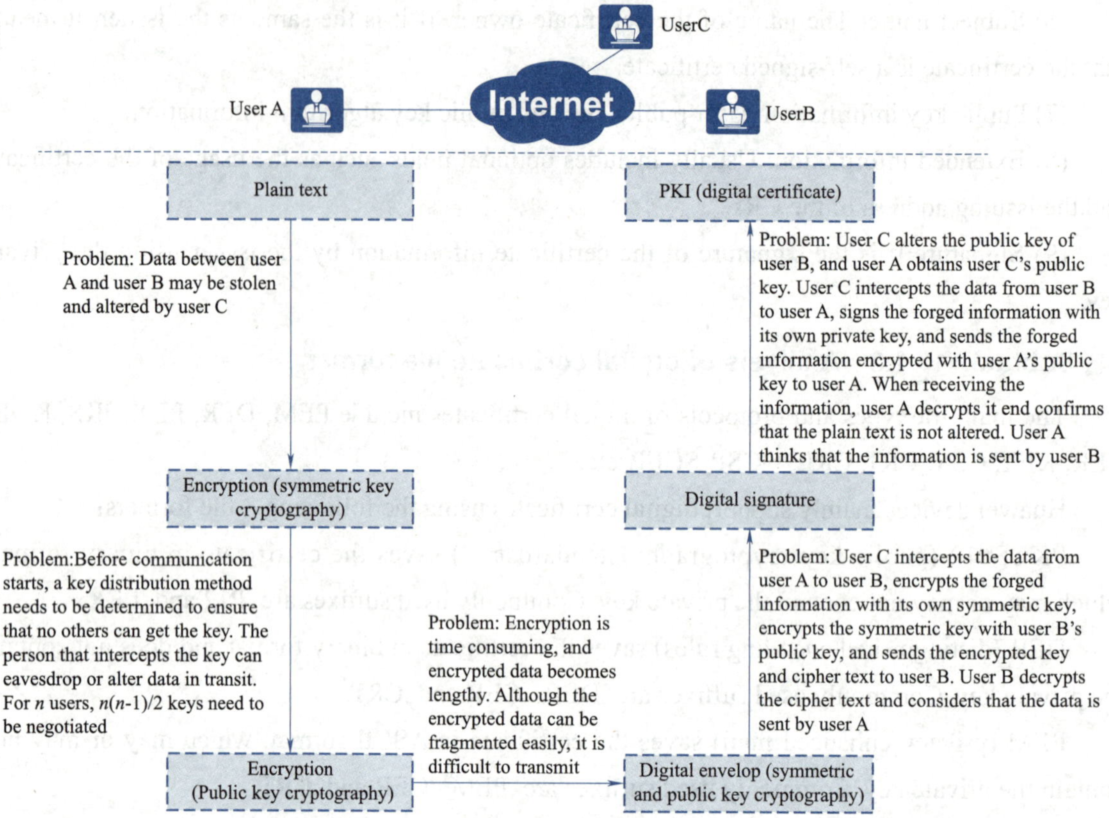

Figure 15-2 The necessity of PKI

Subtask 15.2.1 Understand the basic concepts of PKI

Public key infrastructure (PKI) is a certificate management platform that follows established standards. It uses public key technology to provide security services for all network applications. PKI technology is the core of information security technology, as well as the key and basic technology of e-commerce.

The core technology of PKI revolves around the entire life cycle of the digital certificates, including application,issuance and using of digital certificates. During this entire life cycle, PKI uses symmetric key encryption technology, public key encryption technology, digital envelopes technology, and digital certificates technology.

Subtask 15.2.2 Analysis of PKI architecture

As shown in Figure 15-3, a PKI system consists of four parts: an end entity, a certificate certification authority, a certificate registration authority, and a certificate/CRL repository.

End entity (EE): Also known as PKI entity, it is the end user of PKI products or services, which can be individuals, organizations, devices (such as routers, firewalls), or processes running in computers.

CA (certificate authority): CA is the trust foundation of PKI and a trusted entity used to issue and manage digital certificates. It is an authoritative, trustworthy and impartial third-party organization, usually served by a server, such as Windows Server 2008.

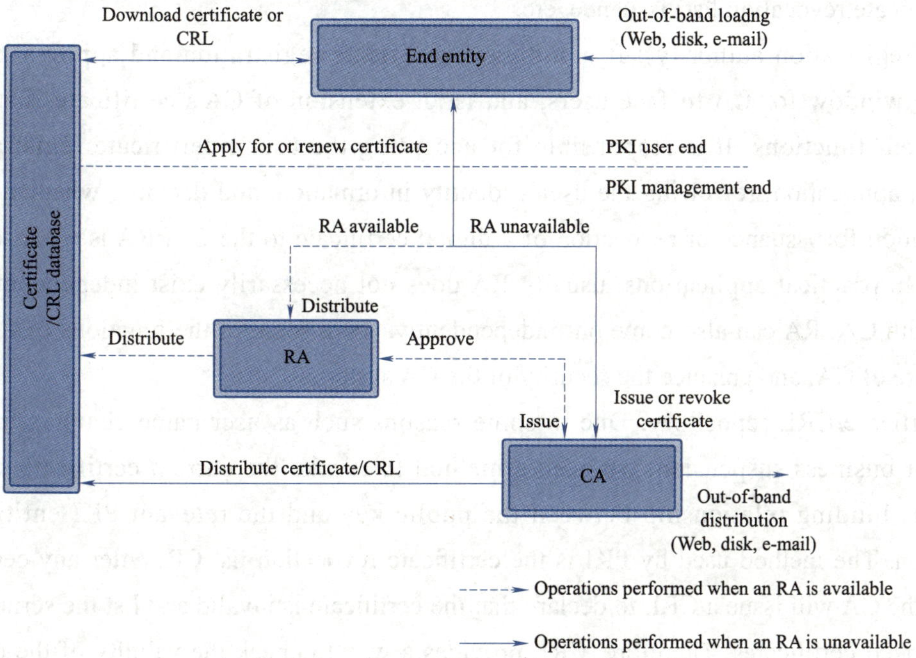

Figure 15-3 PKI system architecture

(1) CA usually adopts a multi-level hierarchical structure. According to the level of the certification authority, it can be divided into root CA and subordinate CA.

Root CA is the first certification authority in the public key system, and it is the origin of trust. The root CA can issue certificates for other CAs, as well as other computers, users, and services. For most certificate-based applications, authentication using certificates can be traced back to the root CA through the certificate chain. The root CA usually holds a self-signed certificate.

The subordinate CA must obtain a certificate from the superior CA. The superior CA can be the root CA or a subordinate CA that has been authorized by the root CA to issue subordinate CA certificates. The upper-level CA is responsible for issuing and managing the certificates of the lower-level CA, and the lower-level CA directly faces users. For example, CA2 and CA3 are subordinate CAs and hold CA certificates issued by CA1; CA4, CA5, and CA6 are subordinate CAs and hold CA certificates issued by CA2.

(2) When a PKI entity trusts a CA, it will pass the trust label through the certificate chain. The certificate chain is a collection of a series of certificates from the user‘s certificate to the root certificate. When the communicating PKI entity receives the certificate to be verified, it will

successively verify the legitimacy of its issuer along the certificate chain.

(3) The core function of CA is to issue and manage digital certificates, including: certificate issuance, certificate renewal, certificate revocation, certificate query, certificate archiving, CRL(certificate revocation list) issuance, etc.

RA (registration authority): It is a digital certificate registration and approval authority. RA is the window for CA to face users, and is an extension of CA's certificate issuance and management functions. It is responsible for accepting the user's certificate registration and revocation application, reviewing the user's identity information, and deciding whether to submit an application for issuance or revocation of a digital certificate to the CA. RA is a part of the CA function. In practical applications, usually RA does not necessarily exist independently, but is merged with CA. RA can also come out independently, share some of the functions of CA, reduce the pressure of CA, and enhance the security of the CA system.

Certificate/CRL repository: Due to some reasons such as user name change, private key leakage or business suspension, we need a method to revoke the current certificate, that is, to revoke the binding relationship between the public key and the relevant PKI entity identity information. The method used by PKI is the certificate revocation list CR. After any certificate is revoked, the CA will issue a CRL to declare that the certificate is invalid and list the serial numbers of all revoked certificates. Therefore, CRL provides a way to check the validity of the certificate. The certificate/CRL repository is used to store and manage information such as certificates and CRLs, and provide query functions. The certificate/CRL repository can be constructed using LDAP (lightweight directory access protocol) server, FTP (file transfer protocol) server, HTTP (hypertext transfer protocol) server or database, etc. Among them, the LDAP specification simplifies the cumbersome X.500 directory access protocol, supports TCP/IP, and has been widely used in the PKI system for certificate information issuance, CRL information issuance, CA policies, and various aspects related to information issuance. If the scale of the certificate is not too large, you can also choose to set up HTTP, FTP and other servers to store the certificate and provide download services to users.

Subtask 15.2.3 Understand PKI life cycle

The core technology of PKI revolves around the entire life cycle of local certificate application, issue, storage, download, installation, verification, update, and revocation, as shown in Figure 15-4.

Certificate application: The certificate application is the registration of the certificate. In the process, a PKI entity introduces itself to the CA and obtains a certificate. Under normal circumstances, a PKI entity generates a pair of public and private keys. The public key and its own identity information (included in the certificate registration request message) are sent to the CA to generate a local certificate, and the private key is saved by the PKI entity itself to digitally sign and decrypt the cipher text sent by the opposite entity. There are two ways for a PKI entity to apply for a local certificate from a CA:

(1) Online application: PKI entities support SCEP (simple certificate enrollment protocol) or CMPv2 (certificate management protocol version 2) to send certificate registration requests to CA to apply for local certificates.

(2) Offline application: Refers to the PKI entity using the PKCS#10 format to print out the local certificate registration request message and save it to a file, and then send the file to the CA for certificate application through out-of-band methods (such as Web, disk, email, etc.) .

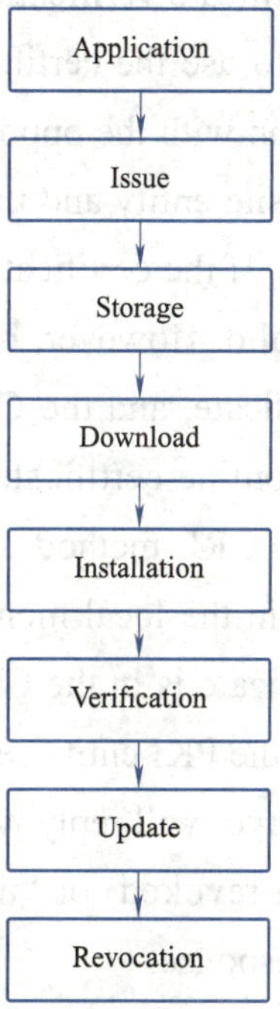

Figure 15-4 The life cycle of PKI

Certificate issuance: When a PKI entity applies for a local certificate from a CA, if there is an RA, the RA will first review the identity information of the PKI entity. After the review is passed, the RA sends the application information to the CA, and the CA generates a local certificate based on the public key and identity information of the PKI entity, and sends the local certificate information to the RA. If there is no RA, the CA directly verifies the identity information of the PKI entity.

Certificate storage: After CA generates a local certificate, CA/RA will publish the local certificate to the certificate/CRL repository, and provide users with download services and directory browsing services.

Certificate download: The PKI entity downloads the issued certificate from the CA server through the SCEP or CMPv2 protocol, or downloads the issued certificate through LDAP, HTTP, or out-of-band methods. The certificate can be your own local certificate, a CA/RA certificate or the local certificate of other PKI entities.

Certificate installation: After the PKI entity downloads the certificate, it needs to be installed , that is, the certificate will be imported into the device's memory, otherwise, the certificate will not take effect. The certificate can be your own local certificate, CA/RA certificate, or the local certificate of other PKI entities. When applying for a certificate through the SCEP protocol, the PKI entity first obtains the CA certificate and automatically imports the CA certificate into the device memory, then obtains the local certificate and automatically imports the local certificate into the device memory.

Certificate verification: After the PKI entity obtains the certificate of the opposite entity, when it needs to use the certificate of the opposite entity, such as establishing a secure tunnel or secure connection with the opposite end, it usually needs to verify the legitimacy of the local certificate of the opposite entity and the CA (Whether the certificate is valid or whether it is issued by the same CA, etc.). If the certificate of the certificate issuer is invalid, all certificates issued by the CA are no longer valid. However, before the CA certificate expires, the device will automatically update the CA certificate, and the CA certificate will expire under abnormal conditions. PKI entities can use CRL or online certificate status protocol (OCSP) to check whether the certificate is valid. When using the CRL method, the PKI entity first searches for the CRL in the local memory. If there is no CRL in the local memory, it needs to download the CRL and install it in the local memory. If the certificate is in the CRL, it means that the certificate has been revoked. When using the OCSP method, the PKI entity sends a request for certificate status information to the OCSP server, and the OCSP server will reply with "valid" (the certificate has not been revoked), "expired" (the certificate has been revoked) or "unknown" [The server(OCSP)cannot determine the requested certificate status] response.

Certificate update: When a certificate expires or the key is leaked, the PKI entity must replace the certificate. The update can be achieved by re-applying, or it can be updated automatically using the SCEP or CMPv2 protocol.

Certificate revocation: Due to changes in user identity, user information, and user public key, the user needs to revoke his digital certificate, which is to revoke the binding relationship between the public key and user identity information. In PKI, CA mainly uses CRL or OCSP protocol to revoke certificates, while PKI entities revoke their own certificates through out-of-band applications.

Task 15.3 Analysis of PKI working mechanism

For a network which uses PKI, the purpose of configuring PKI is for the designated PKI entity to apply for a local certificate from the CA, and the device verifies the validity of the certificate. The PKI working process is shown in Figure 15-5.

(1) The PKI entity requests a CA certificate from the CA, and the CA certificate is the CA server certificate.

(2) When the CA receives the CA certificate request from the PKI entity, it will reply its own CA certificate to the PKI entity.

(3) After the PKI entity receives the CA certificate, it will install the CA certificate.

When a PKI entity applies for a local certificate through the SCEP protocol, the PKI entity uses the configured Hash algorithm to calculate the CA certificate to obtain a digital fingerprint, and compares it with the digital fingerprint of the CA server configured in advance. If they are consistent, the PKI entity accepts CA certificate, otherwise the PKI entity will discard the CA certificate.

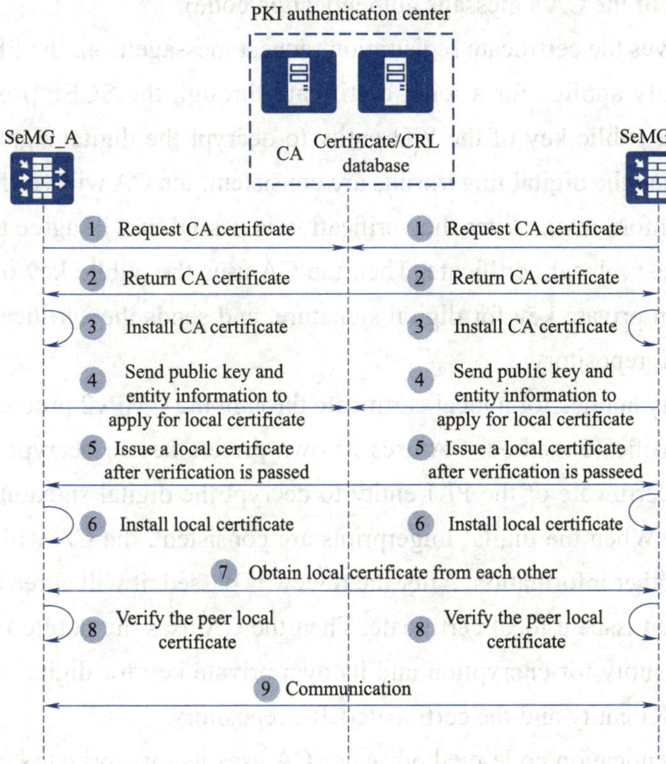

Figure 15-5 The PKI working process

(4) The PKI entity sends a certificate registration request message (including the public key and PKI entity information in the configured key pair) to the CA.

When a PKI entity applies for a local certificate through the SCEP protocol, the PKI entity uses the public key of the CA certificate to encrypt the certificate registration request message and digitally sign its own private key. If the CA requires verification of the challenge code, the certificate registration request message must carry the challenge code (consistent with the CA's challenge password).

When a PKI entity applies for a local certificate through the CMPv2 protocol, the PKI entity can use additional certificates (local certificates issued by other CAs) or message authentication codes for identity authentication.

① Additional certificate method: The PKI entity uses the public key of the CA certificate to encrypt the certificate registration request message and digitally signs the private key corresponding to the additional certificate of the PKI entity.

② Message authentication code method: PKI entity use the public key of the CA certificate to encrypt the certificate registration request. And the certificate registration request must include the reference value and secret value of the message authentication code (consistent with the reference value and secret value of the CA's message authentication code).

(5) The CA receives the certificate registration request message from the PKI entity.

When a PKI entity applies for a local certificate through the SCEP protocol, the CA uses its private key and the public key of the PKI entity to decrypt the digital signature and verify the digital fingerprint. When the digital fingerprints are consistent, the CA will verify the identity of the PKI entity and other information. After the verification is passed, it will agree to the application of the PKI entity and issue a local certificate. Then the CA uses the public key of the PKI entity for encryption and its own private key for digital signature, and sends the certificate to the PKI entity and the certificate/CRL repository.

When a PKI entity applies for a local certificate through the CMPv2 protocol:

① Additional certificate method: CA uses its own private key to decrypt and uses the public key in the additional certificate of the PKI entity to decrypt the digital signature and finally verify the digital fingerprint. When the digital fingerprints are consistent, the CA will review the identity of the PKI entity and other information. After the review is passed, it will agree to the application of the PKI entity, and then issue a local certificate. Then the CA uses the public key in the additional certificate of the PKI entity for encryption and its own private key for digital signature, and sends the certificate to the PKI entity and the certificate/CRL repository.

② Message authentication code method: After CA uses its own private key to decrypt, it will verify the reference value and secret value of the message authentication code. When the reference value and the secret value are consistent, the CA will review the identity of the PKI entity and other information. After the review is passed, it will agree to the application of the PKI entity and issue a local certificate. Then the CA uses the public key of the PKI entity for encryption, and sends the certificate to the PKI entity and the certificate/CRL repository.

(6) The PKI entity has received the certificate information sent by the CA.

When a PKI entity applies for a local certificate through the SCEP protocol, the PKI entity uses its own private key to decrypt, and uses the CA's public key to decrypt the digital signature and then verify the digital fingerprint. When the digital fingerprints are consistent, the PKI entity accepts the certificate information, and then installs the local certificate.

When a PKI entity applies for a local certificate through the CMPv2 protocol:

① Extra certificate method: The PKI entity uses the private key which corresponding to the additional certificate for decryption, and uses the CA's public key to decrypt the digital signature, and then verifies the digital fingerprint. When the digital fingerprints are consistent, the PKI entity accepts the certificate information, and then installs the local certificate.

② Message authentication code method: PKI entity uses its own private key to decrypt and verify the reference value and secret value of the message authentication code. When the reference value and the secret value are consistent, the PKI entity accepts the certificate information, and then installs the local certificate.

(7) When PKI entities communicate with each other, each needs to obtain and install the local certificate of the opposite entity. The PKI entity can download the local certificate of the peer through HTTP/LDAP and other methods. In some special scenarios, such as IPSec, PKI entities will send their local certificates to the peer.

(8) After the PKI entity installs the local certificate of the opposite entity, it will verify the validity of the local certificate of the opposite entity through CRL or OCSP. When the local certificate of the peer entity is valid, the public key of the peer certificate can be used for encrypted communication between PKI entities. If the PKI certification center has an RA, the PKI entity will also download the RA certificate. The RA will review the local certificate application of the PKI entity, and after the review is passed, the application information will be sent to the CA to issue the local certificate.

Subtask Certificate application scenario

1. Log in to the Web via HTTPS

As shown in Figure 15-6, the administrator can securely log in to the web interface of the HTTPS server through HTTPS, and manage the device through the web interface.

In order to improve the security when establishing an SSL connection between the two parties, we specify for the HTTPS client on the device that a local certificate will be issued by a CA trusted by the web browser. In this way, the web browser can verify the legitimacy of the local certificate, avoid possible active attacks, and ensure the safe login of the administrator.

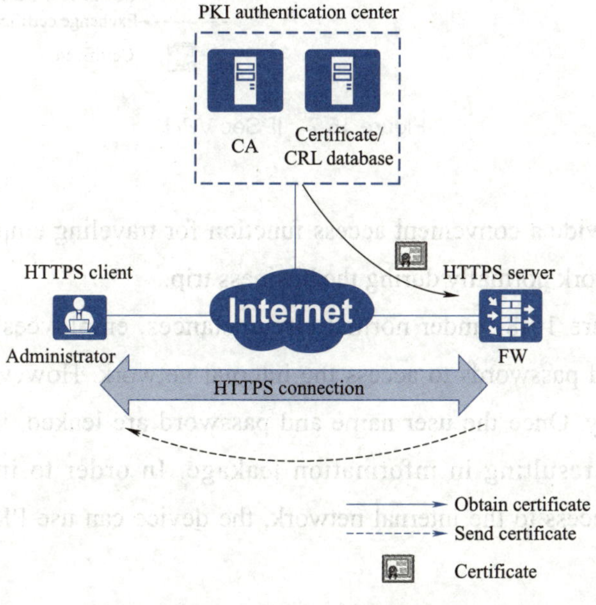

Figure 15-6 Log in to the Web via HTTPS

2. IPSec VPN

As shown in Figure 15-7, the device serves as the egress gateway of Network A and Network B. In Network A and Network B, intranet users communicate with each other through the public network. Because the public network is an insecure network, in order to protect data security, the device uses IPSec technology to establish an IPSec tunnel with the peer device. Normally, IPSec uses pre-shared key to negotiate IPSec. However, when IPSec adopts the pre-shared key method in a large network, there will be problems of insecure key exchange and heavy configuration workload. In order to solve the above problems, PKI-based certificates can be used for identity authentication between devices to complete the establishment of IPSec tunnels.

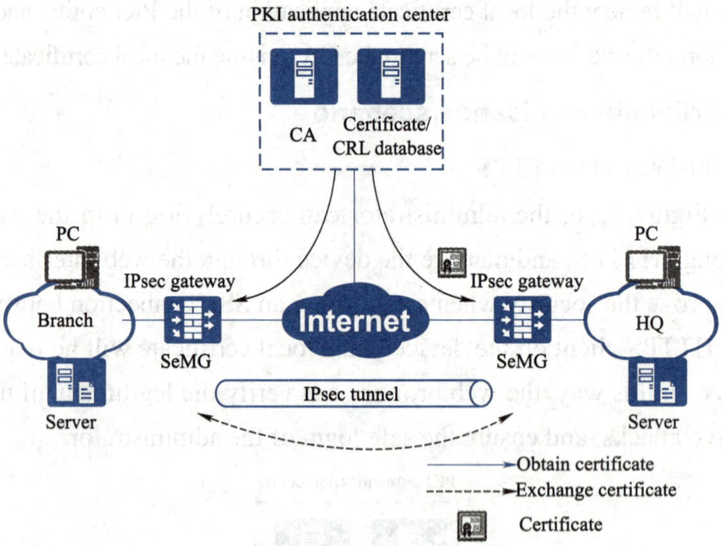

Figure 15-7 IPSec VPN

3. SSL VPN

SSL VPN can provide a convenient access function for traveling employees, so that they can access the internal network normally during the business trip.

As shown in Figure 15-8, under normal circumstances, employees who are on a business trip use user names and passwords to access the internal network. However, this security method has poor confidentiality. Once the user name and password are leaked, illegal users may access the internal network, resulting in information leakage. In order to improve the security of traveling employees' access to the internal network, the device can use PKI certificate methods to authenticate users.

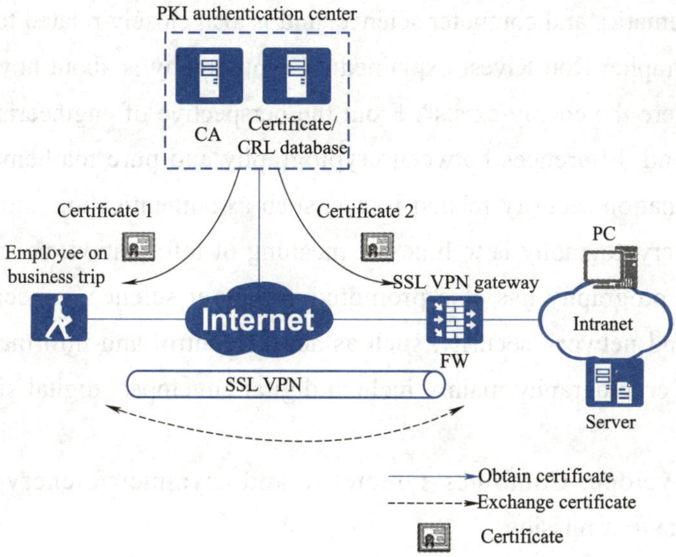

Figure 15-8 SSL VPN

Project 16 Encryption technology application

Through encryption technology, the security of data transmission in the network can be guaranteed, and the data will not be stolen or modified; through signature technology, the integrity of the data can be guaranteed, and the identity of the data sender can be proved. Through the PKI certificate authentication technology, the legitimacy of the public key can be verified, so that users can be guaranteed to access a safe and legal network.

By using these technologies, companies can prevent illegal users from accessing the corporate network. A secure channel will be established between enterprise branches, thus ensuring the security of enterprise data.

Task 16.1 Application of cryptography

Cryptography is a technical science that studies the preparation and deciphering of ciphers. Among them, the science of studying the objective laws of cipher changes and compiling cipher to keep communication secrets is called cryptography. The discipline used in decrypting ciphers to obtain communication information is called deciphering. The above two are collectively referred to as cryptography.

Cryptography is the study of how to secretly transmit information. In modern times, it specifically refers to the mathematical study of information and its transmission. It is often regarded as a branch of mathematics and computer science, and is also closely related to information theory. The famous cryptographer Ron Rivest explained: "Cryptography is about how to communicate in the environment where the enemy exists". From the perspective of engineering, this is equivalent to the similarities and differences between cryptography and pure mathematics. Cryptography is the core of information security related issues, such as authentication and access control. The primary purpose of cryptography is to hide the meaning of information, not to hide the existence of information. Cryptography has also promoted computer science, especially the techniques used in computer and network security, such as access control and information confidentiality. The applications of cryptography mainly include digital envelopes, digital signatures and digital certificates.

(1) Digital envelope: Combines symmetric and asymmetric encryption to ensure the confidentiality of data transmission.

(2) Digital signature: Use hash algorithm to ensure the integrity of data transmission.

(3) Digital certificate: It notarizes the public key through a third party (CA) to ensure the non-repudiation of data transmission.

Combined with actual network application scenarios, digital envelopes, digital signatures, and digital certificates have corresponding scenarios.

(1) VPN: Many VPN technologies use encryption and decryption technologies to ensure data confidentiality, such as IPSec VPN and SSL VPN.

(2) IPv6: In order to prevent the device from being impersonated by attackers, you can configure the SEND (secure neighbor discovery) function authorized by the router on the device, and then use digital certificate technology to select legitimate gateway devices.

(3) HTTPS login: The administrator can safely log in to the web interface of the HTTPS server through HTTPS, and manage the device through the web interface. In order to improve the security when establishing an SSL connection between the two parties, you can specify a local certificate issued by a CA trusted by the web browser for the HTTPS client on the device. In this way, the web browser can verify the legitimacy of the local certificate, avoid possible active attacks, and ensure the safe login of the administrator.

(4) System login authorization: Digest the user password before storing it, and compare authorization when the user logs in.

Task 16.2 Master the classification and configuration of VPN

Subtask 16.2.1 Understand the definition of VPN

VPN (virtual private network) establishes a private data channel through a shared public network. It connects each network or terminal that needs to access this virtual network through a channel. So as to form a dedicated network with a certain degree of security and a certain guarantee of service quality.

There are two key points in this definition.

(1) Virtual: Users no longer need to have actual dedicated long-distance data lines, but use long-distance data lines on the Internet to establish their own private networks.

(2) Private network: Users can develop a network that best meets their needs.

Traditional VPN networking mainly adopts two methods: dedicated line VPN and encrypted VPN based on client devices. Private line VPN means that users lease digital data network (DDN) circuits, ATM permanent virtual circuits (PVC), frame relay (FR) PVCs, etc. to form a two-layer VPN network. The backbone network is maintained by telecom operators, and customers are responsible for managing their own sites and routes. Encrypted VPN based on client devices implements all functions by client devices, and the members of the VPN are interconnected through an untrusted public network. The first method has a relatively high cost and poor scalability; the second method has higher requirements for user equipment and personnel.

In the IETF draft, the understanding of IP-based VPN is: "Use the IP mechanism to simulate a private wide area network." It is the technology of simulating a point-to-point dedicated line on the public data network through the tunnel technology. The so-called "virtual" means that users no longer need to have actual dedicated long-distance data lines, but use the Internet's long-distance data lines to establish their own private networks. The so-called private network means that users can develop a network that best meets their needs.

With the continuous development of IP data communication technology, IP-based VPN technology has gradually become the main choice in the VPN market. Because IP VPN adopts IP network to carry, and the operator's network is gradually improved, so the cost is low, and its service quality is sufficient to meet customer needs, and it also has better scalability and manageability. In general, more and more users are choosing IP VPN, and operators are also building IP VPN to attract more users.

Subtask 16.2.2 Master VPN classification

1. Divided by type of business use

According to the type of business use, VPN can be divided into remote access virtual network (Access VPN), enterprise internal virtual network (intranet VPN) and enterprise extended virtual network (Extranet VPN). These three types of VPNs correspond to the extranet formed by the traditional remote access network, the corporate intranet corporate network, and the corporate network of related partners.

2. Divided by implementation level

According to the level of implementation, VPNs can be divided into two-layer VPN (L2VPN), three-layer VPN (L3VPN), and higher layers such as SSL VPN, as shown in Figure 16-1.

From a conceptual point of view, SSL VPN is a new type of VPN technology that uses the SSL (security socket layer) protocol to achieve remote access. The SSL protocol is a web application-based security protocol proposed by Netscape. It includes: server authentication, client authentication (optional), data integrity on the SSL link, and data confidentiality on the SSL link. For internal and external applications, the use of SSL can ensure the authenticity, integrity and confidentiality of information. At present, the SSL protocol is widely used in various browser applications, and similarly, it can also be used in C/S applications that use the TCP protocol to transmit data such as Outlook. Because the SSL protocol is built into browsers such as IE, there is no need to install a client for SSL VPN that uses the SSL protocol for authentication and data encryption. Compared with the traditional IPSEC VPN, SSL VPN has the following characteristics: simple deployment, client free, low maintenance cost, and strong network adaptability. The difference between these two types of VPNs is similar to the difference between the C/S architecture and the B/S architecture.

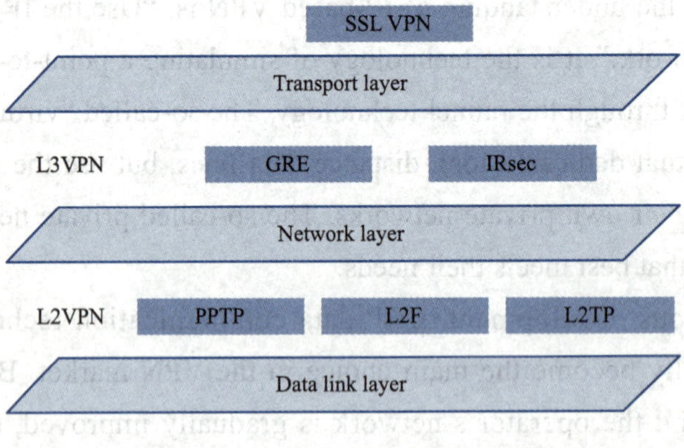

Figure 16-1 Classification of VPN

The L3VPN mainly refers to the VPN technology that works at the network layer of the protocol stack. Taking the IPSec VPN technology as an example, the IPSec header and the IP header work at the same level. When encapsulating the message, it is either encapsulated in an IPinIP manner, or the IPSec header and the IP header encapsulate the data payload at the same time. In addition to IPSec VPN technology, the other main three-layer VPN technology is GRE VPN. GRE VPN came into being earlier, and the implementation mechanism is relatively simple. GRE VPN can realize the encapsulation of any network protocol on another network protocol. Compared with IPSec, it cannot guarantee data security and can only provide a limited and simple security mechanism.

Similar to the L3VPN, the L2VPN refers to the VPN technology working at the data link layer of the protocol stack. Layer 2 VPNs mainly include PPTP (point-to-point tunneling protocol), L2F (layer 2 forwarding), and L2TP (layer 2 tunneling protocol).

3. Divided by VPN application scenarios

According to VPN application scenarios, VPNs can be divided into Site-to-Site VPN and Client-to-Site VPN.

Site-to-Site VPN is used to establish a connection between two local area networks. Available VPN technologies: IPSec, L2TP, L2TP over IPSec, GRE over IPSec, IPSec over GRE.

Client-to-Site VPN refers to a personal-to-site VPN, which is used to establish a connection between the client and the corporate intranet. Available VPN technologies: SSL, IPSec, L2TP, L2TP-over-IPSec.

Subtask 16.2.3 Implementation of L2TP VPN

L2TP (layer 2 tunneling protocol) is a tunneling protocol set up for the transparent transmission of PPP messages between users and the enterprise's server. It provides tunnel transmission support for data packets of PPP link layer. PPP defines an encapsulation mechanism for multi-protocols which across the second layer of point-to-point links. In particular, the user obtains the second layer connection to the network access server (NAS) through one of the following technologies (such as dial-up POTS, ISDN, ADSL, etc.), and then runs PPP on this connection. In such a configuration, the second-layer end point and the PPP session end point are in the same physical device (such as NAS).

There are three main usage scenarios for L2TP VPN: NAS-Initiated VPN, LAC auto-dial and Client-Initiated VPN. At present, we mainly introduce the most commonly used Client-Initiated VPN dial-up scenarios.

Client-Initiated VPN dial-up scenarios are generally used for employees on business trips using PCs, mobile phones and other mobile devices to access the headquarters server, and it is also the most commonly used L2TP dial-up method.

As shown in Figure 16-2, employees who on a business trip use their own PCs, mobile phones or other terminals to dial the LNS (L2TP network server) of the headquarters. In this process, an L2TP tunnel is constructed to encapsulate the packets that visit the headquarters. Encapsulation format: First, the private network IP header is encapsulated into the PPP protocol, then the PPP frame is encapsulated into the L2TP header, and finally the UDP and public network IP headers are encapsulated and forwarded to the public network. Since the target IP encapsulated in the public network IP header is the public network IP of the LNS, the message will be forwarded to the LNS by the public network, and the LNS will decapsulate the received message until the private network IP header is exposed. Finally, LNS checks the routing table and forwards it to the target host. In the Client-Initiated VPN dial-up scenario, the endpoints of the PPP and L2TP tunnels are the same. One end is LNS and the other end is Client.

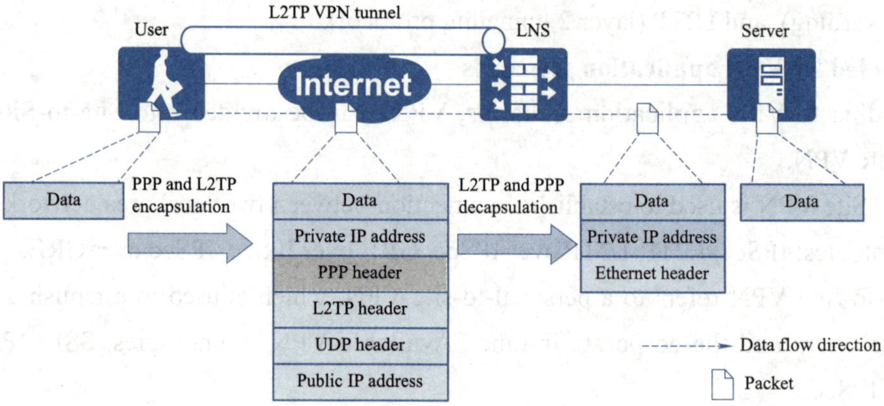

Figure 16-2 Client-Initiated VPN dial

In Client-Initiated VPN, each access user establishes a tunnel with the LNS; each tunnel carries only one L2TP session and PPP connection.

Figure 16-3 shows the process of establishing a call for an L2TP tunnel.

(1) When the access user dials to the LNS, it triggers the establishment of an L2TP tunnel between the access user and the LNS.

(2) After the L2TP tunnel is successfully established, an L2TP session is established on the basis of the tunnel.

(3) LNS authenticates users.

(4) Establish a PPP connection between the user and the LNS.

(5) On the basis of PPP connection, users can access intranet resources through LNS.

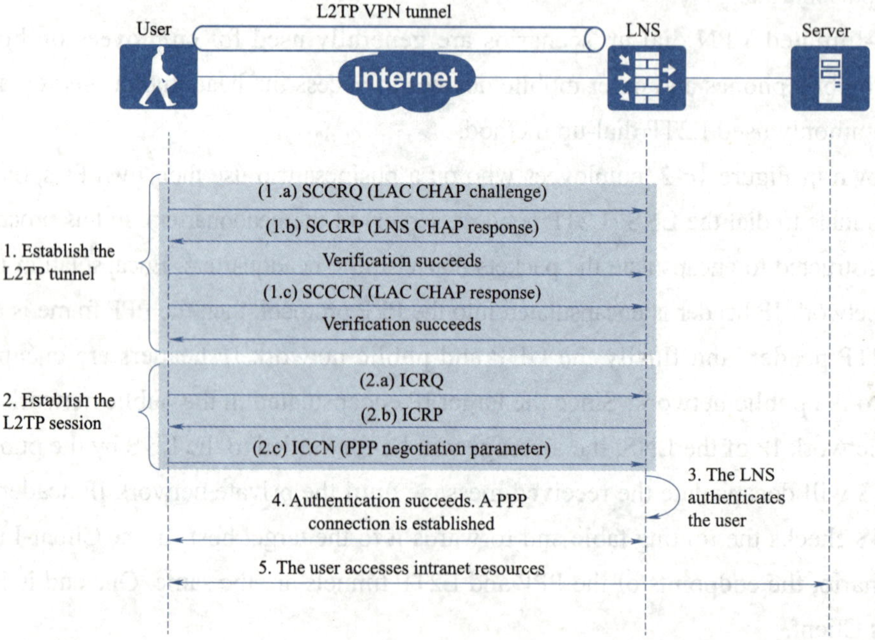

Figure 16-3 Client-Initiated VPN tunnel and session establishment process

Subtask 16.2.4 Implementation of GRE VPN

General routing encapsulation (GRE) is a three-layer VPN encapsulation technology. GRE can encapsulate the packets of certain network layer protocols (such as IPX, Apple Talk, IP, etc.) so that the encapsulated packets can be transmitted in another network (such as IPv4). This solves the problem of message transmission across heterogeneous networks. The channel through which heterogeneous packets are transmitted is called a tunnel.

When PC_A accesses PC_B through a GRE tunnel, the packet forwarding process on FW_A and FW_B is shown in Figure 16-4.

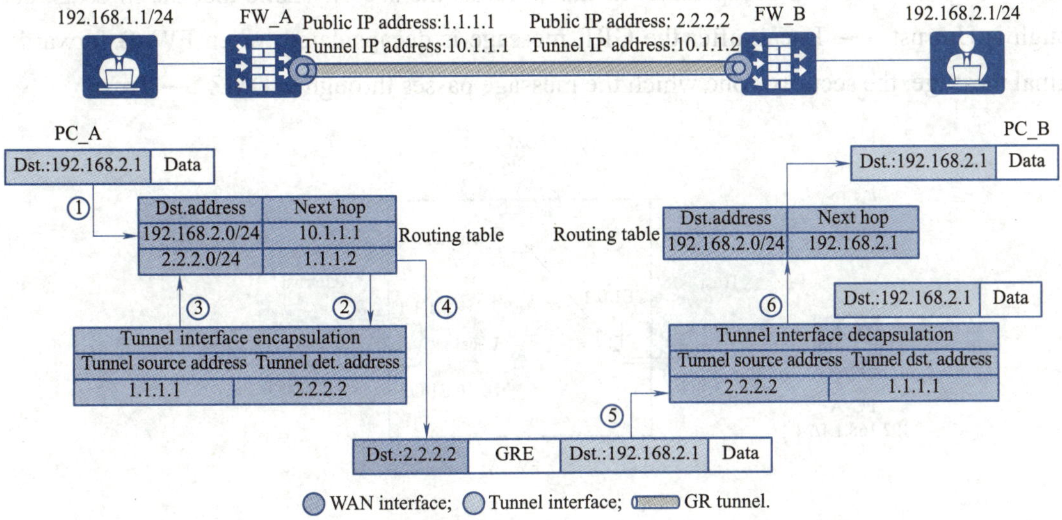

Figure 16-4 GRE VPN

(1) The original message that PC_A accesses PC_B enters FW_A, it first matches the routing table.

(2) According to the search result of the routing table, FW_A sends the packet to the Tunnel interface for GRE encapsulation, and at the same time adds a GRE header and a new IP header in the outer layer.

(3) According to the destination address (2.2.2.2) of the new IP header in the GRE packet, l FW_A searches the routing table again.

(4) FW_A forwards packets according to the route search results.

(5) After FW_B receives a GRE message, it will first determine whether the message is a GRE message. The encapsulated GRE message will have a new IP header. There is a Protocol field in the new IP header, which identifies the inner protocol type. If the value of the Protocol field is 47, it means that the packet is a GRE packet. If it is a GRE packet, FW_B will send the packet to the Tunnel interface for decapsulation, remove the new IP header and GRE header, and restore it to the original packet; if it is not, the packet will be processed as a normal packet.

(6) According to the destination address of the original message, FW_B searches the routing table again, then forwards the message according to the route matching result.

The data flowing through the firewall needs to consider security policy issues. When considering security policies for GRE traffic, the data flow before and after encapsulation and decapsulation should be fully considered, as shown in Figure 16-5. When the original message sent from PC_A enters the tunnel interface, the security zone that the message passes through is Trust → DMZ. After the original message is encapsulated by GRE, when FW_A forwards the message, the security zone that the message passes through is Local → Untrust. When the message reaches FW_B, FW_B will decapsulate. In this process, the security zone that the message passes through is Untrust → Local; after the GRE message is decapsulated, when FW_B forwards the original message, the security zone which the message passes through is DMZ → Trust.

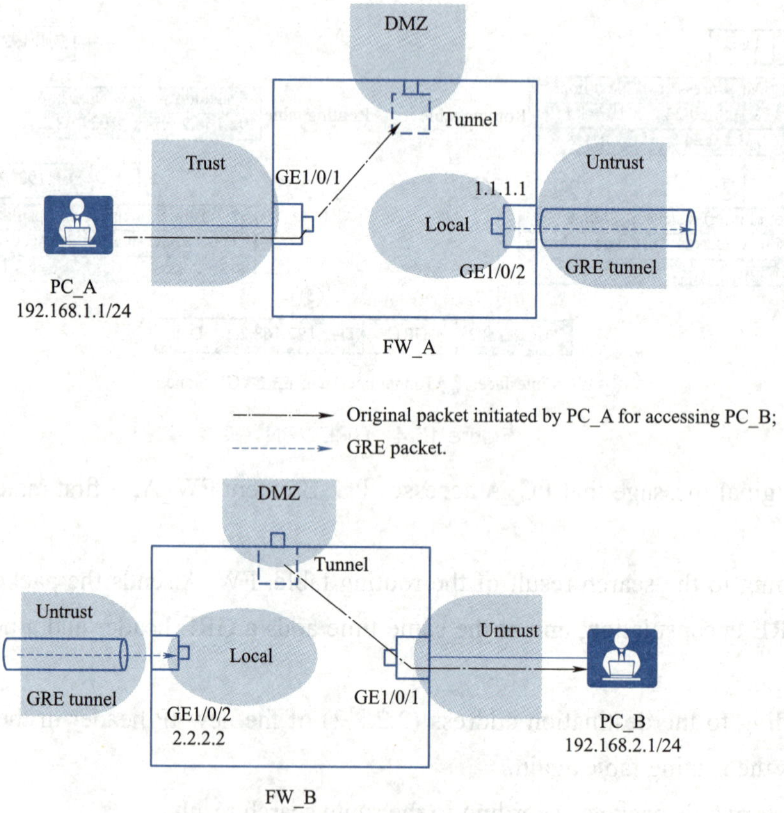

Figure 16-5 GRE VPN security policy

Subtask 16.2.5 Implementation of IPSec VPN

The IPSec (IP Security) protocol suite is a series of security protocols formulated by the IETF. It provides a cryptographic-based, interoperable, and high-quality security protection mechanism for end-to-end IP message interaction. IPSec VPN is a network layer VPN established by IPSec tunnel technology.

As shown in Figure 16-6, IPSec implements the security protection of IP packets through two security protocols: Authentication header (AH) and encapsulating security payload (ESP). AH is a message header verification protocol, which mainly provides data source verification, data integrity verification, and anti-message replay functions. It does not provide encryption. ESP is an encapsulated security payload protocol, which mainly provides encryption, data source verification, data integrity verification and anti-message replay functions. The security functions provided by the AH and ESP protocols depend on the authentication and encryption algorithms they use. The keys used by the IPSec encryption and authentication algorithms can be manually configured or dynamically negotiated through the Internet key exchange (IKE) protocol.

Security protocol	ESP				AH			
Encryption	DES	3DES	AES	SM1/SM4				
Authentication	MD5	SHA1	SHA2	SM3	MD5	SHA1	SHA2	SM3
Key exchange	IKE (ISAKMP and DH)							

Figure 16-6 IPSec protocol system

The successful establishment of SA (security association) between IPSec peers (two endpoints running the IPSec protocol) is the prerequisite for IPSec to transmit data securely, as shown in Figure 16-7.

SA is the agreement on certain elements between the communicating IPSec peers, such as: what kind of security protocol is used between the peers; the characteristics of the data stream that needs to be protected; the encapsulation mode of the data transmitted between the peers; the encryption used by the protocol , Authentication algorithm; What kind of key exchange and IKE protocol are used between peers, and SA life cycle, etc.

SA is uniquely identified by a triplet, which includes security parameter index (SPI), destination IP address, and security protocol number (AH or ESP).

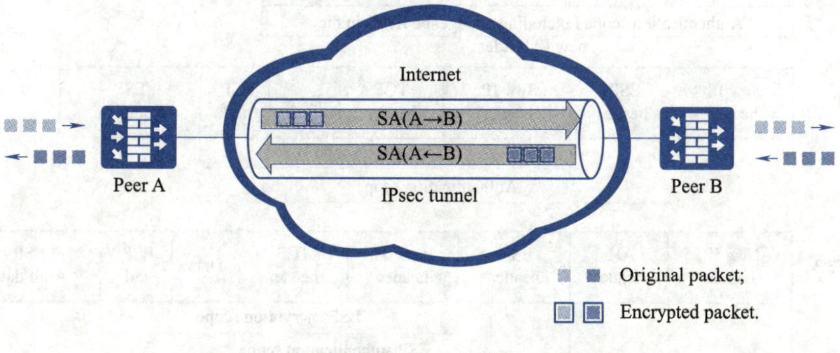

Figure 16-7 IPSec security association

1. Encapsulation modes supported by IPSec

IPSec VPN supports two encapsulation modes: transmission mode and tunnel mode.

As shown in Figure 16-8, in the transmission mode, the AH header or ESP header is inserted between the IP header and the transport layer protocol header to protect the TCP/UDP/ICMP load. The transmission mode does not change the packet header, so the source and destination addresses of the tunnel must be consistent with the source and destination addresses in the IP packet header. Therefore, the transmission mode is only suitable for communication between two hosts or between a host and a VPN gateway.

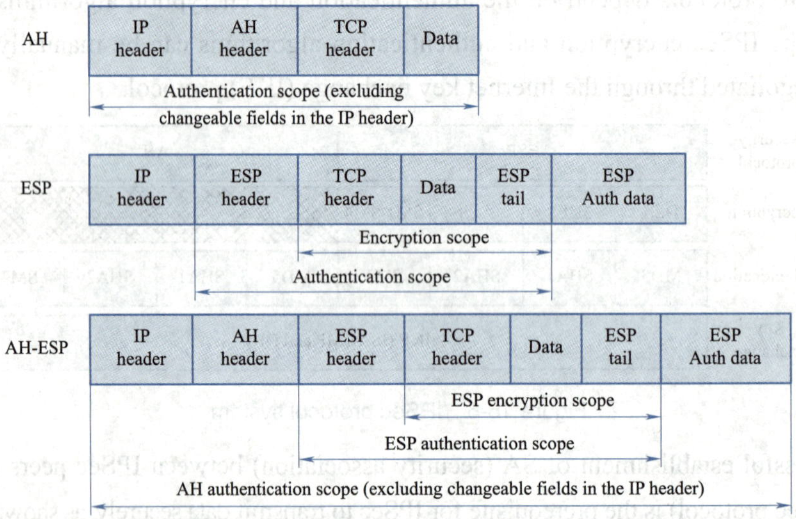

Figure 16-8 Transmission modes supported by IPSec

As shown in Figure 16-9, in tunnel mode, before the AH or ESP header is inserted into the original IP header, a new packet header is generated and placed before the AH or ESP header to protect the IP header and loading balance. The tunnel mode is mainly used for communication between two VPN gateways or between a host and a VPN gateway.

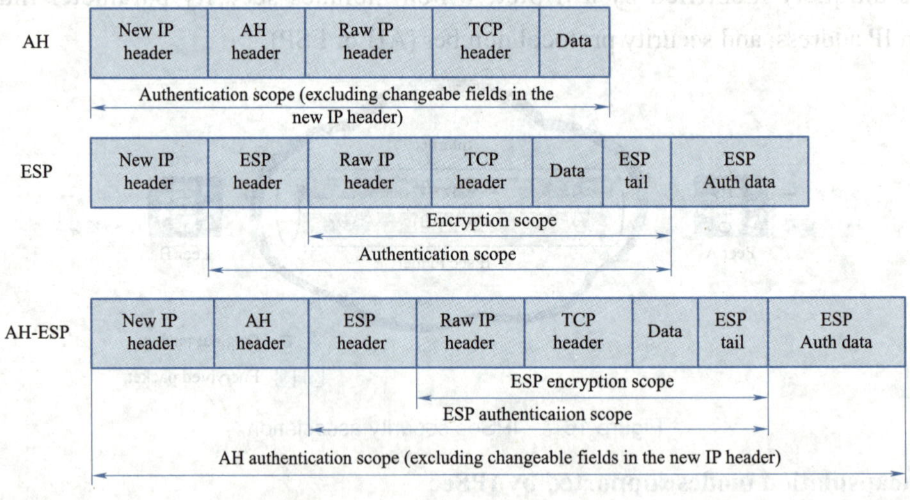

Figure 16-9 IPSec tunnel mode

The difference between transmission mode and tunnel mode is:

(1) In terms of security, the tunnel mode is better than the transmission mode. It can completely verify and encrypt the original IP datagram. In tunnel mode, the internal IP address, protocol type and port can be hidden.

(2) In terms of performance, the tunnel mode has an extra IP header, so it will take up more bandwidth than the transmission mode.

2. IKE SA

Before using IPSec to protect an IP packet, an SA must be negotiated. The SA of IPsec can be configured manually. However, when there are too many nodes in the network, manual configuration will be very difficult, and it will be difficult to ensure data security. At this time, you can use IKE to automate the process of SA establishment and key exchange.

The IKE protocol is built on the framework which defined by ISAKMP (Internet security association and key management protocol). It is an application layer protocol based on UDP, which provides keys for data encryption and simplifies the use and management of IPSec. Overall, it greatly simplifies the configuration and maintenance of IPSec.

First, an IKE SA is established between the peers to complete identity verification and key information exchange. Then, under the protection of IKE SA, a pair of IPSec SA is generated through negotiation between the peers based on the configured AH/ESP security protocol and other parameters. After that, the data between the peers will be encrypted and transmitted in the IPSec tunnel.

In order to serve IPSec SA, IKE SA provides IPSec with automatic key negotiation and establishment of IPSec security association services, as shown in Figure 16-10.

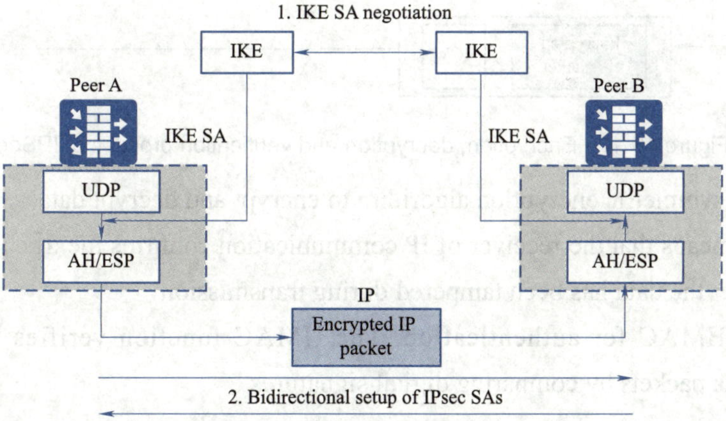

Figure 16-10 IKE SA negotiation

3. Encryption, decryption and verification process of IPSec

The encryption, decryption and verification process of IPSec is shown in Figure 16-11: First, the IPSec sender encrypts the IP message through the encryption algorithm and key that negotiated by the IPSec SA into cipher text.Secondly, the ciphertext will be calculated using the authentication algorithm and key which negotiated by the security association, and the hash value will be added to the message, then sent to the receiver together.After receiving the message, the receiver verifies the message first, if the verification is passed, the receiver decrypts the message into plain text

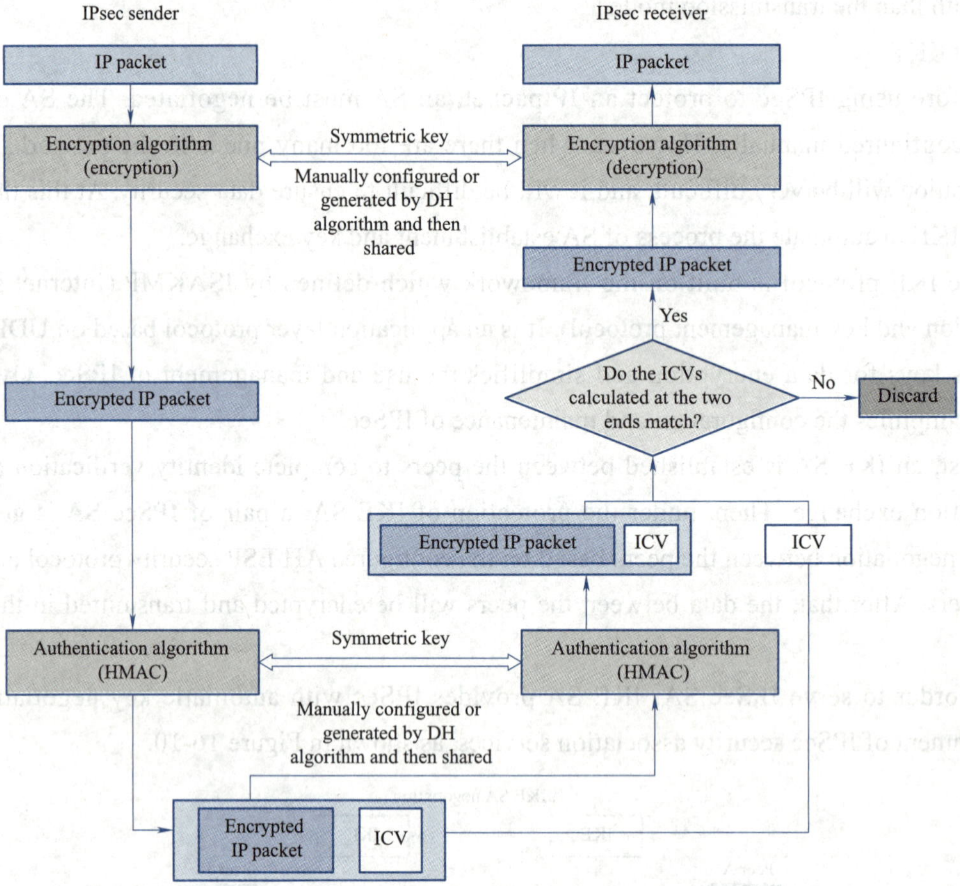

Figure 16-11 Encryption, decryption and verification process of IPSec

IPSec uses a symmetric encryption algorithm to encrypt and decrypt data.

Verification means that the receiver of IP communication confirms the true identity of the data sender and whether the data has been tampered during transmission

IPSec uses HMAC for authentication. The HMAC function verifies the integrity and authenticity of data packets by comparing digital signatures.

Subtask 16.2.6 Implementation of SSL VPN

SSL (secure sockets layer) is a secure protocol that provides secure connections for TCP-based application-layer protocols. SSL is a set of Internet data security protocol developed by

Netscape, the current version is 3.0. As shown in Figure 16-12, it has been widely used for identity authentication and encrypted data transmission between Web browsers and servers. The location of the SSL protocol is between the TCP/IP protocol and various application layer protocols, and is intended to provide security support for data communication.

The SSL protocol can be divided into two layers. SSL record protocol: It is built on a reliable transmission protocol (such as TCP). It provides support for basic functions such as data encapsulation, compression, and encryption for high-level protocols. SSL handshake protocol: It is built on top of the SSL record protocol. It is used to authenticate the identity of the communicating parties, negotiate encryption algorithms, and exchange encryption keys before the actual data transmission starts.

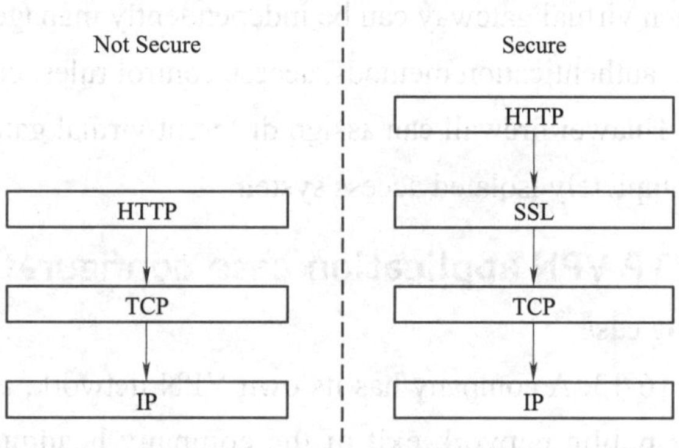

Figure 16-12 SSL VPN

In order to more finely control resource access permissions for mobile office users, SSL VPN divides intranet resources into four types: Web resources, file resources, port resources, and IP resources. Each type of resource has a corresponding access method. For example, if mobile office users want to access the internal Web server of the enterprise, they need to use the Web proxy service provided by SSL VPN; if they want to access the internal file server, they need to use the file sharing service.

Web proxy: It implements clientless page access, which reflects the ease of use of SSL VPN. It is an important feature of SSL VPN that distinguishes it from other VPNs. It forwards the page request (using https protocol) from the remote browser to the web server. Then, it will send the server's response back to the terminal user, and provide detailed URL control to control the user's access to a specific page. According to different implementation methods, Web proxy is divided into two types: Web rewriting and Web link.

File sharing: Sharing resources of different system servers (such as Windows systems that support the SMB protocol, and Linux systems that support the NFS protocol) are provided to users in the form of web pages for access. The file sharing service supports the SMB protocol (Windows) and NFS protocol (Linux) file protocol.

Port forwarding: Used for applications that cannot be accessed using web technology, such as C/S.

Network extension: After using the network extension service, the remote client will obtain an internal network IP address, just like in the internal network, the remote client can access any internal network resources. At the same time, for different access methods, remote clients will have different Internet and local subnet access permissions. There are full route mode (Full Tunnel), separate mode (Split Tunnel) and manual mode (Manual Tunnel).

In addition, Huawei firewalls also support user authentication functions. Virtual gateways can verify client identities. Each virtual gateway can be independently managed, and they can configure their own resources, users, authentication methods, access control rules, etc. When an enterprise has multiple departments, the Huawei firewall can assign different virtual gateways to each department or user group to form a completely isolated access system.

Task 16.3 L2TP VPN application case configuration

1. Background of the case

As shown in Figure 16-13: A company has its own VPN network, and a VPN gateway (USG firewall) is placed at the public network exit of the company headquarters. The case requires traveling personnel to be able to communicate with the company's internal business server through the L2TP tunnel.

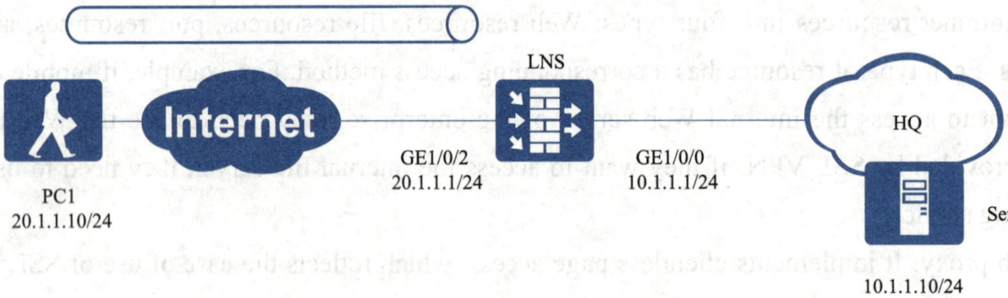

Figure 16-13 L2TP VPN configuration topology

2. Configuration process

The configuration process of L2TP VPN is shown in Figure 16-14.

3. Configuration steps

Step 1: Turn on the L2TP function. As shown in Figure 16-15. Choose "Network" → "L2TP" → "L2TP", in "Configure L2TP", select the "Enable" option after L2TP, and click "Apply" button.

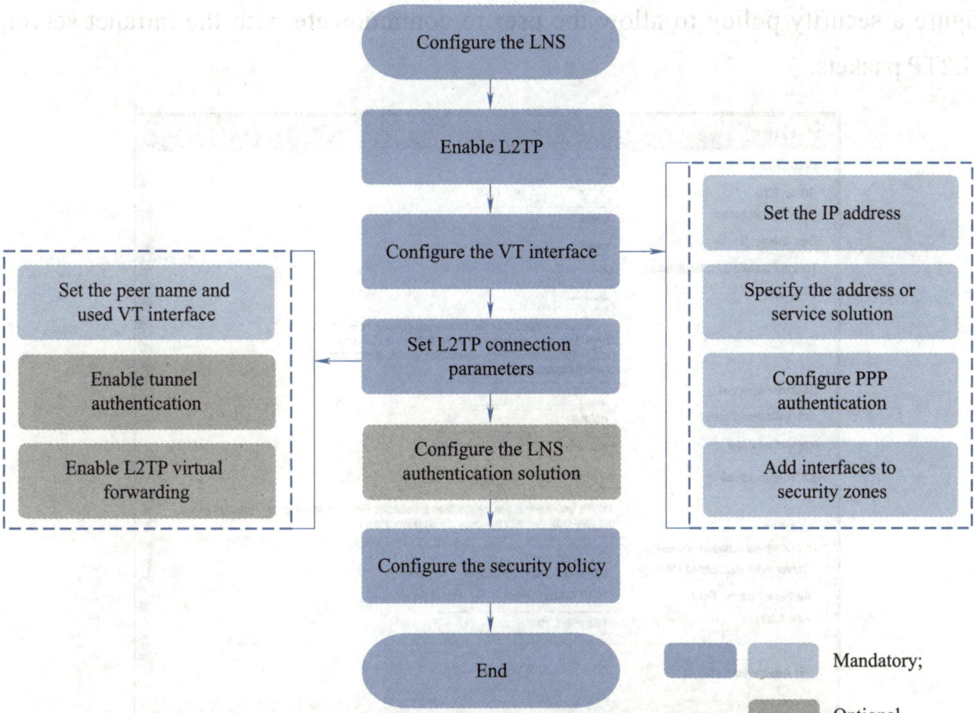

Figure 16-14 Configuration process of L2TP VPN

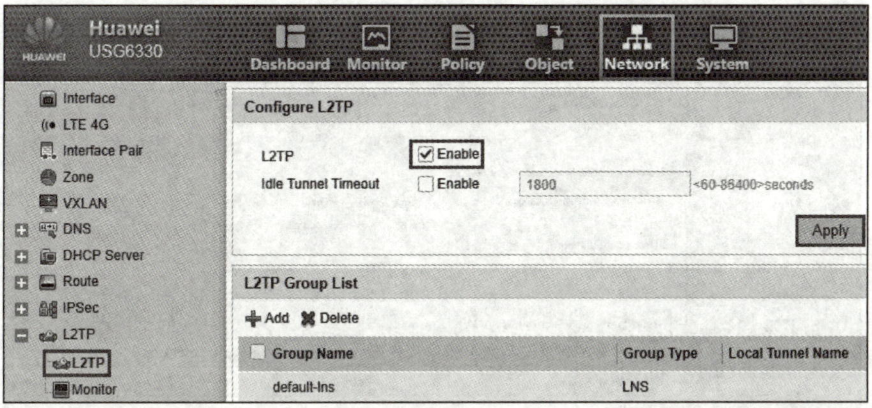

Figure 16-15 Enable L2TP function

Step 2: Configure L2TP parameters. As shown in Figure 16-16. In the "L2TP Group List", click "New" to set the parameters of the L2TP group.

Step 3: Set dial-up user information. As shown in Figure 16-17. Choose "Object" → "User". Select the "default" authentication domain, in the "User Management" option, click "New", and select "New User" to configure the dial-up user name and password. The dial-up user name is "user001" and the password is "Admin@123". After the L2TP configuration is complete, we need to configure a security policy to allow the user to communicate with the intranet server, and to release L2TP packets.

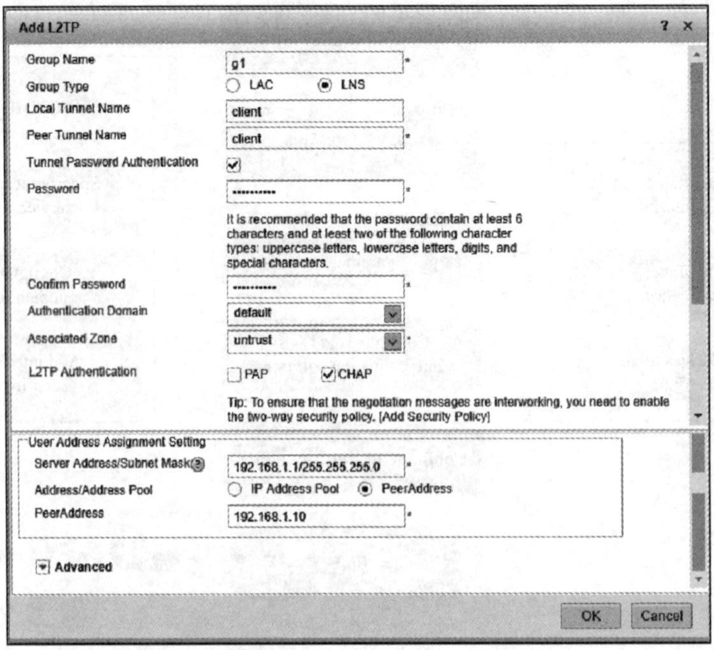

Figure 16-16 Configuring L2TP parameters

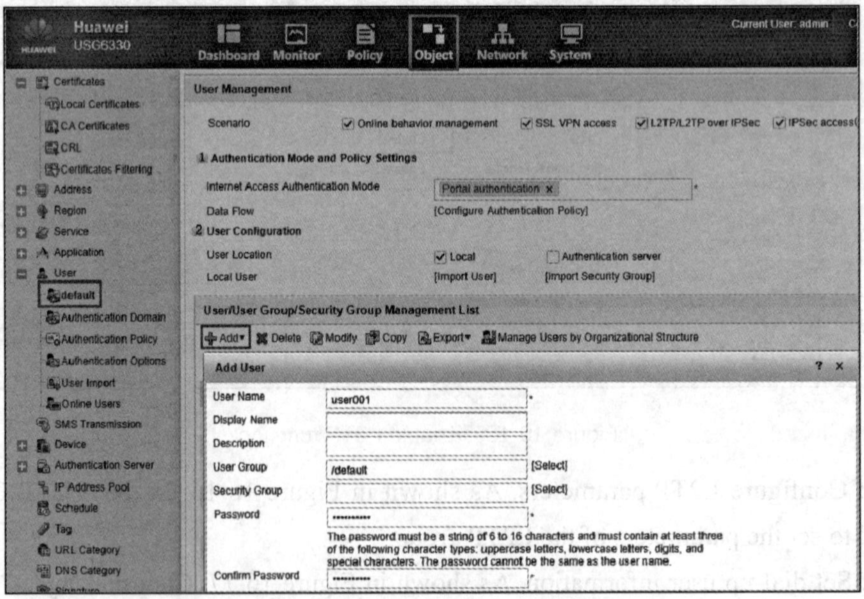

Figure 16-17 Set dial-up user information

Step 4: VPN client setting and verification. As shown in Figure 16-18, the VPN client settings must be consistent with those on the firewall.

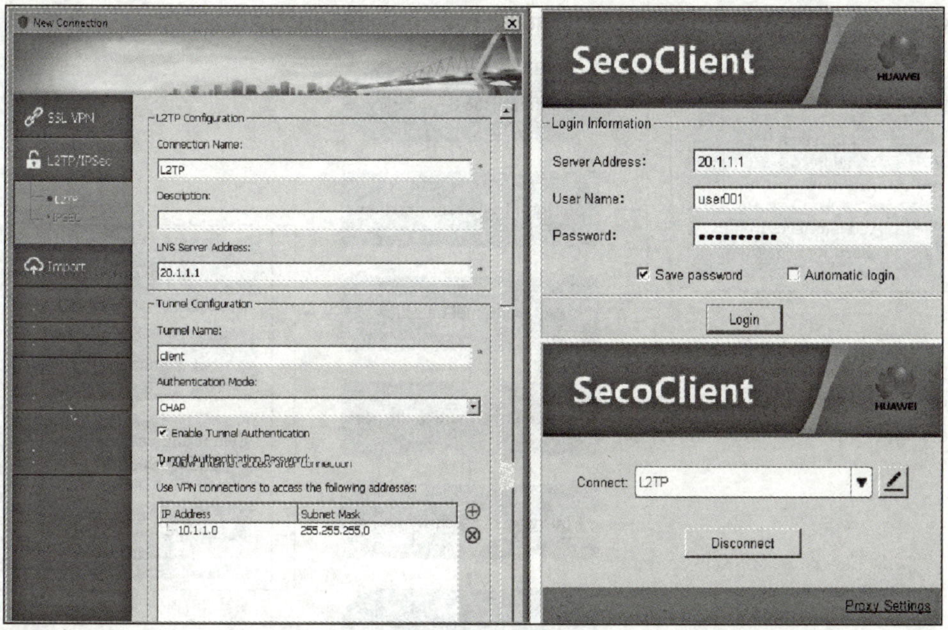

Figure 16-18 VPN client setting and verification

Task 16.4 GRE VPN application case configuration

1. Background of the case

As shown in Figure 16-19, there are two subnets, Network 1 and Network 2, running the IP protocol. They establish a GRE tunnel between FW_ A and FW_ B to achieve mutual access.

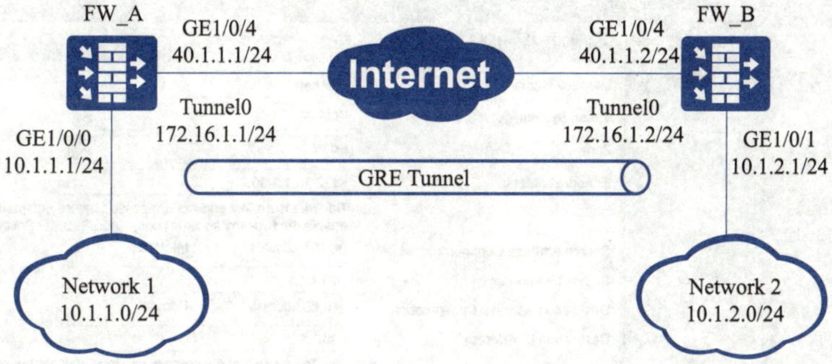

Figure 16-19 GRE VPN configuration topology

2. Configuration process

The GRE VPN configuration process is shown in Figure 16-20.

3. Configuration steps

Step 1: GRE interface configuration. As shown in Figure 16-21. Choose “Network” → “ GRE” → “ GRE”. Click “Add” to configure various parameters of the GRE interface.

Step 2: Routing configuration. As shown in Figure 16-22. Choose “Network” → “Routing” →

"Static Route". Click "Add" to configure a static route to network 2.

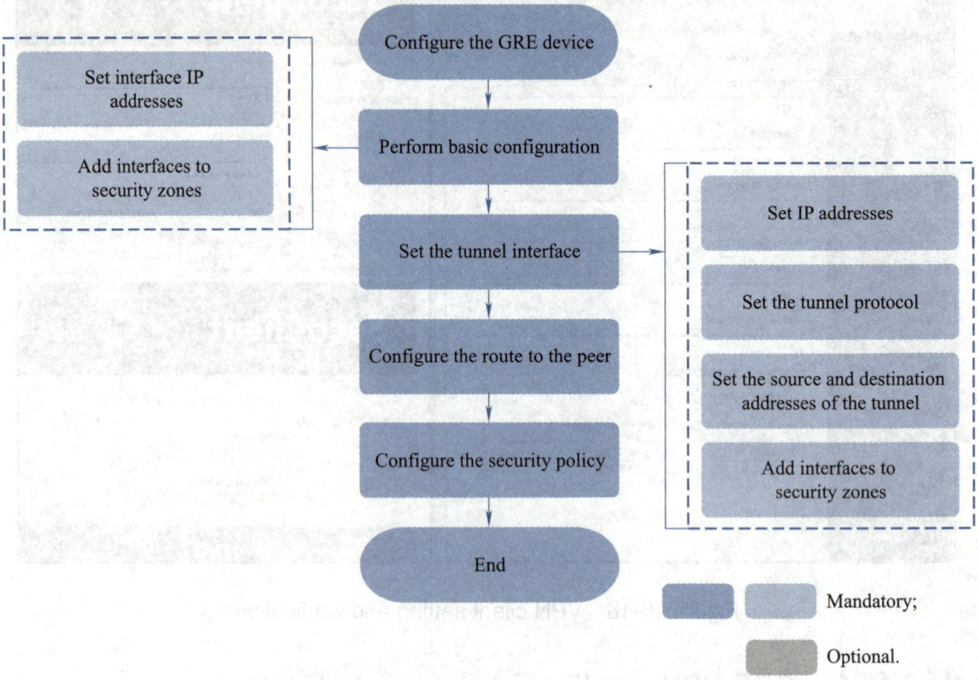

Figure 16-20 GRE VPN configuration process

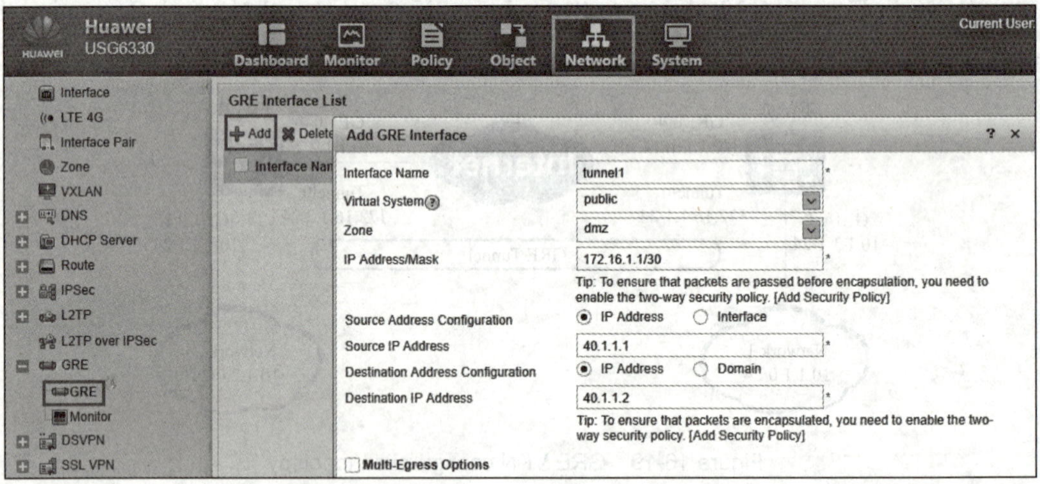

Figure 16-21 GRE interface configuration

Step 3: Verify the results. The PC in Network 1 and the PC in Network 2 can ping each other. Check FW_A "Network" → "Routing" → "Routing Table", you can see that the destination address is 10.1.2.0/24, and the outgoing interface is the route of tunnel1.

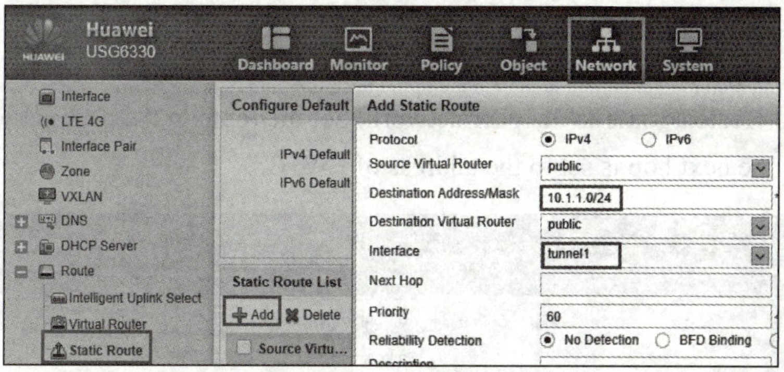

Figure 16-22 GRE routing configuration

Task 16.5 Point-to-point IPSec VPN application case configuration

1. Background of the case

IPSec VPN configuration topology as shown in Figure 16-23.

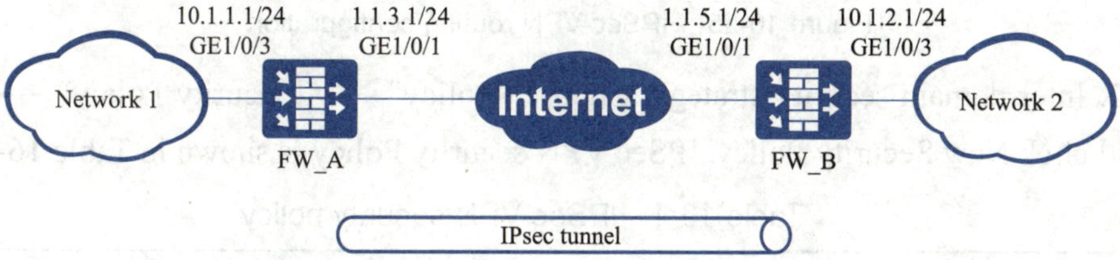

Figure 16-23 IPSec VPN configuration topology

2. Configuration process

Figure 16-24 shows the IPSec VPN configuration process.

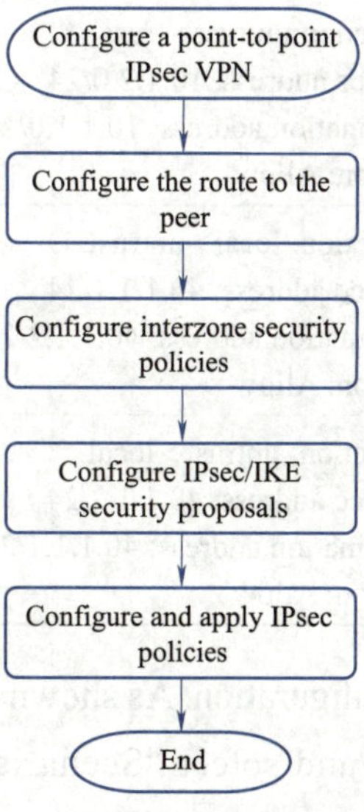

Figure 16-24 IPSec VPN configuration process

3. Configuration steps

Step 1: Routing configuration. As shown in Figure 16-25. Choose "Network" → "Route" → " Static Route" to create a route to the peer network. Set the route to the peer network. Taking FW_A as an example, the next hop is set to the address of FW_B.

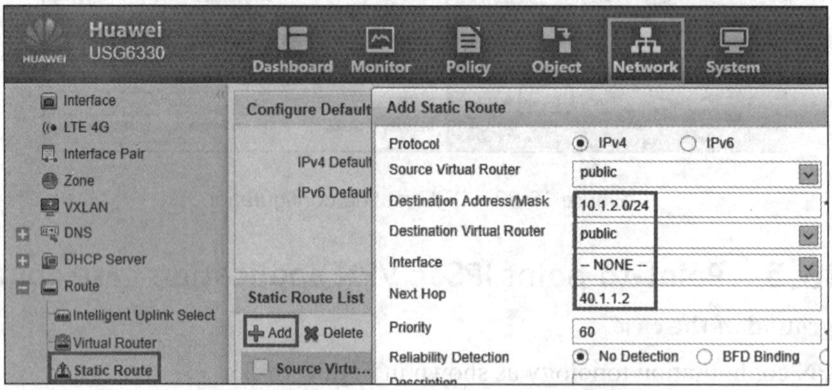

Figure 16-25 IPSec VPN routing configuration

Step 2: Inter-domain security strategy. Choose "Policy" → "Security Policy" → "Security Policy", and click New Security Policy. IPSec VPN security Policy is shown in Table 16-1.

Table 16-1 IPSec VPN security policy

IPsee		FW_A	FW_B
Allow Network 1 and Network 2 to communicate	ipsec 1	Direction: trust > untrust Source address:10.1.1.0/24 Destination address: 10.1.2.0/24 Action: Allow	Direction: trust > untrust Source address: 10.1.2.0/24 Destination address: 10.1.1.0/24 Action: Allow
	ipsec 2	Direction:untrust > trust Source address: 10.1.2.0/24 Destination address: 10.1.1.0/24 Action: Allow	Direction: untrust > trust Source address: 10.1.1.0/24 Destination address: 10.1.2.0/24 Action: Allow
Allow IKE negotiation packets to pass through	ipsec 3	Direction: local > untrust Source address: 40.1.1.1/24 Destination address: 40.1.1.2/24 Action: Allow	Direction: local > untrust Source address: 40.1.1.2/24 Destination address: 40.1.1.1/24 Action: Allow
	ipsec 4	Direction: untrust > local Source address: 40.1.1.2/24 Destination address: 40.1.1.1/24 Action: Allow	Direction: untrust > local Source address: 40.1.1.1/24 Destination address: 40.1.1.2/24 Action: Allow

Step 3: IPSec policy parameter configuration. As shown in Figure 16-26. Choose "Network" → "IPSec" → "IPSec", click "New", and select "Scenario" as "Site-to-site". Set IPSec related parameters in "Basic Configuration".

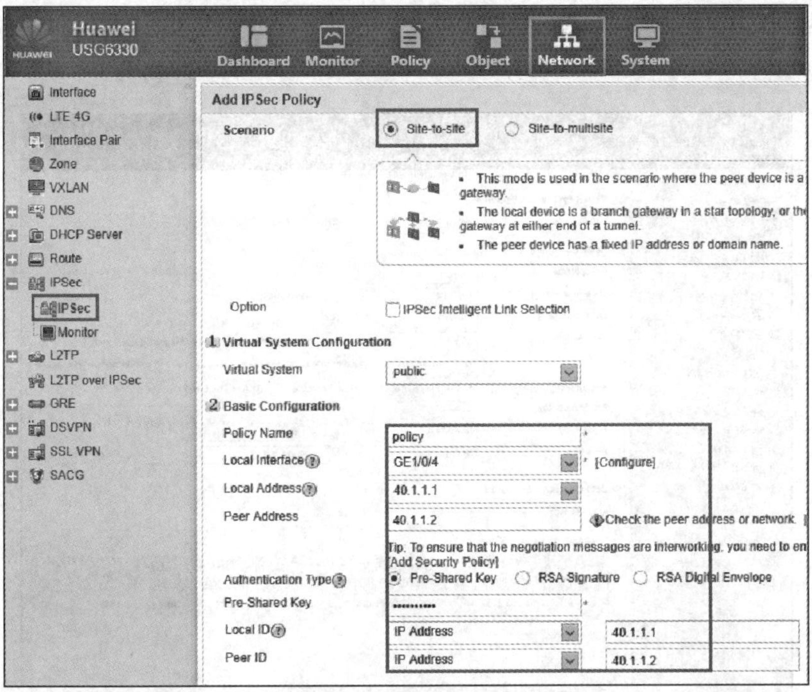

Figure 16-26 IPSec policy parameter configuration

Step 4: Encrypted data stream configuration. As shown in Figure 16-27. Click “New” in “Data Flow to Be Encrypted” on the configuration interface to create a new encrypted flow.

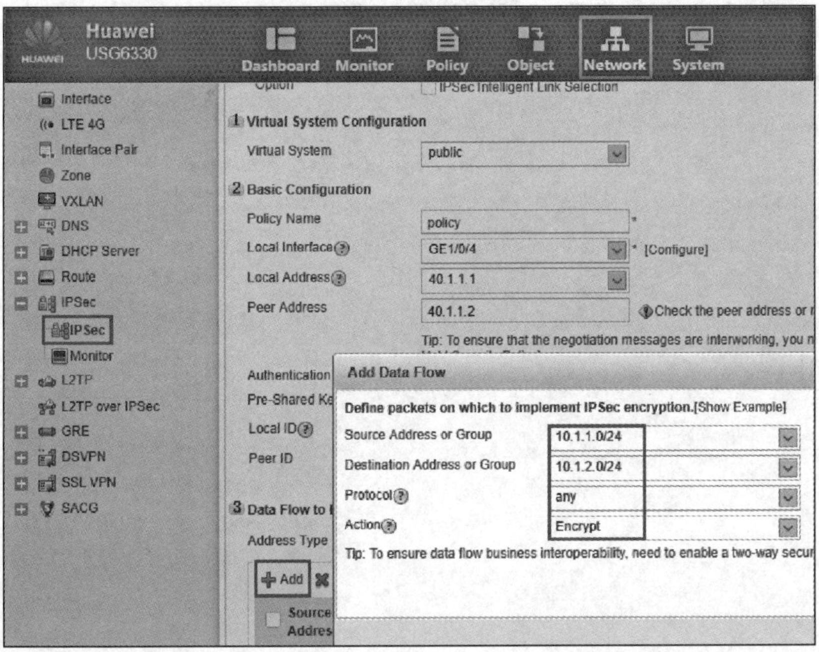

Figure 16-27 Encrypted data stream configuration

Step 5: Apply IPSec policy. As shown in Figure 16-28. After the configuration is complete, click "Apply" at the bottom of the configuration interface to save and apply the IPSec policy.

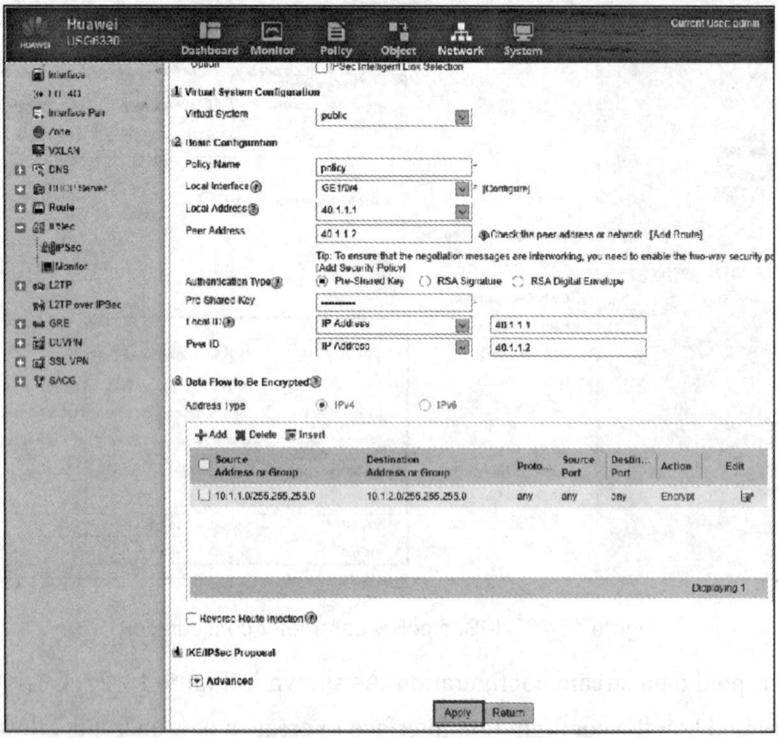

Figure 16-28 Apply IPSec policy

Section 5

Fundamentals of security operations and analysis

With the in-depth understanding of information security, it is not difficult to find that information security network security is a three-dimensional prevention system, including physical security, security technology, security management and other aspects. Security technology is an aspect of realizing network security, which aims at preventing network attacks and maintaining network services through equipment configuration or security equipment. Security management is a series of activities and processes to guide, standardize and manage information security.

This section will focus on safe operation and focus on how to carry out safe operation in four projects:

Fundamentals of security operations and analysis: Introduce the basic knowledge of safety operation, including background and content.

Data monitoring and analysis: Introduce the technical means of data monitoring and analysis, and describe the process of data collection.

Electronic evidence collection: Introduce related tools of electronic evidence collection and describe the process of electronic evidence collection.

Case study of information study: Through the case study of information security, the knowledge of information security norms, security threats, security technologies, security operations, etc. is digested and applied, so as to achieve the purpose of applying what you have learned.

Project 17 Overview and core elements of cybersecurity operation

In today's information age, safe operation has gradually become a hot topic. Information security incidents emerge one after another, which greatly affects the normal operation of enterprises. Up to now, they have caused immeasurable losses to enterprises all over the world, and safe operation is the necessary condition to ensure the continuous and safe operation of enterprises' businesses in all aspects. This project will introduce the basic concepts of safe operation, and briefly introduce the conditions required for safe operation.

Task 17.1 Understand safe operation

Subtask 17.1.1 Understand the concept of safe operation

For enterprises, in the initial stage of building information security, the main content is to purchase a series of security equipment and deploy them at each protocol layer. In order to ensure the daily stable operation of the business.

With the occurrence of security issues, such as information leakage incidents, natural disasters, etc., practitioners have gradually realized that only deploying security equipment cannot achieve effective security operations. Safe operation must be an effective combination of resources, processes and management, so as to achieve the purpose of protecting the safe, continuous and stable operation of the enterprise business

Subtask 17.1.2 Master the basic requirements for safe operation

Safe operation involves many requirements. Figure 17-1 shows the basic operating conditions for safe operation.

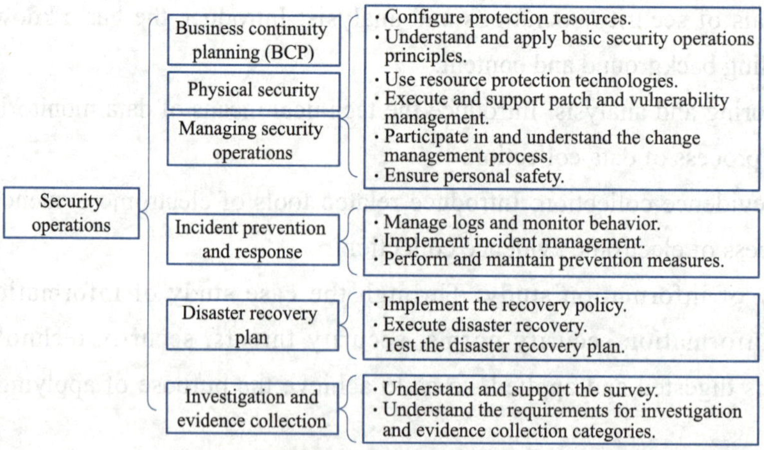

Figure 17-1 Basic conditions for safe operation

(1) Business continuity plan (BCP): It contains the risk assessment of the various processes of the organization, as well as various strategies to minimize the impact in the event of a risk. It is used to maintain the continuous operation of the business in the event of a critical event.

(2) Physical security: The purpose of physical security work is to prevent physical threats, including perimeter security and internal security.

(3) Safety management operations:

Protect the allocation of resources: Manage the allocation of various assets, including physical assets, cloud assets, virtual assets and data assets. Ensure that all systems are in a consistent safe state and maintain this state throughout their life cycle.

Use resource protection technology: use media management methods and asset management methods to protect the configuration and management of resources throughout the life cycle.

The basic security operation principles should be understood and applied: it is the security principle that we need to consider when we conduct responsibility management and authority management for organizational personnel.

Patch support and vulnerability management: Patch management can ensure that the appropriate patches are applied. Vulnerability management helps to verify that the system is free from interference from known threats.

Participate in and understand the process of change management: Change management helps reduce unforeseen interruptions caused by unauthorized changes and ensure that changes cannot cause interruptions. For example, configuration changes.

Participate in solving personal safety issues: implement safety controls to ensure the safety of corporate personnel

(4) Incident prevention and response:

Manage logs and monitoring behavior: Logs, monitoring and auditing procedures can help organizations prevent accidents and can also respond effectively when accidents occur.

Implement incident management: The main purpose of incident response is to minimize the impact on the organization when a security incident occurs.

Perform and maintain prevention measures: Prevent the occurrence of security incidents by configuring corresponding equipment and plans.

(5) Disaster recovery plan: When a disaster causes business interruption, the disaster recovery plan should be able to be implemented almost fully automatically and begin to provide support for the recovery operation.

Implement a recovery strategy: Guide how to restore business after a disaster。

Execute disaster recovery.

Test the disaster recovery plan process

(6) Investigation and evidence collection: When the threats and damage caused by some information security incidents are serious enough to require access by law enforcement agencies, investigators must carefully investigate to ensure that the correct steps are performed.

Understand and support the survey.

Understand the requirements for investigation and evidence collection categories.

Task 17.2 Understand the content of safe operation

Subtask 17.2.1 Understand business continuity plans

The goal of business continuity planning (BCP) is to provide quick, calm, and effective response under emergency situations, so that companies can more easily recover from disruptive events immediately, as shown in Figure 17-2.

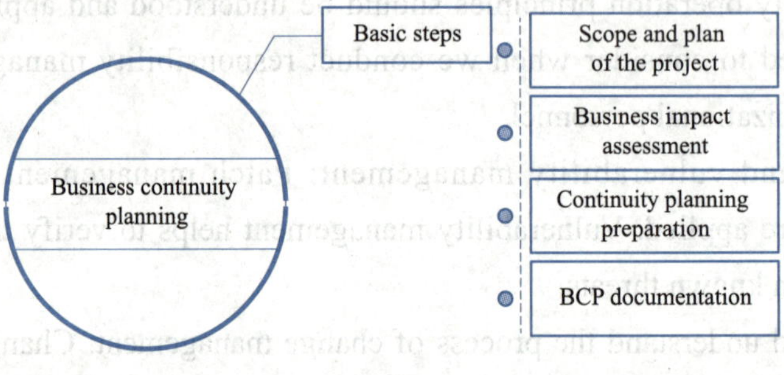

Figure 17-2 Business continuity planning

1. Scope and plan of the project

When formulating any process or plan, it must be based on the actual business scale and nature of the specific organization. At the same time, it must conform to the corporate culture and comply with relevant laws. Figure 17-3 shows the specific requirements for the unrated project scope at the initial stage of planning:

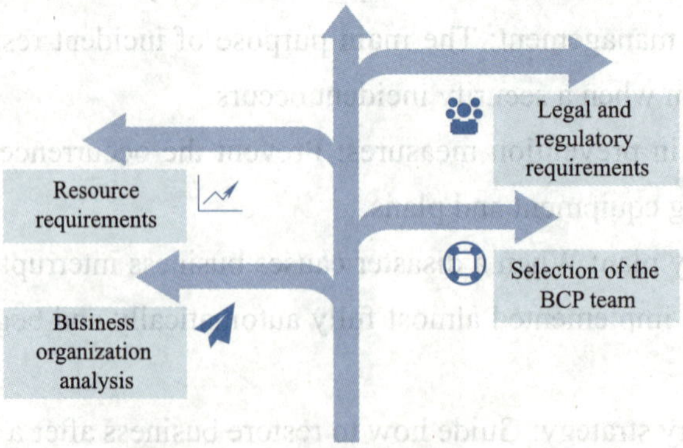

Figure 17-3 Scope and plan of the project

(1) Business organization analysis: Used to determine all relevant departments and personnel which are involved in the business continuity planning process. The following key areas need to be considered when analyzing:

Operation department that provides core service business.

Support service department, such as IT department and others that maintain the system of the operation department.

Senior executives and corporate decision makers.

(2) Selection of the BCP team: According to the above analysis of business organization, business continuity has important relationships with the operation department, service department, and the senior level of the enterprise. Therefore, the BCP formulation and maintenance team must have the participation of people from these departments, including but not limited to the following:

Representatives of each core business operation department.

Representatives of the support department.

IT representatives with technical expertise in the field set up by BCP.

A security representative who understands the BCP process.

Legal representative who understands relevant laws.

Representatives of senior management.

(3) Resource requirements: Development, testing, training, maintenance and implementation of BCP require a lot of manpower, time and materials.

(4) Legal and regulatory requirements: The state and local governments have issued corresponding laws and regulations for business continuity. These laws and regulations not only indicate the standards for the continuity of business operations, but also protect the vitality of the national economy.

2. Business impact assessment

Business impact assessment identifies the resources that can determine the sustainable development of the organization and assesses the threats to these resources. It also evaluates the actual likelihood of each threat and its impact on the business. The business impact assessment includes the following parts:

(1) Determining priority: In the event of a disaster, determining the priority of the business is critical for recovery operations. The priority of the business can be quantitatively analyzed using maximum tolerable downtime (MTD).

(2) Risk identification: In this part, the organization should identify possible risks, including natural risks and man-made risks. At this stage, only a qualitative analysis will be done to pave the way for subsequent evaluation.

(3) Possibility assessment: Assess the likelihood of the above-mentioned risks that may

threaten the organization

(4) Impact assessment: According to the risk and the possibility of the occurrence of the risk, the impact on the organization will be oriented or quantitatively assessed. It includes, but is not limited to, reputation, public influence, and resource loss.

(5) Resource prioritization: Prioritize business continuity plan resources that allocated to various risks.

3. Continuity planning preparation

The last two stages are mainly used to determine the working process of the BCP and the limited sequence of protecting business assets. In the stage of preparing the continuity plan, it focuses on the development and realization of the continuity strategy. In this stage, the following tasks are involved:

(1) Strategy development: Determine the mitigation measures to be taken for each risk based on the results of the business impact assessment.

(2) Plan realization: Use specific resources to develop a plan which based on strategy, and make it reach the predetermined goal.

(3) Preparation and processing: Provide necessary resources and protective measures for the formulation, maintenance and implementation of business continuity plans. These resources include people, buildings/equipment and infrastructure.

(4) Training and education: All relevant personnel involved in the BCP will be trained on the business continuity plan, so that they can understand the tasks and deal with emergencies calmly and orderly.

(5) Plan approval: After the design of the business continuity plan is completed, it is necessary to obtain approval from the top of the organization.

4. BCP documentation

In the event of an emergency, the BCP documentation helps provide guidance to BCP personnel in dealing with threats. At the same time, the document records the modification history. This will provide experience for subsequent handling of similar incidents or modification of documents. The content of the document should include the following:

(1) The goals of the continuity plan.

(2) Materiality statement.

(3) Declaration of organizational responsibilities.

(4) Declaration of urgency and time limit.

(5) Risk assessment.

(6) Acceptable risk/risk mediation.

(7) Major record plan.

(8) Guiding principles for responding to emergencies.

(9) Maintenance.

(10) Tests and exercises.

Subtask 17.2.2 Understanding incident response management

Not all threats can be prevented. The business continuity plan provides guidance on the process of handling emergencies and responding to threats as soon as possible, so as to minimize the impact of the incident on the organization, as shown in Figure 17-4.

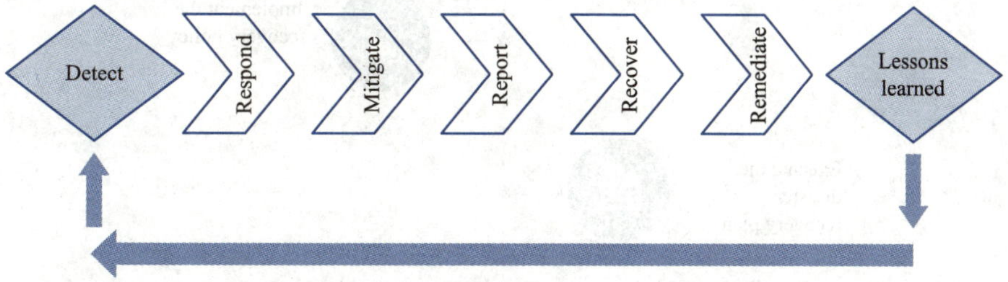

Figure 17-4 Incident response management

(1) Detect: The staff detects the occurrence of security incidents through data monitoring and data analysis, such as collecting logs.

(2) Respond: After detecting and verifying the incident, activate the response program. The computer security incident response team needs to assist in investigating the incident, evaluating the damage, collecting evidence, reporting the incident and restoring the program, repairing and drawing lessons, and helping to do root cause analysis as soon as possible. Responding to safe time as soon as possible can reduce damage even more.

(3) Mitigate: Mitigation is also a way of emergency response to contain the impact of the incident, such as isolating the problem by interrupting the connection between the infected host and the corporate network.

(4) Report: When an incident occurs, it needs to be reported internally to the organization, and sometimes to the outside world. Minor security incidents do not need to be reported to the senior management of the organization, but in order to adjust the response policy in a timely manner to contain the impact, major incidents must be notified to senior executives.

(5) Recover: Restore the system to a normal state. If evidence needs to be collected, it needs to be performed before the system is restored.

(6) Remediate: At this stage, the root cause of the incident will be analyzed and system vulnerabilities will be repaired to prevent similar incidents from happening again.

(7) Lessons learned: We should summarize the incident, learn lessons, and apply the output of this stage to the subsequent inspection and maintenance stage of the business continuity plan.

Subtask 17.2.3 Understand disaster recovery plans

When a disaster causes business interruption, the disaster recovery plan starts to guide the work of the personnel responsible for emergency response, so as to restore the business to a normal operating state.

The content of the disaster recovery plan is shown in Figure 17-5.

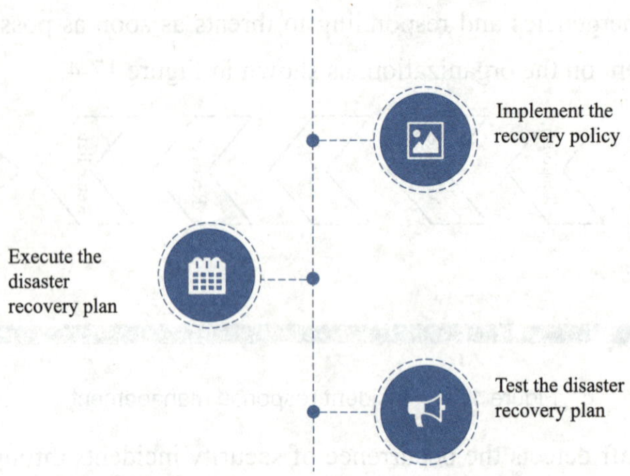

Figure 17-5 Disaster recovery plan

The types of disasters include:

(1) Natural disasters: earthquakes, floods, fires, etc.

(2) Man-made disasters: terrorist acts, power outages, etc.

(3) Public facilities failures and infrastructure failures: software/hardware failures, demonstrations, vandalism, etc.

Recovery strategy: Back up core data and facilities to improve system resilience and fault tolerance, so as to maintain high business availability and improve service quality.

(1) Backup storage strategy.

(2) Site recovery strategy.

(3) Mutual assistance agreements.

Test the disaster recovery plan: The disaster recovery plan must be regularly tested to ensure that the content of the plan is feasible and meets the changing needs of the organization. The types of tests include:

(1) Read-through test.

(2) Structured testing.

(3) Simulation test.

(4) Parallel testing.

(5) Completely interrupted testing.

Subtask 17.2.4 Understand investigation and evidence collection

Before the corresponding plan is implemented, it is necessary to confirm that the attack has occurred, and after being attacked, it is necessary to investigate the security incident and collect evidence. The purpose of the investigation is to find out what happened and the extent of the damage, as shown in Figure 17-6.

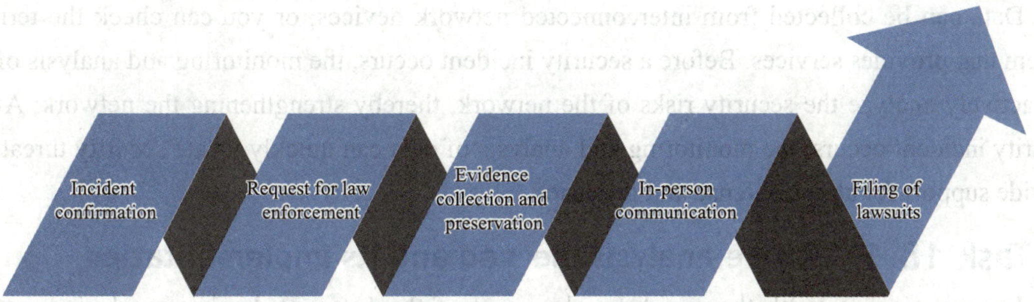

Figure 17-6 Investigation and evidence collection process

The method of investigation must comply with laws and regulations.

(1) Operational investigation: Mainly used to investigate the organization's computing facilities. For example, whether there are performance problems, configuration problems, etc. Operational investigations are mainly used to analyze problems, but do not require particularly rigorous evidence.

(2) Criminal investigation: It is an investigation conducted by law enforcement officers for illegal acts.

(3) Civil investigations: Civil investigations usually do not involve the work of internal employees and external legal counsel teams.

(4) Supervisory investigation: Supervisory investigations are carried out by government agencies when the organization may violate the law.

(5) Electronic discovery: that is, electronic forensics. For details, refer to the follow-up project Electronic evidence collection.

In order to successfully report a crime, sufficient evidence must be provided to confirm the crime. The evidence must be admissible evidence:

(1) The evidence must be related to the established facts;

(2) The evidence must be related to the incident;

(3) Evidence must be obtained through legal means.

The types of evidence are divided into:

(1) Physical evidence: refers to objective evidence, such as objects;

(2) Document evidence: written content, such as computer logs, etc.;

(3) Verbal evidence: witnesses and testimony.

Project 18 Data monitoring and analysis

This project mainly introduces how to obtain effective information through technical means, and analyze the obtained information to locate security risks and threats.

Data can be collected from interconnected network devices, or you can check the terminal system that provides services. Before a security incident occurs, the monitoring and analysis of data can actively analyze the security risks of the network, thereby strengthening the network; After a security incident occurs, the monitoring and analysis of data can quickly locate security threats and provide support for attack defense and forensics.

Task 18.1 Active analysis method and its implementation

In recent years, with the rapid development of the Internet, hackers and virus attacks have occurred frequently. According to statistics, about 95% of successful web attacks are due to failure to patch known vulnerabilities. The reason for the increase in attacks is due to the wider application of the Internet and the increasing number of easy-to-use hacking tools. The most important and fundamental reason is the vulnerability of computer systems that can be penetrated, which is also called security vulnerability. In the actual computer network system, vulnerability is almost omnipresent, so it is particularly important to evaluate the vulnerability.

Active analysis refers to the security assessment of the network status before an attack occurs. To achieve the purpose of promptly correcting the exposed problems, strengthening the network, and improving the security of the network.

The security assessment method is shown in Figure 18-1.

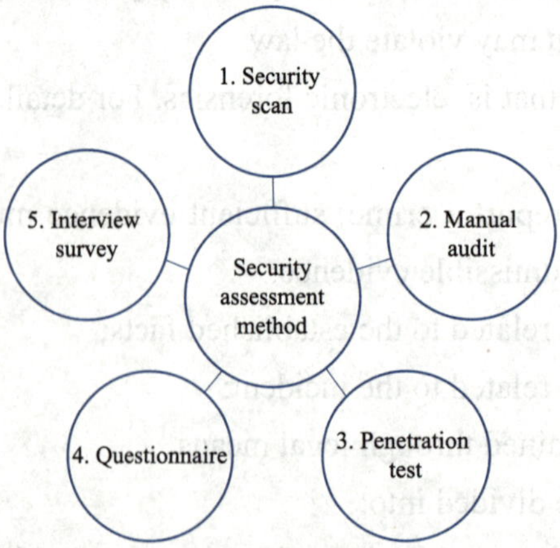

Figure 18-1 Security assessment method

(1) Security scan: Fully understand the current network security vulnerabilities of the target

system. In order to discover related vulnerabilities, it is necessary to scan the target system with a scanning analysis and evaluation tool.

(2) Manual audit: Manual inspection of target systems including host systems, business systems, databases, network equipment, security equipment, etc.

(3) Penetration test: Simulate the attack technology and vulnerability discovery technology that hackers may use. In this way, the security of the target system can be deeply probed to find the most vulnerable part of the system.

(4) Questionnaire: Check the business, assets, threats, vulnerability and other aspects of network system administrators, security administrators, technical leaders, etc.

(5) Interview survey: Confirm the results of the questionnaire survey to obtain detailed management status and listen to users' thoughts and opinions.

Based on the results of the security assessment and with reference to the current network and system status, we can propose practical solutions for the improvement or upgrade of the customer's information security architecture. By helping customers implement security reinforcement work, we can improve network security and ensure the best network performance.

Subtask 18.1.1 Security scan

The purpose of security scan is to fully understand the current network security vulnerabilities of the target system. In order to discover related vulnerabilities, it is necessary to scan the target system with a scanning analysis tool.

The content of the security scan includes: the open port number of the system, the existence of weak passwords, SQL injection vulnerabilities, cross-site scripting vulnerabilities, etc.

After the security scan is completed, we need to use the output scan tool to generate the results.

Common scanning tools are shown in Table 18-1.

Table 18-1 Common scanning tools

Tool type	Tool name	Use
Port scanning tool	SuperScan	Powerful port scanning software: ① It checks whether the IP is online through ping; ② It can check the service category provided by the target computer; ③ It can check whether the target computer is online within a certain range, and it can also check the port status
	Nmap	It is a network scanning and sniffing toolkit under Linux. Basic functions: ① It can detect whether a group of hosts are online; ② It can scan the host port to check the network services provided; ③ It can also infer the operating system used by the host
Vulnerability scanning Tool	Sparta	It is a vulnerability scanning tool integrated in Kali. It can discover the services and open ports supported in the system
Application scanning tool	Burp Suite	Burp Suite is an integrated platform for penetrating web applications. It contains many tools and designed many interfaces for these tools to facilitate the process of attacking applications

1. Using demonstration of SuperScan

As shown in Figure 18-2: Use SuperScan to scan the web server in the experimental environment.

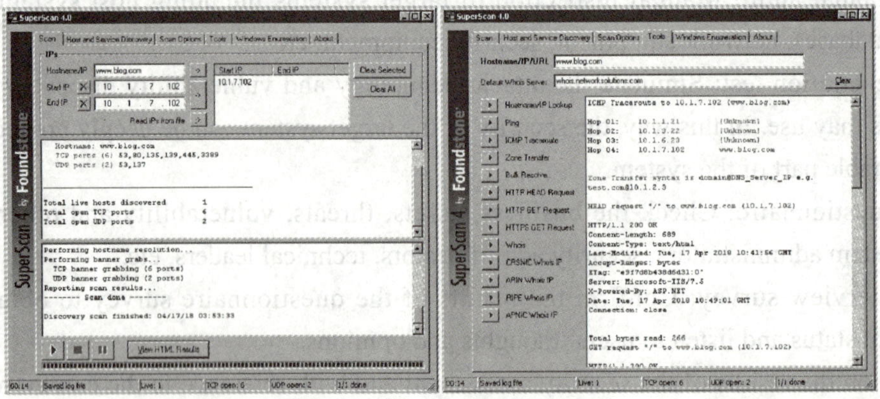

Figure 18-2 Scanning with SuperScan

In the "Scan" menu bar, click "View HTML Result" to view the scan result, as shown in Figure 18-3.

2. Using demonstration of Nmap

Select "Information Gathering" in "Applications" and click "nmap", as shown in Figure 18-4. According to the parameter rules of the Nmap tool, information is collected on the target system. As shown in Figure 18-5, it scans the open TCP ports of the host.

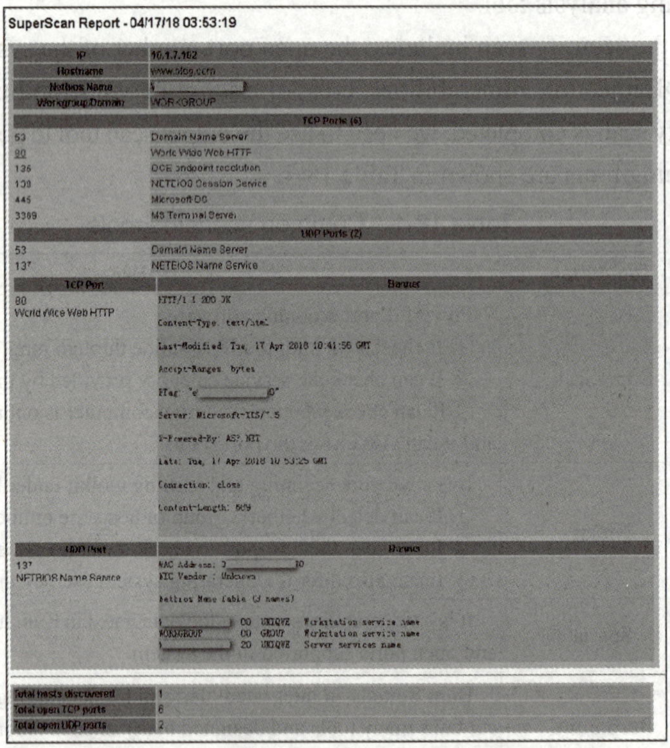

Figure 18-3 Scanning results using SuperScan

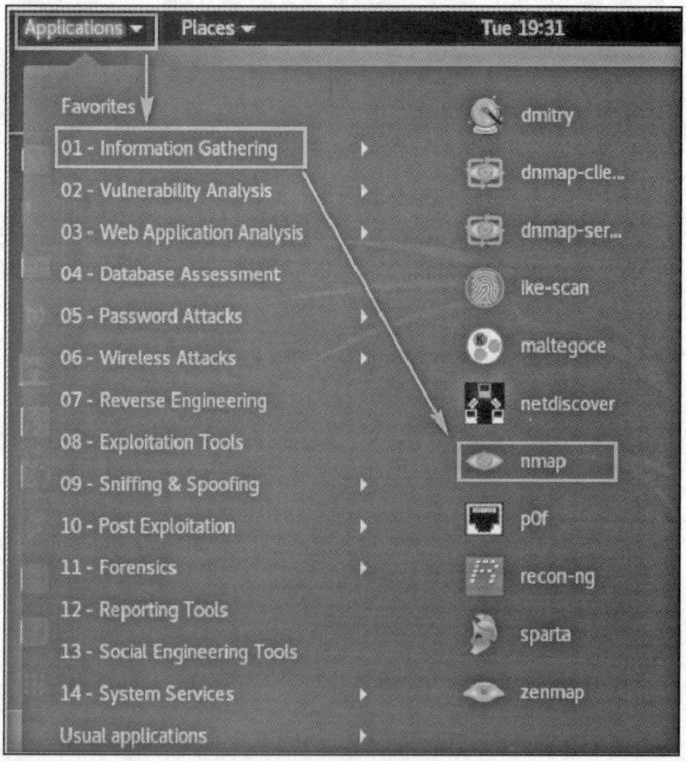

Figure 18-4 Nmap tool

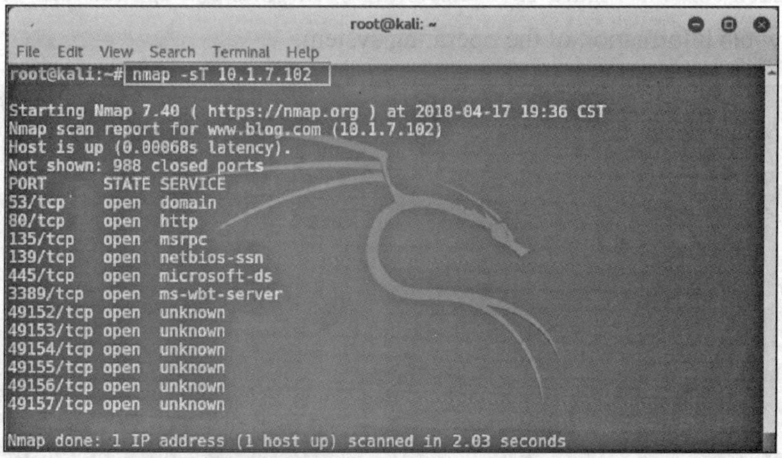

Figure 18-5 Scanning result of Nmap

3.Using demonstration of Sparta

Sparta is a graphical interface tool, it is easy to operate, and it integrates the port scanning function and exhaustive cracking function.

Select “Vulnerability Analysis” in “Applications” and click “sparta”, as shown in Figure 18-6.

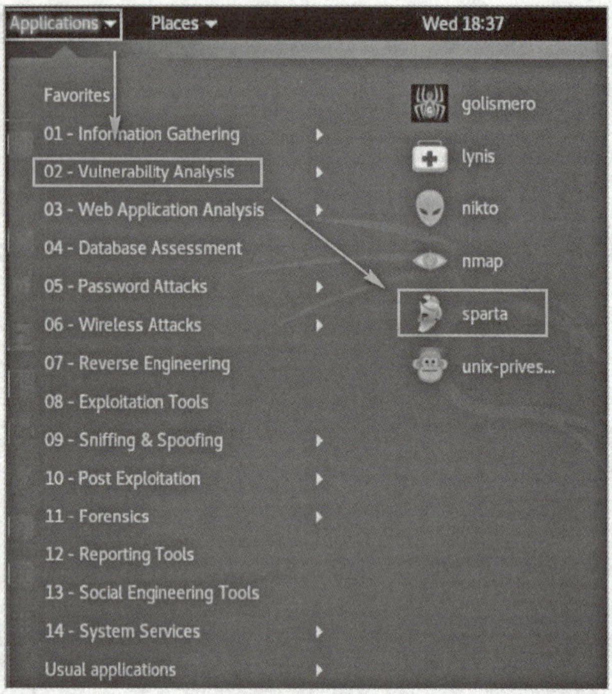

Figure 18-6 Sparta tool

Add the address network segment or host address to be scanned in the operation interface, and then scan the added content. Figure 18-7 shows the open ports and applications of the target host, and you can view the information of the operating system.

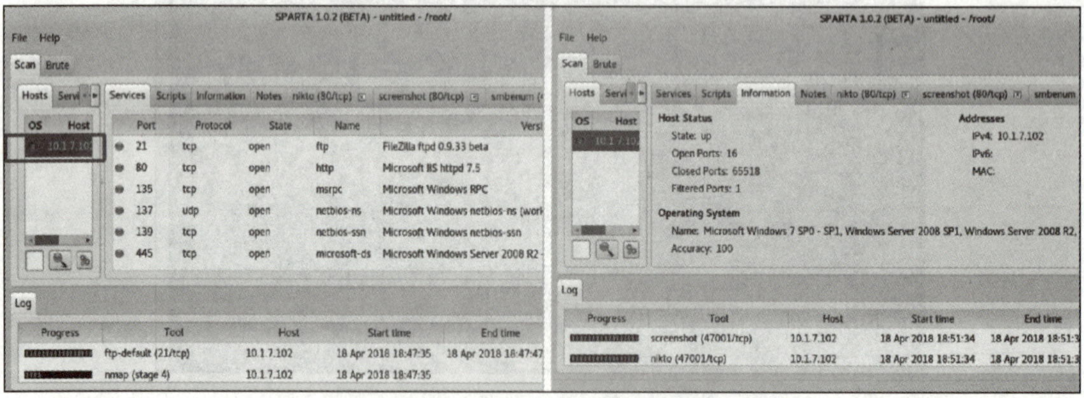

Figure 18-7 Scanning result of Sparta

4.Using demonstration of Burp Suite

Burp Suite (see Figure 18-8) will send a request to the application and verify the vulnerability through the payload. It has a good scanning effect for the following two types of vulnerabilities:

(1) Client-side vulnerabilities, such as XSS, Http header injection, operation redirection;

(2) Server-side vulnerabilities, such as SQL injection, command line injection, and file

traversal.

Before using Burp Suite, in addition to correctly configuring Burp Proxy and setting it as a browser proxy, we also need the domain and URL module path to be scanned in Burp Target's site map. When these domains or URL paths exist in Burp Target's sItemap, we can perform full scans or branch scans on the specified domains or URLs.

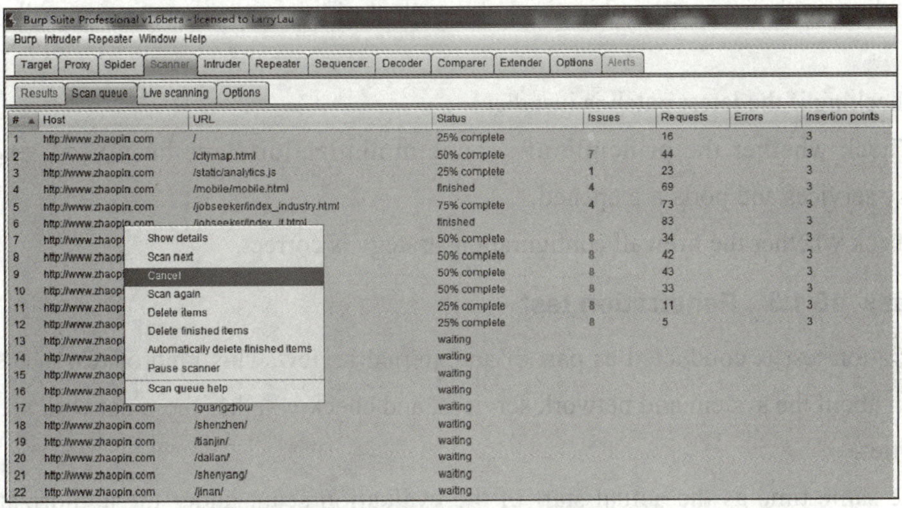

Figure 18-8 Burp Suite tool

As shown in Figure 18-19, in the Results interface, the system automatically displays the details of the vulnerabilities that have been scanned in the queue.

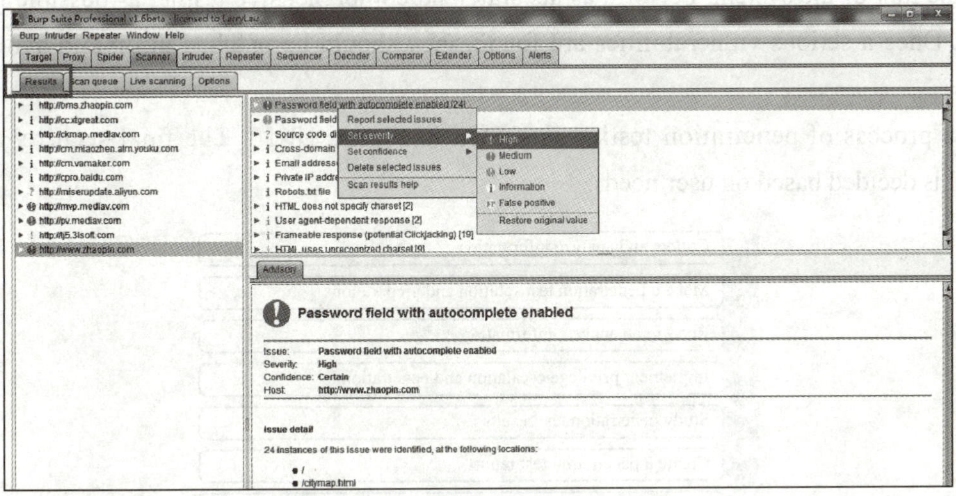

Figure 18-9 Burp Suite scan results

Subtask 18.1.2 Manual audit

Manual audit is a supplement to instrumental assessment. It does not need to install any software on the target system to be evaluated, and it has no effect on the operation and status of the system to be evaluated. It is very useful on important hosts where software is not allowed to be checked.

In the manual audit work, security experts conduct manual checks on target systems including host systems, business systems, databases, network equipment, and security equipment.

Depending on the inspection objectives, the content of the inspection may cover the following aspects:

(1) Checking if the latest patch is installed.

(2) Check whether the principle of service minimization have been used, and whether unnecessary services and ports are opened.

(3) Check whether the firewall configuration strategy is correct.

Subtask 18.1.3 Penetration test

Penetration test is conducted as part of an external review. This kind of test needs to obtain information about the system and network services, and check whether these network services have vulnerabilities.

At the same time as the actual start of the evaluation scan, under the authorization of the evaluated party, a penetration test is performed according to the key IP requested by the customer. It completely simulates the attack technology and vulnerability discovery technology that hackers may use, makes in-depth detection of the security of the target system, and discovers the most vulnerable link of the system. Perform as accurate and comprehensive testing as possible for these key IPs. Once a serious vulnerabilities are found, they should be patched in time to avoid future troubles.

The process of penetration testing is shown in Figure 18-10. The final security solution proposal is decided based on user needs.

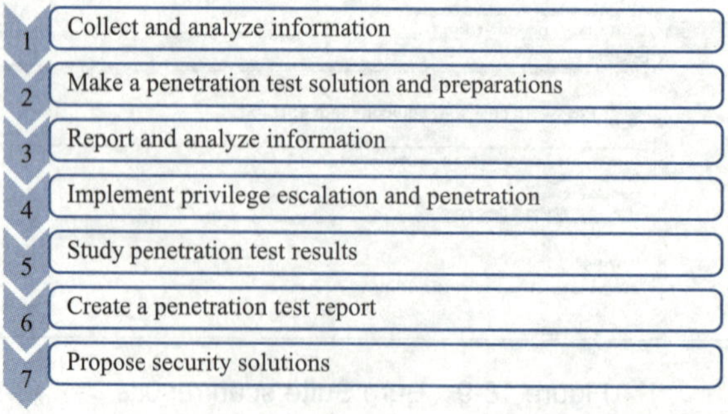

Figure 18-10 Penetration testing process

Subtask 18.1.4 Questionnaire

In the questionnaire stage, network system administrators, security administrators, technical leaders, etc. are inspected in terms of business, assets, threats, and vulnerabilities.

The subjects of investigation include: network system administrators, security administrators, technical leaders, etc.

The contents of the investigation include: business, assets, threats, vulnerabilities (management aspects).

The principles of questionnaire design: completeness, specificity, conciseness and consistency.

Figure 18-11 shows a sample questionnaire.

Operational operation

1. Does the network administrator recommend changes to the configuration of the network system?

   A. Yes　　B. No　　C. Others______

2. Are there regular audits and checks on important servers and routers running on the network?

   A. Yes　　B. No　　C. Others______

3. Are the backbone routers strictly using fixed terminals for configuration and management?

   A. Yes　　B. No　　C. Others______

4. Has the intrusion detection device running on the network system been selected to detect and control system intrusion events in a timely manner?

   A. Yes　　B. No　　C. Others______

5. Have you considered future resource and capacity requirements when designing a new application system?

   A. Yes　　B. No　　C. Others______

6. Are the administrators of important hosts monitoring the usage of major system resources, including processor, main memory, file storage, and communication system bandwidth?

   A. Yes　　B. No　　C. Others______

Figure 18-11 Sample questionnaire

Subtask 18.1.5 Interview survey

Interview survey refers to confirming the results of questionnaire surveys, obtaining detailed management status and listening to users' thoughts and opinions.

Interviewees include: security administrators, technical leaders, network system administrators, etc.

The content of the interview includes: confirming the results of the questionnaire, obtaining detailed information about the status of management implementation, and listening to users' thoughts and opinions.

Finally, the security management access record needs to be output, as shown in Table 18-2.

Table 18-2 Security management interview record

No.	Category	Investigate subject	Results
1	Security strategy	Does the company have a systematic, unified security strategy?	
2	Security strategy	Is the security policy detailed to the collective requirements of all aspects of information security?	
3	Security strategy	Has the security policy been developed by a dedicated person for maintenance?	
4	Security strategy	Are you reviewing your published security policy, checking and finding out where it is inappropriate or outdated?	
5	Security organization	Has the company set up a dedicated information security department or working group responsible for security management?	
6	Security organization	Does the company have relevant departments or leaders to coordinate the implementation of safety management in all departments?	
7	Security organization	Does the company explicity require individuals to be responsible for the safety of personally-managed assets? Is there a clear disciplinary rule for employees who violate security policies and security procedures?	

Task 18.2 Passive acquisition method and its implementation

Subtask 18.2.1 Passive acquisition

Passive acquisition means that when an attack occurs, it collects data in time, analyzes the methods used in the attack, and analyzes the problems in the network, so as to make timely remediation and reduce losses.

The data collection method is shown in Figure 18-12.

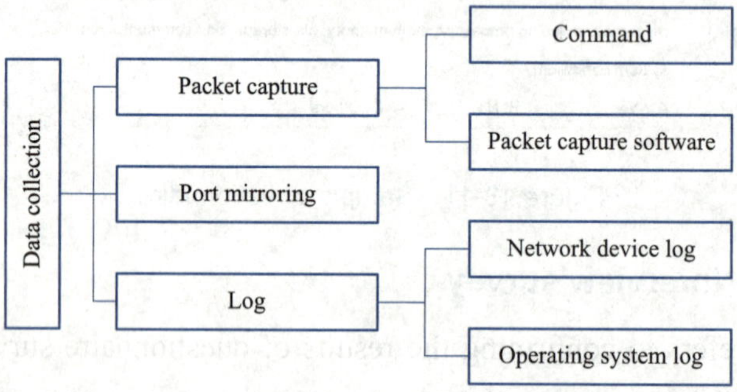

Figure 18-12 Data collection method

Subtask 18.2.2 Grab a bag

Packet capture can obtain all message information and directly analyze and use it. You can use command line or packet capture software to capture packets.

As shown in Figure 18-13, this is to use the command line to view the OSPF neighbor establishment process.

```
<Switch> debugging ospf event
Apr 12 2018 12:03:16.370.1-05:00 SW3 RM/6/RMDEBUG:
 FileID: 0x7017802d Line: 1136 Level: 0x20
 OSPF 1: Nbr 10.1.10.34 Rcv HelloReceived State Down -> Init.
Apr 12 2018 12:03:16.380.1-05:00 SW3 RM/6/RMDEBUG:
 FileID: 0x7017802d Line: 1732 Level: 0x20
 OSPF 1: Nbr 10.1.10.34 Rcv 2WayReceived State Init -> 2Way.
Apr 12 2018 12:03:16.390.1-05:00 SW3 RM/6/RMDEBUG:
 FileID: 0x7017802d Line: 1732 Level: 0x20
 OSPF 1: Nbr 10.1.10.34 Rcv AdjOk? State 2Way -> ExStart.
Apr 12 2018 12:03:16.400.1-05:00 SW3 RM/6/RMDEBUG:
 FileID: 0x7017802d Line: 1845 Level: 0x20
 OSPF 1: Nbr 10.1.10.34 Rcv NegotiationDone State ExStart -> Exchange.
Apr 12 2018 12:03:16.410.1-05:00 SW3 RM/6/RMDEBUG:
 FileID: 0x7017802d Line: 1957 Level: 0x20
 OSPF 1: Nbr 10.1.10.34 Rcv ExchangeDone State Exchange -> Loading.
Apr 12 2018 12:03:16.430.1-05:00 SW3 RM/6/RMDEBUG:
 FileID: 0x7017802d Line: 2359 Level: 0x20
 OSPF 1: Nbr 10.1.10.34 Rcv LoadingDone State Loading -> Full.
```

Figure 18-13 Use the command line to view the OSPF neighbor establishment process

As shown in Figure 18-14, the command line is being used to view the specific content of OSPF packet information.

```
<Switch> debugging ospf packet
Apr 12 2018 12:05:42.930.2-05:00 SW3 RM/6/RMDEBUG:  Source Address: 10.1.10.34
Apr 12 2018 12:05:42.930.3-05:00 SW3 RM/6/RMDEBUG:  Destination Address:
224.0.0.5
Apr 12 2018 12:05:42.940.1-05:00 SW3 RM/6/RMDEBUG:  Ver# 2, Type: 1 (Hello)
Apr 12 2018 12:05:42.940.2-05:00 SW3 RM/6/RMDEBUG:  Length: 48, Router:
34.34.34.34
Apr 12 2018 12:05:42.940.3-05:00 SW3 RM/6/RMDEBUG:  Area: 0.0.0.0, Chksum:
4dcf
Apr 12 2018 12:05:42.940.4-05:00 SW3 RM/6/RMDEBUG:  AuType: 00
Apr 12 2018 12:05:42.940.5-05:00 SW3 RM/6/RMDEBUG:  Key(ascii): * * * * * * *
*d
Apr 12 2018 12:05:42.940.6-05:00 SW3 RM/6/RMDEBUG:  Net Mask: 255.255.255.0e
Apr 12 2018 12:05:42.940.7-05:00 SW3 RM/6/RMDEBUG:  Hello Int: 10, Option: _E_
Apr 12 2018 12:05:42.940.8-05:00 SW3 RM/6/RMDEBUG:  Rtr Priority: 1, Dead Int:
40
Apr 12 2018 12:05:42.940.9-05:00 SW3 RM/6/RMDEBUG:  DR: 10.1.10.34
Apr 12 2018 12:05:42.940.1-05:00 SW3 RM/6/RMDEBUG:  BDR: 10.1.10.33
Apr 12 2018 12:05:42.940.2-05:00 SW3 RM/6/RMDEBUG:  # Attached Neighbors: 1
Apr 12 2018 12:05:42.940.3-05:00 SW3 RM/6/RMDEBUG:  Neighbor: 33.33.33.33
```

Figure 18-14 Use the command line to view the specific content of OSPF packet information

As shown in Figure 18-15, this is using Wireshark to view the specific content of OSPF message information.

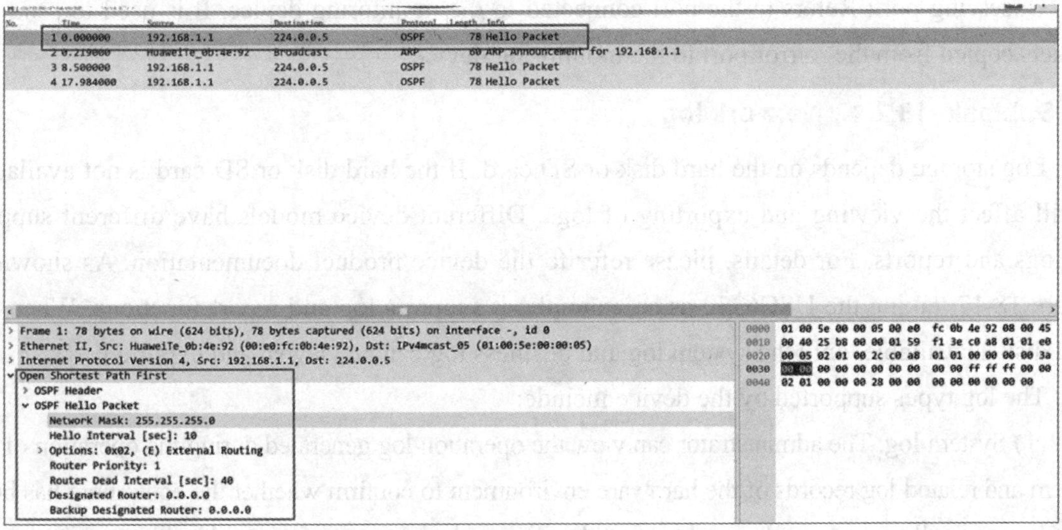

Figure 18-15 Use wireshark packet capture tool to view the specific content of OSPF message information

Subtask 18.2.3 Port mirroring

Port mirroring means that the device replicates the packets flowing through the mirrored port, and transmits the packets to the designated observing port for analysis and monitoring. Sometimes it is necessary to configure monitoring equipment to analyze port mirroring through special software. In this way, data reports can be obtained to determine threats or vulnerabilities, as shown in Figure 18-16.

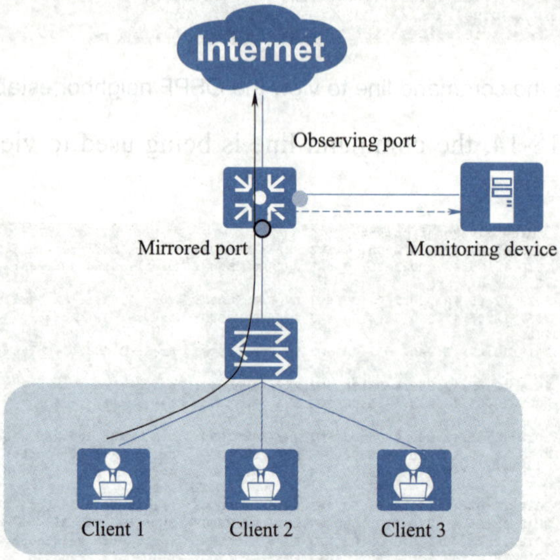

Figure 18-16 Port mirroring

Mirrored port: Refers to the port being monitored. The messages received and sent by the mirror port will be copied to the port connected to the monitoring device.

Observing port: Refers to the port connected to the monitoring device. It is used to send the packets copied from the mirror port to the monitoring device.

Subtask 18.2.4 Network log

Log storage depends on the hard disk or SD card. If the hard disk or SD card is not available, it will affect the viewing and exporting of logs. Different device models have different support for logs and reports. For details, please refer to the device product documentation. As shown in Figure 18-17, taking the USG6330 as an example, it supports log and report functions. When the hard disk unavailable, only the system log and business log can be viewed and exported.

The log types supported by the device include:

(1) System log: The administrator can view the operation log generated during the operation of the system and related log records of the hardware environment to confirm whether the equipment has been operating normally, which can be used for fault location and analysis when a problem occurs.

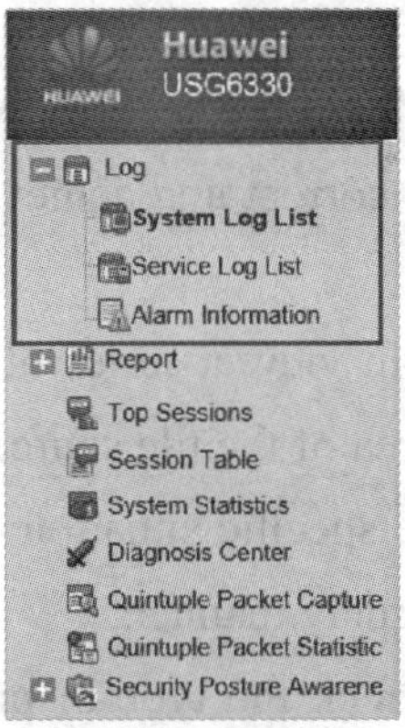

Hard disk unavailable

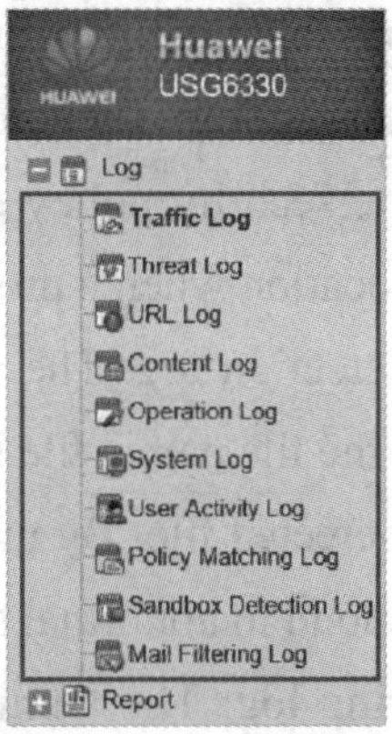

Hard disk available

Figure 18-17 Port mirroring

(2) Service log list: By viewing the business log, the administrator can obtain relevant information in the network, and can locate and analyze the fault in time.

(3) Alarm information: View the alarm information generated by the device through the web interface, including alarm level, occurrence, alarm source, and description information.

(4) Traffic log: Through the traffic log, the administrator can learn the characteristics of the traffic in the network, understand the current network bandwidth occupation and the effectiveness of the security policy and bandwidth policy configuration.

(5) Threat log: Through the threat log, the administrator can view the records of the detection and defense of AV, intrusion, DDoS, zombies, Trojans, worms, APT and other network threats. In this way, we can understand the threats that have occurred and the threats that are happening, and make strategic adjustments or proactive defenses in time.

(6) URL log: The administrator can view the permission, warning or blocking of the user's access to the URL, and understand the user's URL access and the reason for the permission, warning or blocking.

(7) Content log: Through the content log, the administrator can view the alarms and blockades generated when users transmit files or data, send and receive e-mails, and visit websites. In this way, the administrator can understand the user's security risk behavior and the reason for being alerted or blocked.

(8) Operation log: Through the operation log display information, you can view the records of all administrators' login and logout, device configuration and other operations, understand the history of device management, and enhance device security.

(9) User activity log: The administrator can view the user's online records, such as login time, online time/freezing time, and login IP address. In this way, the administrator can understand the current user activities in the network, find abnormal user logins or abnormal network access

behaviors, and respond in a timely manner.

(10) Policy matching log: The administrator can learn the security policy hit by the traffic to determine whether the security policy configuration is correct and achieve the desired effect. This log is used for fault location when a problem occurs.

(11) Sandbox detection log: The administrator can view a series of information detected by the sandbox, such as the file name, file type, security area of the file source, and destination address security area of the detected file. By understanding the specific situation of sandbox detection, the administrator can respond to abnormal situations in a timely manner.

(12) Mail filtering log: The administrator can check the type of protocol used to send and receive mail, the number and size of attachments in the mail, and the reason why the mail was blocked. Then take reasonable countermeasures.

(13) Audit log: By viewing the audit log, the administrator can know the user's FTP behavior, HTTP behavior, e-mail sending and receiving behavior, QQ online and offline behavior, administrators can also search for keywords and the effectiveness of the audit policy configuration.

The log format supported by the firewall is shown in Table 18-3.

Table 18-3 The log format supported by the firewall

Format	Using scenario
Binary format	When the session log is output in binary format, it occupies less network resources. But it cannot be viewed directly on the FW, it needs to be output to the log server to be viewed
Syslog format	When the session log, packet loss log, and system log are output in Syslog format, the log information is presented in text format
Netflow format	For session logs, FW also supports the function of outputting to the log server for viewing in Netflow format, which is convenient for the administrator to analyze the IP packet flow information in the network
Dataflow format	The business log is output in dataflow format and viewed on the log server

1. System log

FW outputs system logs through the information center. The information center is the information hub of the system software modules on the FW. It categorizes the output information of the system in detail, and can effectively filter the information.

According to the severity or urgency of the information, the information is divided into 8 levels. The more serious the information, the smaller the value of its severity level:

(1) Emergency (0): The device has a fatal exception and the system cannot return to normal. The device must be restarted. For example, the device restarts due to an abnormal program, or an error is detected in the use of memory, etc.

(2) Alert (1) : There is a major abnormality in the equipment, and immediate measures need to

be taken. For example, the memory usage of the device reaches the limit.

(3) Critical (2) : There is a major abnormality in the equipment, and measures need to be taken to deal with or analyze the cause. For example, the memory usage of the device exceeds the low limit, the temperature exceeds the low temperature warning line, BFD detects that the device is unreachable, and detects an error message (the message is generated internally by the device), etc.

(4) Error (3) : There is an error in the operation or abnormal process of the equipment, which will not affect the follow-up business, but it needs attention and cause analysis. Such as: wrong command, wrong user password, detection of wrong protocol message (the message is obtained by other devices).

(5) Warning (4) : The abnormal point of the abnormal operation of the equipment, or the process that may cause business failure, they need to be paid attention to. For example, the routing process is closed, a packet detected by BFD is lost, and an incorrect protocol packet is detected.

(6) Notice(5) : The key operation information for the normal operation of the equipment. Such as the execution of the shutdown command, the discovery of adjacent nodes, the normal jump of the protocol state machine, etc.

(7) Informational (6) : General operating information for the normal operation of the equipment.

(8) Debugging (7) : General information about the normal operation of the device, and the user does not need to pay attention.

Take the firewall USG6000 as an example, the steps to view the system log are shown in Figure 18-18.

(9) Choose "Monitor" → "Log" → "System Log" to view the system log information of the firewall.

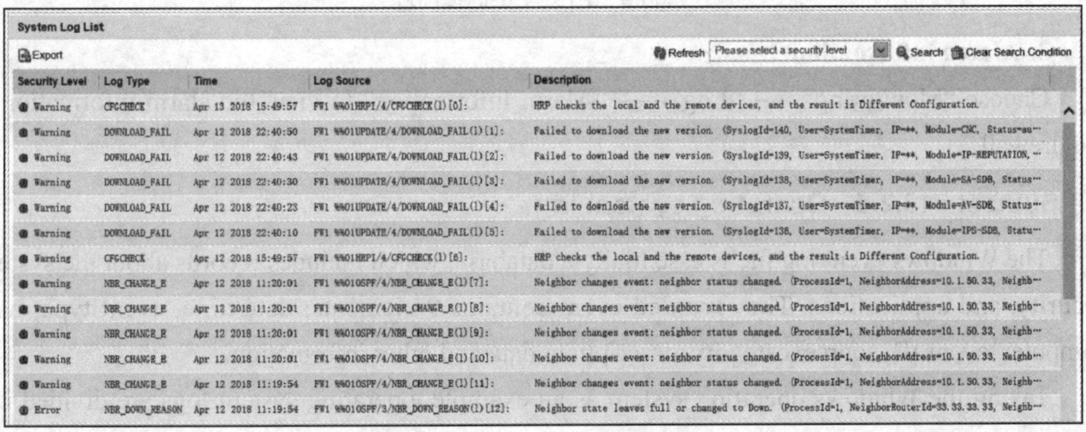

Figure 18-18 System log

2. Business log

FW supports the output of logs of safety business functions. The administrator can view these log information through the web interface and log server.

Business logs on FW include threat logs, content logs, policy hit logs, mail filtering logs, URL filtering logs, and audit logs.

The business log on the FW can be output to the web interface, or directly to the log host, or through the information center. Administrators can learn about business operation and network status by viewing business logs.

Take the firewall USG6000 as an example to view service logs, as shown in Figure 18-19.

Choose "Monitor" → "Log" → "Service Log" to view the firewall's business log information.

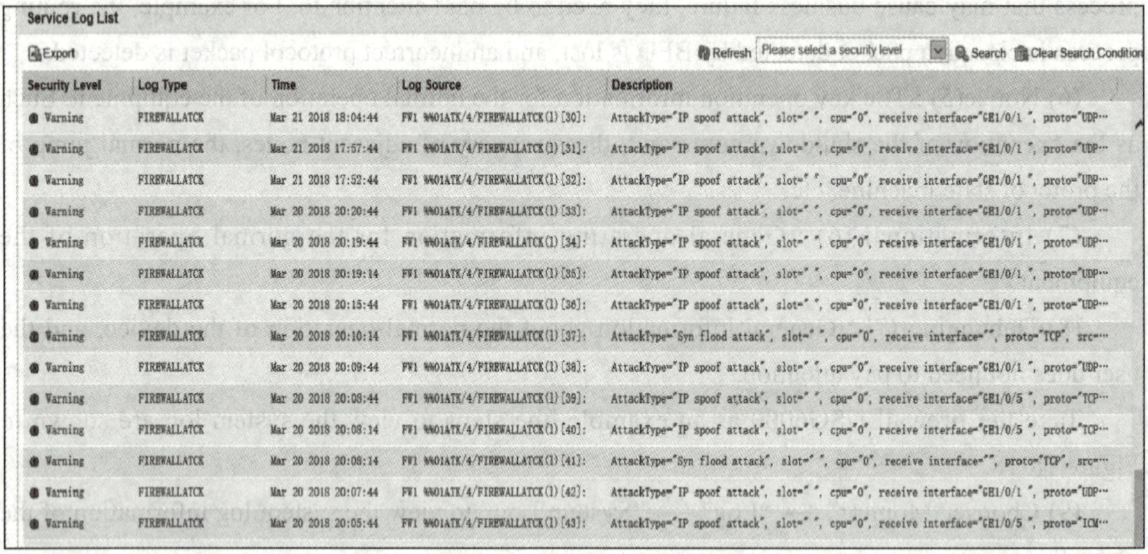

Service Log List

Export Refresh Please select a security level Search Clear Search Condition

Security Level	Log Type	Time	Log Source	Description
Warning	FIREWALLATCK	Mar 21 2018 18:04:44	FW1 %%01ATK/4/FIREWALLATCK(l)[30]:	AttackType="IP spoof attack", slot=" ", cpu="0", receive interface="GE1/0/1 ", proto="UDP…
Warning	FIREWALLATCK	Mar 21 2018 17:57:44	FW1 %%01ATK/4/FIREWALLATCK(l)[31]:	AttackType="IP spoof attack", slot=" ", cpu="0", receive interface="GE1/0/1 ", proto="UDP…
Warning	FIREWALLATCK	Mar 21 2018 17:52:44	FW1 %%01ATK/4/FIREWALLATCK(l)[32]:	AttackType="IP spoof attack", slot=" ", cpu="0", receive interface="GE1/0/1 ", proto="UDP…
Warning	FIREWALLATCK	Mar 20 2018 20:20:44	FW1 %%01ATK/4/FIREWALLATCK(l)[33]:	AttackType="IP spoof attack", slot=" ", cpu="0", receive interface="GE1/0/1 ", proto="UDP…
Warning	FIREWALLATCK	Mar 20 2018 20:19:44	FW1 %%01ATK/4/FIREWALLATCK(l)[34]:	AttackType="IP spoof attack", slot=" ", cpu="0", receive interface="GE1/0/1 ", proto="UDP…
Warning	FIREWALLATCK	Mar 20 2018 20:19:14	FW1 %%01ATK/4/FIREWALLATCK(l)[35]:	AttackType="IP spoof attack", slot=" ", cpu="0", receive interface="GE1/0/1 ", proto="UDP…
Warning	FIREWALLATCK	Mar 20 2018 20:15:44	FW1 %%01ATK/4/FIREWALLATCK(l)[36]:	AttackType="IP spoof attack", slot=" ", cpu="0", receive interface="GE1/0/1 ", proto="UDP…
Warning	FIREWALLATCK	Mar 20 2018 20:10:14	FW1 %%01ATK/4/FIREWALLATCK(l)[37]:	AttackType="Syn flood attack", slot=" ", cpu="0", receive interface="", proto="TCP", src=…
Warning	FIREWALLATCK	Mar 20 2018 20:09:44	FW1 %%01ATK/4/FIREWALLATCK(l)[38]:	AttackType="IP spoof attack", slot=" ", cpu="0", receive interface="GE1/0/1 ", proto="UDP…
Warning	FIREWALLATCK	Mar 20 2018 20:08:44	FW1 %%01ATK/4/FIREWALLATCK(l)[39]:	AttackType="IP spoof attack", slot=" ", cpu="0", receive interface="GE1/0/5 ", proto="TCP…
Warning	FIREWALLATCK	Mar 20 2018 20:08:14	FW1 %%01ATK/4/FIREWALLATCK(l)[40]:	AttackType="IP spoof attack", slot=" ", cpu="0", receive interface="GE1/0/5 ", proto="TCP…
Warning	FIREWALLATCK	Mar 20 2018 20:08:14	FW1 %%01ATK/4/FIREWALLATCK(l)[41]:	AttackType="Syn flood attack", slot=" ", cpu="0", receive interface="", proto="TCP", src=…
Warning	FIREWALLATCK	Mar 20 2018 20:07:44	FW1 %%01ATK/4/FIREWALLATCK(l)[42]:	AttackType="IP spoof attack", slot=" ", cpu="0", receive interface="GE1/0/1 ", proto="UDP…
Warning	FIREWALLATCK	Mar 20 2018 20:05:44	FW1 %%01ATK/4/FIREWALLATCK(l)[43]:	AttackType="IP spoof attack", slot=" ", cpu="0", receive interface="GE1/0/5 ", proto="ICM…

Figure 18-19 Business log

3. Alarm information

Choose "Monitoring" → "Log" → "Alarm Information" to view the alarm information of the firewall.

Subtask 18.2.5 Operating system log

The Windows event log file is essentially a database, which includes records about the system, security, and applications. The recorded event contains 9 elements: date/time, event type, user, computer, event ID, source, category, description, data and other information.

Taking the Windows operating system as an example, operating system logs are divided into system logs, application logs, and security logs, as shown in Table 18-4.

Table 18-4 Operating system logs of Windows

Log type	Effect	Log storage location of Vista/Win 7/Win 8 Win 10/Server 2008/Server 2012
System log	Record events generated by operating system components, including: crashes and data of drivers, system components, and application software	%SystemRoot%\System32\Winevt\Logs\System.evtx
Application log	It contains events recorded by application programs or system programs, and mainly records events related to program operation	%SystemRoot%\System32\Winevt\Logs\Application.evtx
Security log	It records system security audit events, including various types of login logs, object access logs, process tracking logs, privilege usage logs, account management logs, policy change logs, and system events	%SystemRoot%\System32\Winevt\Logs\Security.evtx

As shown in Figure 18-20, click "Start", right-click "Computer", select "Manage", and select "System Tools" → "Event Viewer" → "Windows Log" to view Windows log information.

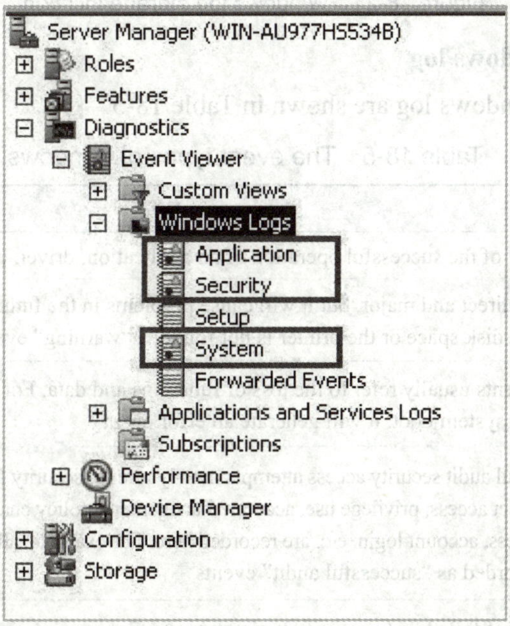

Figure 18-20 Windows log

1. Windows log storage location

Take the security log as an example, select "Properties" to view the specific information of the log, as shown in Figure 18-21.

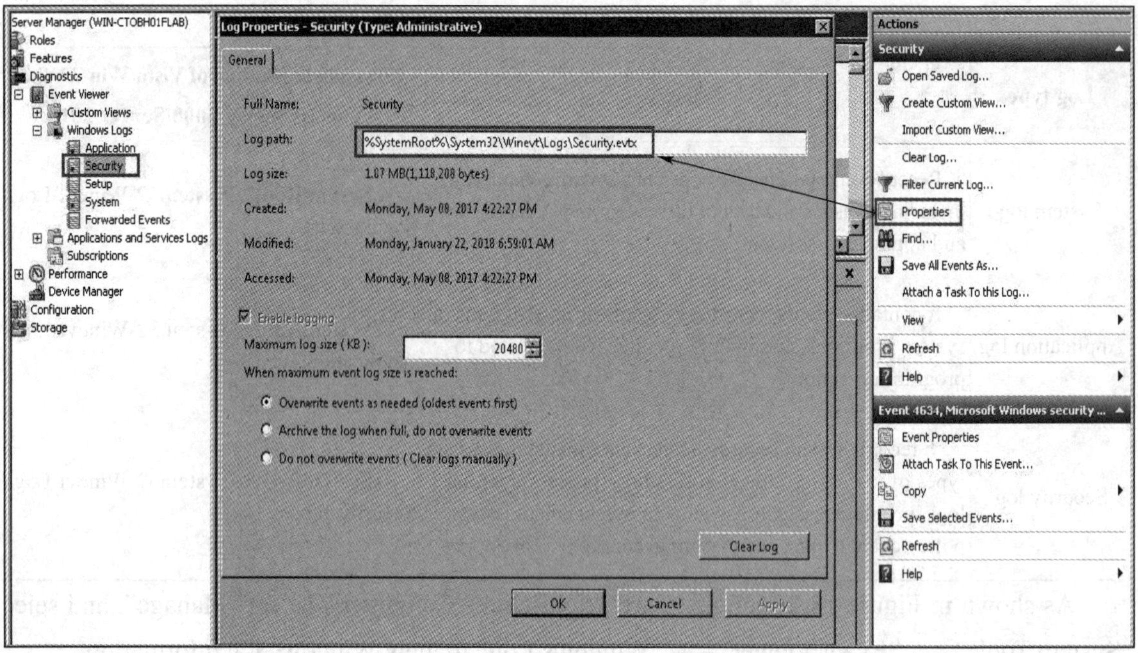

Figure 18-21 Windows log storage location

2. Event type of Windows log

The event types of Windows log are shown in Table 18-5.

Table 18-5 The event types of Windows log

Event type	Effect
Information	An event of the successful operation of an application, driver, or service
Warning	It is not direct and major, but it will cause problems in the future. For example, when there is insufficient disk space or the printer is not found, a "warning" event will be logged
Error	Error events usually refer to the loss of functions and data. For example, if a service cannot be loaded as a system boot, it will generate an error event
Success audit	Successful audit security access attempts mainly refer to security logs. Events such as user login/logout, object access, privilege use, account management, policy changes, detailed tracking, directory service access, account login, etc. are recorded here. For example, all successful logins to the system will be recorded as "successful audit" events
Failure audit	A record of failed audit security login attempts. For example, if a user attempts to access a network drive and fails, the attempt will be recorded as a failed audit event

3. Windows log format

The Windows log consists of two parts: the header field and the description field, as shown in Figure 18-22.

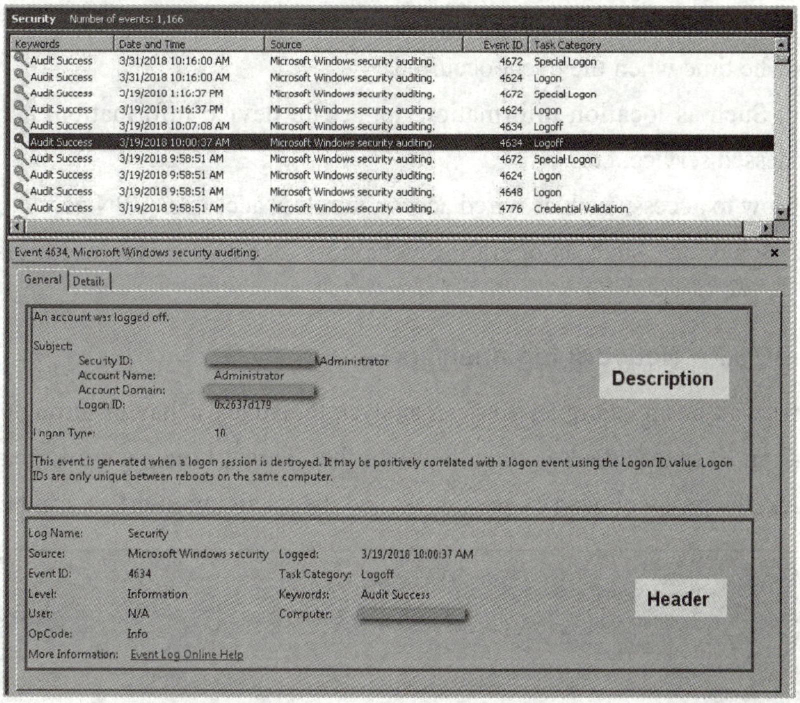

Figure 18-22 Windows log format

Task 18.3 Method and implementation of data analysis

The key to combating cybercrime is how to extract and analyze the "traces" left by criminals in the computer system, and make it an important means and evidence for tracking and arresting criminal suspects. Important events in the network will be recorded in the log, therefore, the analysis of the log is particularly important.

Subtask 18.3.1 Log analysis

The key main points of log analysis are shown in Figure 18-23.

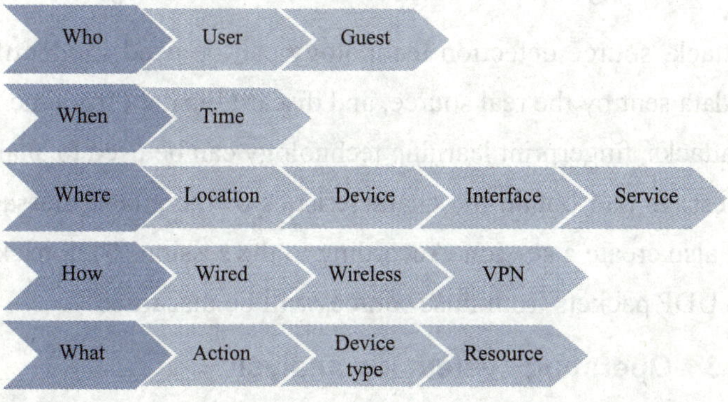

Figure 18-23 The key main points of log analysis

(1) Who: Distinguish between users and visitors.

(2) When: The time when the event occurred.

(3) Where: Such as location information, logged-in device information, accessed interface information, accessed service, etc.

(4) How: How to access, such as wired access, wireless access or VPN access.

(5) What: Behaviors, such as operating behaviors, access devices used, resource services accessed, etc.

Subtask 18.3.2 Network log analysis

Take the firewall as an example, you can analyze the attack behavior through logs. As shown in Figure 18-24, an IP spoof attack is found through the "Threat Log", and information such as the time of the attack, the protocol used by the attack, and the receiving interface can be obtained.

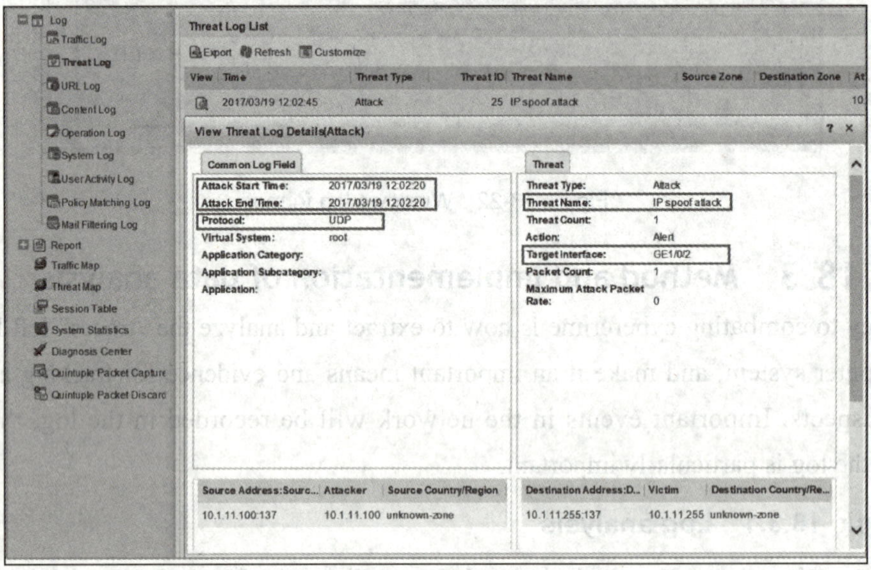

Figure 18-24 Analysis of network device logs

For IP spoof attack, source detection technology can be used to identify the identity of the attacker, release the data sent by the real source, and discard the data from the false source.

For UDP-type attacks, fingerprint learning technology can be used to analyze the characteristic area of the attack message and obtain the characteristics of the attack message to provide a basis for defense. You can also create a session. According to the session, UDP packets from real sources will be released, and UDP packets from false sources will be discarded.

Subtask 18.3.3 Operating system log analysis

Since all events of the operating system are recorded in the log, it is particularly important to quickly filter out key information from a large amount of data. In the Windows operating system, key events can be filtered as needed, as shown in Figure 18-25. The figure shows the log information in the security log with the filtering conditions: the log source is "Application Error", the event level is "Critical" or "Warning", and the event result is "Audit Failure".

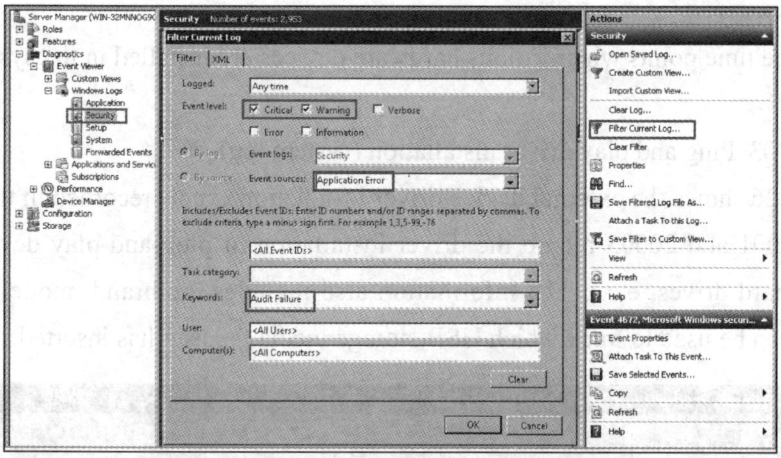

Figure 18-25 Operating system log analysis

For Windows event log analysis, we can quickly obtain the information we care about based on the Event ID according to our own needs. Different Event IDs have different meanings.

1. Analysis of user login and logout events

(1) Use scenario:

Determine which users have logged in to the operating system.

Analyze the user's operation of the operating system.

(2) Event ID:

4624: Successful login (security log).

4625: Login failed (security log).

4634: Logout successful (security log).

4672: Use super user (administrator) to log in (security log).

2. Analysis of modified system time events

(1) Use scenario:

Look for evidence that users modify the system time.

(2) Event ID:

1: Kernel-General (system log).

4616: Change system time (security log).

Event 1 records the user name, original time, and modified time of the time modification event.

Event 4616 not only records the user name, original time, and modified time of the time modification event, but also records the process through which the time modification was executed.

3. Analysis of external settings usage events

(1) Use scenario:

Analyze the time points when various hardware devices are installed in the system.

(2) Event ID:

20001/20003: Plug and play driver installation (system log).

Figure 18-26 shows the external device driver installation events recorded in the system log.

Events 20001 and 20003 record the driver installation of plug-and-play devices (such as U disks, mobile hard drives, etc.). The information also includes the brand, model, and SN of the device, which can be used to track which USB storage media the user has inserted.

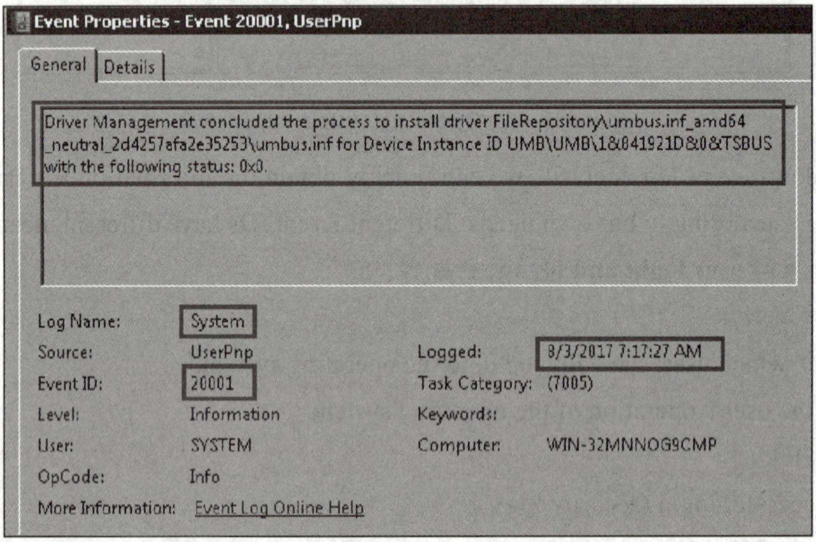

Figure 18-26　Event log of external settings usage

Subtask 18.3.4　Introduction to analysis tools

Log Parser is a log analysis tool produced by Microsoft. It is powerful and easy to use. It can analyze text-based log files, XML files, CSV (comma delimited) files, as well as operating system event logs, registry, file system, and Active Directory. But you need to be able to use SQL proficiently.

Log Parser uses a graphical interface, which is relatively easy to use. You don't even need to memorize cumbersome commands, only need to make settings and write basic SQL statements to get the results intuitively.

Project 19 Electronic evidence collection

With the continuous development of information technology, computers are increasingly involved in people's work and lives. Computer-related legal cases, such as e-commerce disputes, computer crimes, etc., have also continued to appear. In the process of determining or disposing of various disputes and criminal cases, a new form of evidence-electronic evidence, has gradually become one of the new litigation evidence. Because the electronic evidence itself and the forensic process have many characteristics that are different from traditional physical evidence and forensic methods. This raises new research topics for both judicial and computer science fields.

This project will show the technology and tools used in electronic forensics by introducing the process of electronic forensics.

Task 19.1 Understanding electronic evidence collection

Subtask 19.1.1 Understand computer crime

Computer crime refers to the behavior of the perpetrator deliberately breaking into the computer information system in violation of national regulations. Or it refers to the perpetrator using various technologies to damage the functions of the computer information system and related data, application programs, etc.; it also refers to the behavior of making and spreading computer viruses, affecting the normal operation of the computer system and causing serious consequences.

Computer crime is nothing more than the following two methods:

(1) Use computers to store information about criminal activities;

(2) Directly use computers as tools for criminal activities.

In recent years, computer crime cases have been increasing year by year. It has brought immeasurable serious consequences and huge economic losses to the country, even threatened the country's security, and undermined the good social order. Computer crime presents the following characteristics:

(1) Professional means.

(2) Complex and diverse motivations.

(3) The number of members is large and the members are younger.

(4) The potential harm is huge.

(5) Transnational.

(6) Concealed form.

The criminal motives of computer crimes are complex and diverse, as follows:

(1) Fame: Prove your own strength and get respect and recognition from the same kind.

(2) Retaliation: After being suspended, dismissed, demoted or treated unfairly, the perpetrator retaliates against it. The consequences are terrible.

(3) Ignorance: The perpetrator is learning about computers and networks, and unintentionally discovering a weakness or vulnerability that may lead to data destruction or misoperation.

(4) Profit-driven: The perpetrator is hired by others for huge rewards, helping others to break into the target system, stealing or tampering with information.

(5) Political purpose: destruction, stealing intelligence, information warfare.

(6) Mischief: Criminals are extremely boring and possess certain skills, always want to visit all sites of interest.

There are many forms of computer crime. The following are common forms: Trojan horses, hackers, backdoors, viruses, worms, DDos, and internal and external leaks.

Subtask 19.1.2 Understanding electronic evidence

Electronic evidence, also known as digital evidence or computer evidence, refers to an electromagnetic record produced during the operation of a computer or computer system, and its recorded content is used to prove the facts of a case.

There are various forms of electronic evidence, as shown in Figure 19-1, which can be text, graphics, images, animations, audios and videos and other media information. This kind of computer evidence in the form of multimedia covers almost all types of traditional evidence.

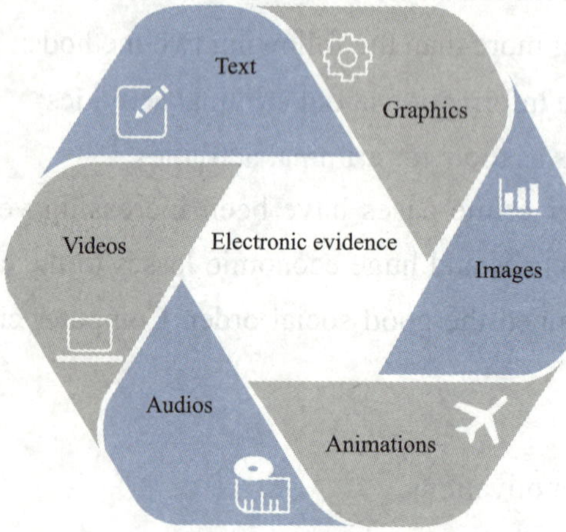

Figure 19-1 Electronic evidence

Common electronic evidence in judicial practice can be divided into three categories, as shown in Figure 19-2:

(1) Electronic evidence related to modern communication technology;

(2) Electronic evidence related to other modern information technologies such as broadcasting technology, television technology, and film technology;

(3) Electronic evidence related to computer technology or network technology.

The characteristics of electronic evidence, as shown in Figure 19-3, include: man-made, high-tech, diversity, fragility and destructibility, dynamic dissemination, vividness and intangibility.

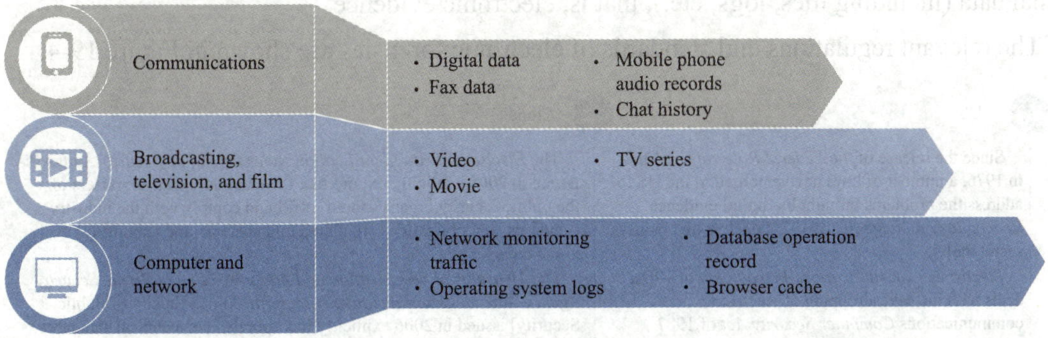

Figure 19-2 Sources of electronic evidence

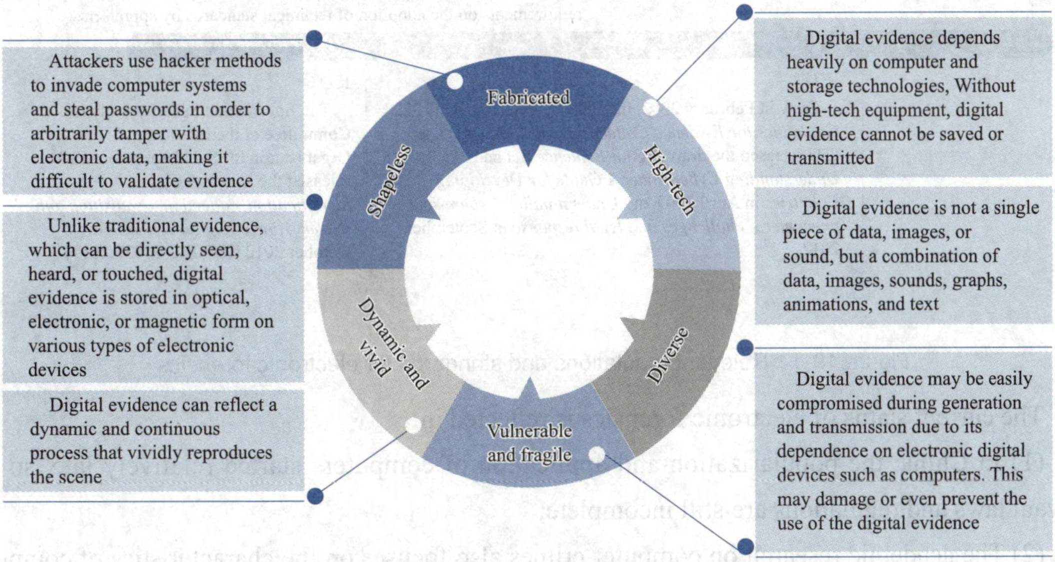

Figure 19-3 Features of electronic evidence

Computer forensics is also called digital forensics or electronic forensics. Computer forensics refers to the process of confirming, protecting, extracting, filing and presenting electronic evidences in court. These electronic evidences are acceptable to the court, reliable and persuasive enough, and exist in computers and related peripherals.

Technically, computer forensics is the process of scanning and cracking the compromised computer system to reproduce the intrusion event.

Computer forensics includes two stages: physical evidence acquisition and information discovery:

(1) Acquisition of physical evidence means that investigators go to the scene of computer crime or intrusion to find and detain relevant computer hardware;

(2) Information discovery refers to finding evidence that can be used to prove or refute from original data (including files, logs, etc.), that is, electronic evidence.

The relevant regulations and standards of electronic forensics are shown in Figure 19-4.

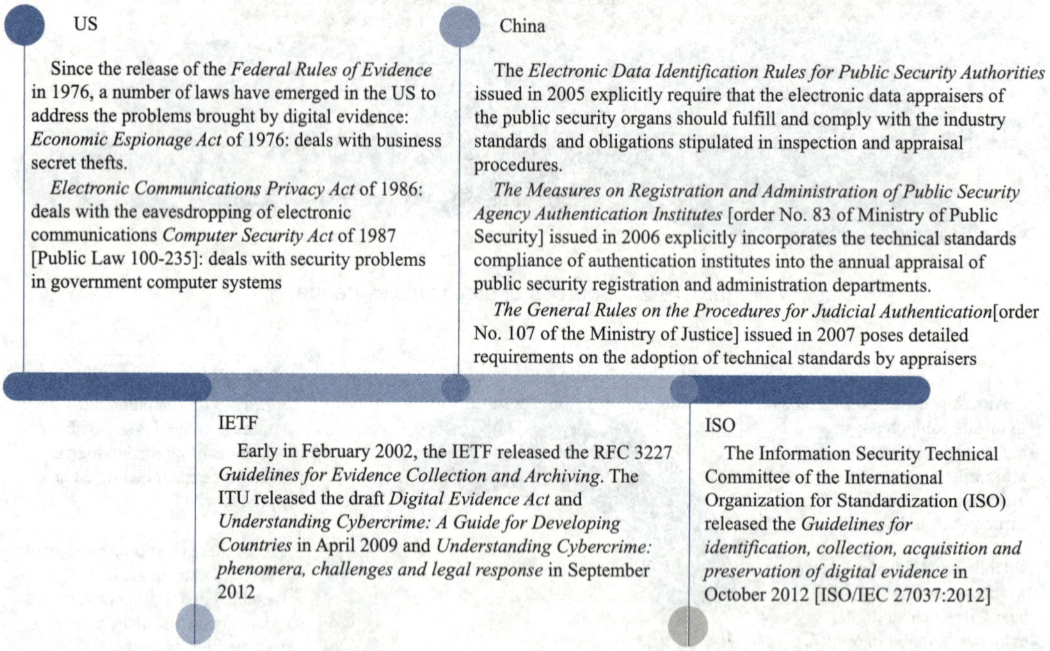

Figure 19-4 Relevant regulations and standards for electronic forensics

The current status of electronic forensics is reflected in:

(1) In China, the popularization and application of computers started relatively late, so the relevant laws and regulations are still incomplete;

(2) The academic research on computer crimes also focuses on the characteristics of computer crimes, preventive countermeasures and their impact on human beings. Forensic technology can no longer meet the requirements of combating computer crimes, protecting network and information security;

(3) It is necessary to independently develop computer forensics tools and software that are suitable for the national conditions of the country and can comprehensively check the computer and network system.

The development trend of electronic forensics is reflected in:

(1) Forensic technology integrates other theories and technologies (such as artificial intelligence, machine learning, neural network and data mining theories and information security technology);

(2) Specialization and automation of forensic tools;

(3) In the process of network protocol design, future forensics needs will be considered, and sufficient information will be reserved for potential forensic activities.

Task 19.2 Electronic forensics process and implementation

Subtask 19.2.1 Principle of obtaining evidence

Due to the intangible, polymorphic, fragile, system-dependent, and easily tampered characteristics of electronic evidence, it is difficult to distinguish the modification, deletion, and coverage of electronic data caused by any human factors or external forces. Therefore, compared with the traditional evidence collection rules and methods, electronic evidence needs to be analyzed and obtained through specific technical means. In the process of obtaining electronic evidence, special attention should be paid to standardizing the process of electronic forensics and following the basic principles of electronic forensics, as shown in Figure 19-5.

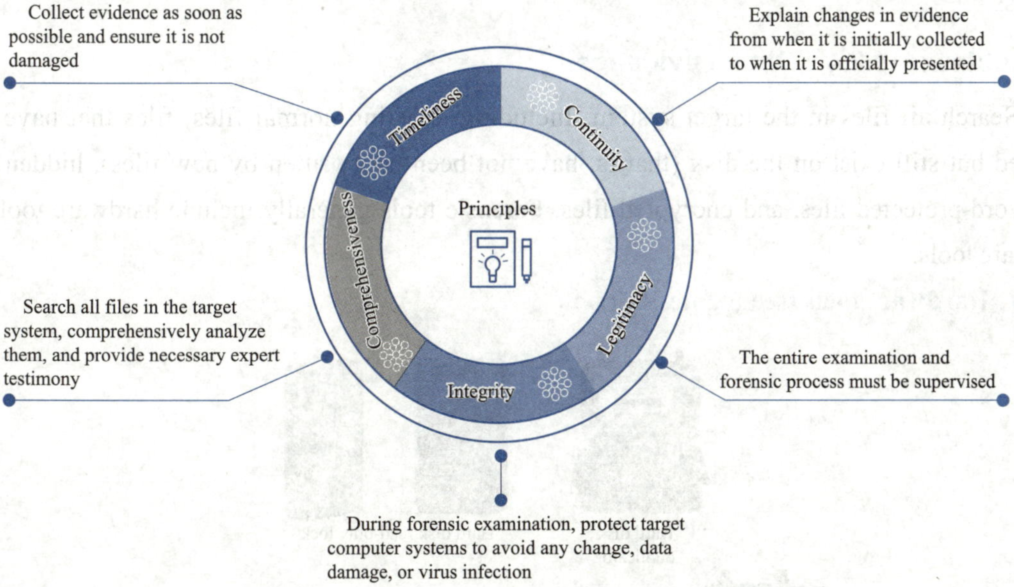

Figure 19-5 Principles of electronic forensics

Subtask 19.2.2 Electronic forensics process

According to the characteristics of electronic evidence, when conducting computer forensics, we must collect evidence as soon as possible and ensure that it has not been damaged in any way. Forensic work is generally carried out in accordance with the steps shown in Figure 19-6.

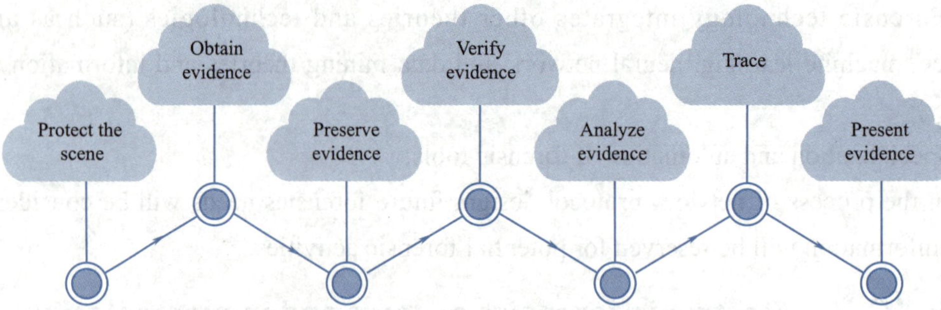

Figure 19-6 The process of electronic forensics

Subtask 19.2.3 Protect the scene

When obtaining evidence, the target computer system must be frozen first, so as not to give the criminal suspect the opportunity to destroy the evidence. Avoid any situation that changes system settings, damages hardware, destroys data, or be infected by viruses.

When the evidence is formally submitted, it must be able to explain the changes from the initial state of the evidence to the presentation of the evidence, and of course it is best if there is no change. The most important thing is that computer forensics must be supervised throughout the process.

Subtask 19.2.4 Obtain evidence

Search all files in the target system. Including: existing normal files, files that have been deleted but still exist on the disk (that is, have not been overwritten by new files), hidden files, password-protected files, and encrypted files. Forensic tools generally include hardware tools and software tools.

1. Hardware tools (see Figure 19-7)

Figure 19-7 Hardware tools

(1) Hard disk duplicator: Because it is not allowed to directly operate the hard disk during the forensic process, the hard disk duplicator can mirror the contents of the hard disk to ensure that the data in the hard disk will not be modified.

(2) Hard disk read-only lock: It is used to block the writing channel of the hard disk, which can effectively ensure that the data in the storage medium will not be modified during the acquisition and analysis process, thereby ensuring the integrity of the data.

(3) Forensic all-in-one appliance: A forensic device based on copy cloning, reading, and destruction.

(4) Forensic tower: It has functions such as mobile phone forensic analysis, mobile phone mirroring, mobile phone chip mirroring, and media forensics.

(5) Media repair device: Supports the repair of various electronic data storage media such as electronic hard disks, mechanical hard disks, mirror files, U disks, and TF cards.

2. Software tools

(1) Picture detection tool: Thumbs Plus is a tool designed to help individuals view and easily browse all the photos on your computer and edit them.

(2) Undelete tool: Hetman Uneraser is a software that can restore deleted files or folders.

(3) CD-ROM tool: Use CD-R Diagnostics, you can see the data that can not be seen under normal circumstances.

(4) Text search tool: dtSearch is a good text search tool, especially with the ability to search Outlook.pst files.

(5) Disk erasing tools: This type of tool is mainly used before using the forensic analysis machine to ensure that the drive of the analysis machine does not contain residual data. Obviously, simple formatting will definitely not work. After booting from the floppy disk, run the DiskScrub program of NTI Company to clear the data of every sector on the hard disk.

(6) Drive imaging program: Disk imaging software that can meet forensic analysis, that is, copy bit by bit to create an image of the entire drive, including SafeBackSnapBack, Ghost, DD, etc.

As shown in Figure 19-8, EnCase is a fully integrated forensic application based on the Windows interface. Its functions include: data browsing, search, disk browsing, data preview, case creation, evidence file creation, case preservation, etc.

3. Electronic forensics related technologies

Electronic forensics related technologies include: network data packet analysis and evidence collection, log forensics technology, honeypot forensics technology, hidden code forensics technology, data mining technology, etc.

With the continuous advancement of technology, some new forensics technologies have also emerged:

(1) Chip forensics: When communication equipment is maliciously damaged, buried, or flooded, making it unusable, chip forensics can be used to extract information from the device.

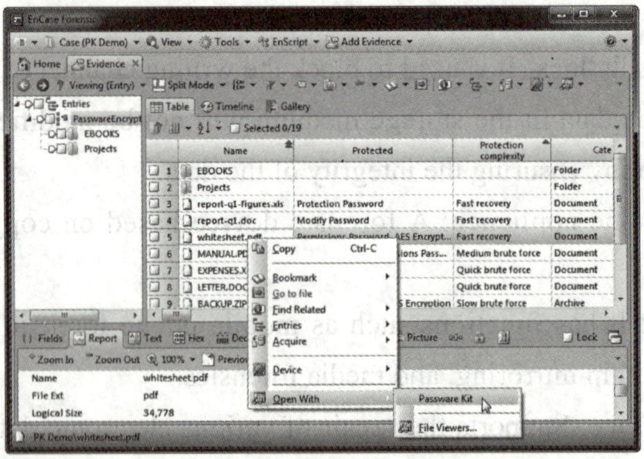

Figure 19-8 EnCase

Use Joint Test Action Group to analyze. It can access the internal registers of the analysis processor through the TAP (test access port) inside the chip. Even when the phone is damaged, it can collect evidence on the chip.

Dynamic simulation: Similar to using a simulator to simulate a mobile phone, restore chat records, group chat discussions and other information. You can extract the files of the chat software, analyze the amount of forwarding, the scope of influence, and the source of information.

(2) Cloud forensics: When data is transferred in the cloud, the integrity of the data will be guaranteed. For deleted data, the cloud service provider can be located and the cloud data can be restored.

(3) Side channel attack forensics: SAC (side channel attack), a method of attacking encrypted devices. It is aimed at side channel information leakage such as time consumption, power consumption, or electromagnetic radiation during operation of encrypted electronic devices

(4) Internet of things forensics: When a networked device is invaded, the Internet of things forensics technology can obtain relevant information. The technologies involved include: Internet of things black box technology, distributed IDS technology sniffing forensics technology and so on.

Subtask 19.2.5 Preserve evidence

Evidence preservation is directly related to the legal effect of evidence. The authenticity and reliability of evidence can only be guaranteed if the preservation technology complies with legal procedures.

Any changes to system settings, hardware damage, data damage or virus infection should be avoided during the forensic process. Ensure that the evidence is still in the original state when presented. The evidence preservation technique is shown in Figure 19-9.

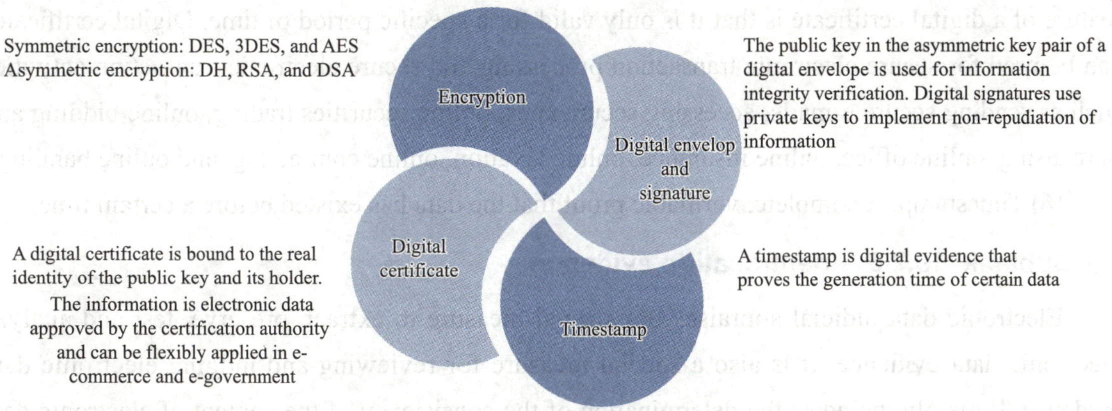

Figure 19-9 Evidence preservation technology

(1) Symmetric encryption: In the symmetric encryption algorithm, only one key is used, and both sender and receiver use this key to encrypt and decrypt data. This requires the decryption party to know the encryption key in advance.

(2) Asymmetric encryption: The asymmetric encryption algorithm requires two keys for encryption and decryption. The two keys are a public key and a private key.

(3) Digital envelope: Symmetrical cryptography and public key cryptography are used in digital envelopes. The information sender first encrypts the information with a randomly generated symmetric password, and then uses the receiver's public key to encrypt the symmetric password. The symmetric password encrypted by the public key is called a digital envelope. When transmitting information, if the information receiver wants to decrypt the information, it must first decrypt the digital envelope with its own private key to obtain the symmetric password, and then use the symmetric password to decrypt the obtained information. This ensures the authenticity and integrity of data transmission.

(4) Digital signature technology: Digital signature technology is to encrypt the summary information with the sender's private key, and transmit it to the receiver together with the original text. The receiver can only use the sender's public key to decrypt the encrypted summary information, and then use the Hash function to generate a summary information based on the received original text, and compare it with the decrypted summary information. If they are the same, it means that the received information is complete and has not been modified during transmission. Otherwise, it means that the information has been modified, so the digital signature can verify the integrity of the information.

(5) A digital certificate is a file that contains the information of the public key owner and the public key that has been digitally signed by the certificate authority. The simplest certificate contains a public key, name, and digital signature of the certificate authority. Another important

feature of a digital certificate is that it is only valid for a specific period of time. Digital certificates can be used for secure electronic transaction processing and secure electronic transaction activities such as sending secure e-mails, accessing secure sites, online securities trading, online bidding and purchasing, online office, online insurance, online taxation, online contracting, and online banking.

(6) Timestamp: A complete, verifiable proof that the data has existed before a certain time.

Subtask 19.2.6 Identification evidence

Electronic data judicial appraisal is a special measure to extract, preserve, test and analyze electronic data evidence. It is also a special measure for reviewing and judging electronic data evidence. It mainly includes: the determination of the consistency of the content of electronic data evidence; the determination of the data content stored in various electronic devices or storage media; the determination of the deleted data content in various electronic devices or storage media; determination of the data content of encrypted files; recognition of computer program function or system status; recognition of authenticity and formation process of electronic data evidence, etc.

The following standards can be used to verify the integrity of the evidence and determine whether it is true.

(1) The relevance standard of electronic data authentication: Relevance is the natural attribute of evidence and the inevitable connection between evidence and the facts of the case. If electronic evidence can have a substantial impact on the facts of the case, the court should rule that the evidence is relevant.

(2) Objectivity standard for electronic data appraisal: Objectivity can also be called authenticity. It should be ensured that the content of electronic evidence remains unchanged during the entire process from the initial acquisition and collection to the submission and use of the evidence as litigation evidence. Then the electronic evidence has objectivity or authenticity.

(3) The legality standard of electronic data authentication: Legality means that only the evidence obtained by legal entities in legal form and obtained by legal means can have the ability to prove. The acquisition, storage, and submission of electronic evidence should be legal, and the process does not constitute a serious infringement of basic rights, such as national interests, social welfare, and personal privacy.

As a special scientific and technological activity, electronic data authentication has its specific principles and requirements.

(1) The principle of legality: This principle requires the standardization and institutionalization of electronic data judicial expertise in business scope, authentication procedures and technical standards. This includes two aspects:

(2) The principle of legal behavior: This is a requirement for the appraiser. Forensic appraisers

of electronic data must first undergo special training, and they should be authenticated in a timely manner to prevent electronic data from changing over time.

(3) The principle of legal status: This is a requirement for the status of electronic data. Electronic data should have multiple backups, and electronic devices should be placed away from high magnetic fields, high temperatures, dust, backlogs, humidity, and other environments. We should try our best to keep the state of the electronic equipment consistent with the original state of the target system or change to a minimum.

(4) The principle of independence: Forensic appraisers of electronic data should independently express their appraisal opinions without interference from outsiders. Then the appraiser will make a scientific judgment based on the actual results obtained. The appraisal result is a manifestation of independent will, and the appraiser only obeys science and the law.

(5) Supervision principle: The judicial appraisal of electronic data needs a supervision mechanism, including two aspects:

① Supervision by investigators: The transfer, storage, unpacking, disassemble, and analysis of electronic data must be supervised and executed by investigators. Then the appraiser and the supervisor jointly signed to assume responsibility.

② Social supervision: The intervention of social supervision should be publicly required to overcome corruption. Electronic data authentication is an activity consistent with justice. The intervention of social supervision can reflect social justice to the greatest extent

Subtask 19.2.7 Analyze evidence

Evidence analysis technology: It refers to finding and matching keywords or key phrases in the data stream that has been obtained to analyze the relevance of events. Specifically:

(1) Password decryption, data decryption;

(2) Analysis of file attributes;

(3) Analysis of digital summaries;

(4) Log analysis technology;

(5) Reverse engineering, etc.

Subtask 19.2.8 Track it

With the upgrading of computer crime technology, the forensic function has been integrated with network security tools, such as intrusion detection. For dynamic forensics, the purpose of tracking is to find the source of the attack.

Forensic tracking technology:

(1) Log analysis:

① Operating system log;

② Firewall logs;

③ Firewall logs;

(2) Set up traps (honeypots).

(3) We can use relevant equipment to track and catch criminal suspects. For example, hackers usually use functions such as data encryption, log cleaning, sniffers, springboards, and bastion hosts to hide or clean up traces of intrusion. Therefore, it is necessary to grasp the information of intrusions and collect evidence. The tracking techniques usually used are as follows:

① Ping, netstat, tracert, nslookup and other commands;

② Data packet analysis;

③ Log analysis;

④ Link testing (also called segment-by-segment tracking) determines the source of attack services by testing the network links between routers. Most of the technical methods we use are to start with the router closest to the victim host and test its input (uplink) link to determine which router is carrying traffic. If a spoofed data packet is detected (by comparing the source IP address of the data packet with its routing table information), it will log in to the superior router and continue to monitor the data packet. If the spreading attack of spoofing is still detected, it will log in to the superior router to detect the spoofed data packet again. Repeat this process until the actual attack source is reached.

⑤ Data packet recording: Data packet recording technology records data packets on key routers of the Internet, and then uses data drilling technology to extract information about the source of the attack business. Although this solution seems obvious and can accurately analyze the attack business (even after the attack has stopped), its biggest disadvantage is the large amount of processing and storage capacity required to keep records. There are also legal and confidentiality issues with record keeping and the need to share this information between ISPs.

⑥ Data packet marking technology inserts tracking data into the tracked IP data packet, thereby marking each router on the network link from the local machine to the target host. The simplest way to implement packet marking is to use the record routing option (specified in RFC791) to store the router address in the option field of the IP header. However, this method increases the packet length processed by each router and may cause additional fragmentation. Moreover, the attacker may try to use fake data to fill in the fields reserved by the route, and use this method to avoid tracking.

⑦ Spam tracking: Shallow mail behavior analysis can realize functions such as connection time analysis, sending IP address, sending time, sending frequency, number of recipients, shallow e-mail header checking, sending behavior detection, etc. You can also use the SMTP MTA host to transparently analyze the source of the mail to determine whether the sender has illegal acts such as anonymity, forgery, and spamming. And you can take a return or delay countermeasure mechanism for the above actions.

Subtask 19.2.9 Show evidence

The process needs to indicate the time, location, equipment, extractor and witnesses of the evidence. Then it will be submitted to the judiciary in a visible form and legal procedures. And you should provide a complete chain of supervision to the judiciary.

Project 20 Case study of information security

After learning basic theoretical knowledge and practicing basic experimental operations, everyone has a certain understanding of information security regulations, security threats, and security operations. Next, we will use specific cases for actual combat training, digest and absorb the knowledge learned and apply it.

Task 20.1 Discussion on operation steps of information security deployment

1. Background

Xiao A is responsible for network operation and maintenance in a company. Since the country promulgated the "Network Security Law", the company has paid more attention to network security. Now it is necessary for Xiao A to sort out the existing network security issues based on previous operation and maintenance experience. But what aspects should Xiao A start with to achieve the company's information security?

2. Information security operation steps

After analysis, Xiao A believes that it can start from the following four aspects, but what is the order of operation?

Establish norms and cultivate awareness → information security planning → information security deployment → information security operation.

3. Establish norms and cultivate awareness

Task : Please talk about the domestic and foreign laws and regulations related to information security or the company's information security regulations that you know about.

List common behaviors with low information security awareness.

Discussion time: 5 minutes.

Achievement: Free answers and supplements by group members.

4. Information security planning

Task : Discuss what are the key aspects of the information security regulations of your company? What aspects need to be considered in the company's information security design?

Discussion time: 5 minutes.

Achievement: Free answers and supplements by group members.

Reference answer:

(1) Information security regulations focus on: Personnel security awareness, network security deployment, security operation procedures and emergency response, etc.

(2) Network security design: Security personnel arrangement, personnel awareness training, network security design and emergency response process planning, etc.

5. Information security deployment

Task : Discuss what security considerations are in the network design of your company?

Discussion time: 5 minutes.

Achievement: Free answers and supplements by group members.

Reference answer: Network security design. Consider from the aspects of terminal security, access control, intrusion prevention and communication security.

6. Information security operation

Task : If an information security incident occurs, such as information leakage, network attacks, etc., how should we deal with it?

Discussion time: 5 minutes.

Achievement: Free answers and supplements by group members.

Reference answer: Consider from the aspects of information security incident prevention, information monitoring and analysis, threat processing and emergency response.

Task 20.2 Network security case discussion

In order to protect the information security of a company, the leader of the company asked the administrator Xiao M to analyze the company's information security situation and give a specific implementation plan.

In the case, we will give specific information security risks as tasks, and we need to give a reasonable solution to each task. According to the number of trainees, each group member will finally give a solution and choose one person to explain the plan.

Grading criteria:

(1) The shortest time to complete the task and the correct answer will get 5 points, and the rest will be sorted backward.

(2) If the question raised by another team cannot be answered or the answer is wrong, the questioning team will add 1 point each time.

(3) The team with the highest cumulative total score in the game is the winning team.

(4) In the event of the same score, one extra question will be answered, and the final interpretation right belongs to the instructor.

(5) The winning team will get a prize.

Time rules for the game:

(1) Discussion time for each task: 15 minutes.

(2) Speech time for representative of each team: 5-10 minutes.

After Xiao M's analysis, he found that the company has the following information security risks:

(1) The employees randomly record the office account number, password and other information on the sticky note of the workstation;

(2) For office convenience, employees often copy important company documents to personal computers for office work;

(3) For the convenience of Internet access, employees freely access the TP-Link router within the company;

(4) A lot of instant messaging software and video software are installed on the employees' office PCs;

(5) Although a firewall is deployed at the company's network exit, the default security policy action is Permit;

(6) The company deploys a Web server to display the company's business, but it often suffers from unknown attacks from the external network, and network administrators are often caught off guard, resulting in the company's homepage cannot be accessed normally.

1. Task one

Poor management of personal accounts and passwords, and random copying of important company information are all signs of inadequate information security awareness. Have you encountered a similar incident? What problems can this cause? How to solve such problems?

Reference answer:

The following dangerous behaviors exist in such incidents:

(1) Password security:

① The complexity of the password is not high enough, and information such as birthday and mobile phone number is often used as the password;

② Multiple accounts use the same password and have not been changed for a long time;

③ When the office equipment is idle, the lock screen is not set.

(2) E-mail security:

① Open the executable file in the anonymous mail;

② Easily trust e-mails such as winning and redeeming prizes, and open the link in the e-mail to enter personal information.

(3) Internet security:

① In order to use the free network, connect to the open unencrypted Wi-Fi;

② Install pirated software and access the push message link.

The solutions that can be taken are:

(1) At the company level:

① Develop detailed and enforceable safety systems and operation manuals;

② Various publicity (safe websites, safety activities, safety brochures, etc.);

③ Regular safety training seminars.

(2) At the personal level:

① The password should be as complex as possible, and it should be changed regularly;

② Don't trust anonymous mail information easily, and don't disclose personal information (mobile phone number, bank card, ID number, etc.) easily;

③ Do not connect to unencrypted Wi-Fi in public areas.

Summary: A weak awareness of information security will cause personal information to be commercialized and sold, which will ultimately lead to personal property losses.

2. Task two

Employees connect equipment to the company network at will, and download non-office software on office computers, which belongs to the company's terminal security management category. Have you encountered a similar incident? What problems can this cause? How to solve this kind of problem?

Reference answer:

The following dangerous behaviors exist in such incidents:

(1) Staff management:

① Failed to authenticate employee's terminal equipment access to the network;

② Failed to conduct security inspections on the connected terminal equipment;

③ No restrictions are imposed on the network resources and business traffic that employees can access.

(2) Visitor management:

① No access authentication for visitors;

② There are no restrictions on the network resources or network areas that visitors can access.

The solutions that can be taken are:

(1) At the Staff management level:

① Certify employees' access to equipment, and only after passing the certification can they access the network;

② Check the software installed on the employees' office equipment. If the installed software does not meet the requirements, it needs to be repaired;

③ Restrict and manage the business traffic of employees to prevent employees from occupying too much bandwidth.

(2) At the visitor level:

① Perform identity authentication for visitors, and only after passing authentication can they access the network;

② Restrict the network resources and areas that visitors can access, so as to prevent visitors from accessing confidential company information.

Summary: User authentication realizes the identification of users, and security check realizes the safe access of equipment. The combination of the two can increase the security of the network. Reasonable network planning is the cornerstone of enterprise information security.

3. Task three

If the default security policy action of the firewall is permit, then the firewall is just a "furnishing" and does not play a role in security defense. Have you ever encountered an incident similar to "Ears and Bells"? What problems can this cause? How to solve this problem?

Reference answer:

The following dangerous behaviors exist in such incidents:

(1) Safety requirements:

① Because they are not clear about the internal business of the enterprise, the administrator has purchased the security equipment without actually using it, but only connected it in series with the network;

② The administrator only deploys a firewall at the exit, and has a one-sided consideration of network security, ignoring the overall protection of the network.

(2) Network deployment:

① There is no redundant backup in network planning, which is prone to single point of failure;

② The network does not have a hierarchical design, which is difficult to maintain, and if there is a failure, it is difficult to troubleshoot;

③ There is no encryption when data is transmitted through the Internet, so the data is "naked".

The solutions that can be taken are:

(1) At the security requirements level:

① Clarify the needs of network security and make full use of security equipment;

② When planning and designing the network, fully consider the overall security of the network, such as: external network attack protection, terminal protection, server protection, access security, etc.

(2) At the network deployment level:

① Redundant design of the network, such as: equipment redundancy, link redundancy and disaster recovery backup, etc.;

② Carry out hierarchical design of the network, such as: network hierarchy division, security zone division, authentication domain division, etc.;

③ When transferring information through the public network, select VPN technology to protect data according to needs.

Summary: A good network design and deployment is not only convenient for administrators to maintain, but when the network fails it also can quickly converge to ensure that services are not affected.

4. Task four

If the company's servers are frequently attacked, and the administrator's failure to respond in a timely manner causes frequent interruption of access, this belongs to the scope of the company's network security operations. Have you encountered a similar incident? What problems can this cause? How to solve this problem?

Reference answer:

The following dangerous behaviors exist in such incidents:

(1) Network reinforcement.

(2) Emergency response:

① Before the network failed or was attacked, we did not evaluate the security of the network;

② There is no emergency drill, and the administrator is not proficient in operation after failure or attack.

The solutions that can be taken are:

(1) At the network reinforcement level:

① Conduct security tests on the network, and strengthen the network according to the test results;

② Emergency drills should be held regularly to improve the emergency response capabilities of administrators and employees.

(2) At the emergency response level:

① Formulate a standardized emergency response process and set up an emergency response team;

② Tracking technology or equipment should be deployed so that when an attack occurs, data can be recovered in time, and evidence can be investigated and collected.

Summary: Security operations involve many aspects, from process planning to network deployment, from personnel arrangements to network operation and maintenance. Only when each link is solidly implemented and coordinated, can the overall security of the network be guaranteed.